ANALYSIS OF ORGANIC MICROPOLLUTANTS IN WATER

Commission of the European Communities

ANALYSIS OF ORGANIC MICROPOLLUTANTS IN WATER

*Proceedings of the Second European Symposium
held in Killarney (Ireland), November 17-19, 1981*

Edited by

A. BJØRSETH

Central Institute for Industrial Research, Oslo

and

G. ANGELETTI

*Directorate-General for Science, Research and Development,
Commission of the European Communities, Brussels*

D. REIDEL PUBLISHING COMPANY
DORDRECHT : HOLLAND / BOSTON : U.S.A.
LONDON : ENGLAND

Library of Congress Cataloging in Publication Data

Main entry under title:

Analysis of organic micropollutants in water.

    At head of title: Commission of the European Communities.
    Includes index.
    1.   Organic water pollutants—Analysis—Congresses.
I. Bjørseth, A., 1941–   . II.   Angeletti, G., 1943–
III. Commission of the European Communities.
TD427.07A527   628.1'61   82–463
ISBN-13:978-94-009-7806-5   e-ISBN-13:978-94-009-7804-1
DOI: 10.1007/978-94-009-7804-1

The Symposium was jointly organized by
- The Commission of the European Communities, Brussels
- National Board for Science and Technology, Dublin
- An Foras Forbartha, Dublin

Publication arrangements by
Commission of the European Communities
Directorate-General Information Market and Innovation, Luxembourg

EUR 7623
Copyright © 1982, ECSC, EEC, EAEC, Brussels and Luxembourg
Softcover reprint of the hardcover 1st edition 1982

Published by D. Reidel Publishing Company
P.O. Box 17, 3300 AA Dordrecht, Holland

Sold and distributed in the U.S.A. and Canada
by Kluwer Boston Inc.,
190 Old Derby Street, Hingham, MA 02043, U.S.A.

In all other countries, sold and distributed
by Kluwer Academic Publishers Group,
P.O. Box 322, 3300 AH Dordrecht, Holland

D. Reidel Publishing Company is a member of the Kluwer Group

FOREWORD

The Commission of the European Communities presents with this volume
the proceedings and the conclusions of the second European Symposium
on the analysis of organic micropollutants in water.

This symposium has been organized within the framework of the
Concerted Action "Analysis of Organic Micropollutants in Water".
This research programme is jointly implemented by the European
Communities and Norway, Portugal, Spain, Sweden, Switzerland and
Yugoslavia within the framework of a COST (Coopération Scientifique
et Technique) agreement. The project, also known as COST Project 64b bis,
aims at coordinating all relevant research in this field in the
participating countries.

An effort is made to improve methods and techniques for the
identification and quantitative determination of organic compounds
present in all types of water.

The symposium permitted to review the results achieved during
the past three years of research in the following areas:

- Sampling and sample treatment
- Gas-chromatography
- Separation of non-volatile compounds, in particular high
  performance liquid chromatography (HPLC)
- Mass-spectrometry
- Data processing
- Specific analytical problems, in particular the analysis
  of organic halogens and phenolic compounds.

The volume gives a rather complete overview of the activities in
this field in Europe.

We are confident that it constitutes a valuable contribution to
solving the important problems posed by the huge number of already
identified or yet unknown organic pollutants in water.

The Commission of the European Communities wishes to express their
sincere thanks to the co-organizers, the National Board for Science
and Technology, Dublin and An Foras Forbartha, Dublin.

Brussels, December 1981

G. ANGELETTI   H. OTT

Directorate-General for Science,
Research and Development,
Commission of the European Communities,
Brussels

C O N T E N T S

SESSION II - GAS-CHROMATOGRAPHY

SESSION V - DATA PROCESSING

SESSION VI - SPECIFIC ANALYTICAL PROBLEMS

A. Organic halogens

ANALYSIS OF ORGANIC MICROPOLLUTANTS IN WATER -
- SOME INTRODUCTORY REMARKS

A. BJØRSETH
Central Institute for Industrial Research
Forskningsveien 1, Blindern
Oslo 3 - Norway

The Commission of the European Communities has a number of projects aimed at Cooperation in Science and Technology, the socalled COST projects. One of these projects, the COST 64b-bis is directed towards the study of "Organic micropollutants in water". This is a subject which has attracted large interest both among the member countries, as well as many non-member countries. However, when we look at the different problems and research interests in the various countries, we realize that they span a very broad research area. There are different environmental problems in the southernmost compared to the northernmost countries in Europe, as well as the easternmost compared to the westernmost ones. We, in the scientific committee of the COST 64b-bis project, have tried to reflect this when composing the program. Over the three days to come we will have a good opportunity to look into several aspects of analysis of micropollutants in water.

If we take a few moments and look back into history, we will realize that water pollution is not a new problem. The laws established over three thousand years ago by Moses detail how human waste should be disposed of by burial outside the immediate confines of the camp and away from running water which may be used for drinking purposes. The large sewer in Rome, the "Cloaca maxima" build by the Emperor Tarquinius was not only constructed for esthetic reasons, but rather for the practical purposes of avoiding diseases and protecting ground water supplies. About year 1250 water pollution legislation was introduced in Venice and in 1501 laws were promulgated in Paris to protect the water of the Seine from the discharges of the growing City. And an unpleasant situation in year 1800 was very nicely described by Samuel Coleridge when he wrote the poem,

> The River Rhine, it is well known
> Doth wash your city of Cologne
> But tell me, nymphs, what power divine
> Shall henceforth wash the River Rhine.

Over the years there has been a great effort to clean most of the polluting discharges from municipal, agricultural and industrial sources, and much has been achieved in terms of removing major pollutants from our environment. In recent years there has also been a large interest for organic micropollutants in water. The reason for this interest is of course the possibilities for some of these micropollutants to bioaccumulate due to their chemical and biological persistence or to have long-term effect.

Furthermore, what is also characteristic for the development in the area of pollution studies are the very strong ties between environmental chemistry and analytical chemistry. In many respects we have seen that the development of analytical methods has revealed new environmental problems. This is true both for organic and inorganic pollutants. On the organic side, I only have to mention the importance of chromatographic techniques, such as gas chromatography and high performance liquid chromatography. To some extent, it is also true that realization of environmental problems has created a demand for new analytical instrumentation. I believe that the development of for instance modern mass spectrometry has been very important for the development of environmental chemistry. On the other hand, the demand for better analytical methods has provided a great push for the development of mass spectrometric instrumentation. It is apparent that some of the manufacturers of mass spectrometric equipment would not have existed without the market created by the environmental protection legislation.

The relation between analytical chemistry and environmental chemistry has several dimensions. First of all we have been working continually to improve the analytical sensitivity. Our interest is frequently focused on compounds which occur in very low concentrations. Furthermore, we are interested in the specificity of the chemical compounds. We cannot use collective parameters only, but we also like to know exactly what kind of chemical species are present. These two analytical aspects, sensitivity and specificity, are then used in environmental interpretations. It is important to know what kind of pathways or fates the chemicals in the environment undergo, if there is any bioaccumulation or biodegradation, what are the environmental or health effects, is it chronic or acute toxicity, are the compounds mutagenic or carcinogenic and so on. The interrelation between these three parameters is illustrated in Figure 1.

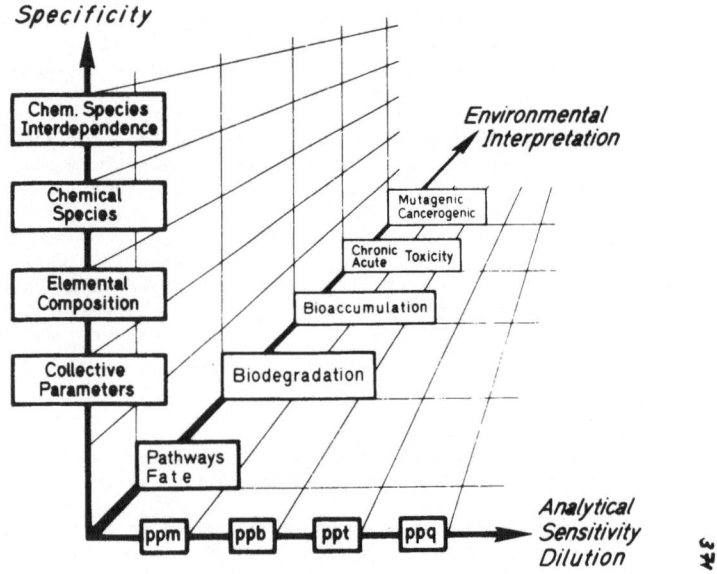

Figure 1.   The interrelation between analytical Sensitivity,
            Specificity and Environmental Interpretation
            (W. Stumm and R. Schwarzenbach, EAWAG, 1980).

As seen the analytical sensitivity may vary over several orders of magnitude. We also have to consider the specificity of our determinations. We can determine the collective parameters, we can determine elemental compositions of the individual chemical species and we can determine the chemical species' interdependence. In our studies of organic micropollutants in water these three dimension aspects are very important.

Recently, we have also seen that studies of organic micropollutants in water put increased requirements on the analytical methods in terms of both qualitative and quantitative aspects. New and improved methods are needed for monitoring and inventory of pollutants as well as studies of transport and fate of organic pollutants in the aquatic environment.

The aim of this Symposium is to try to look into the state-of-the-art of certain areas related to organic micropollutants in water and to discuss their future trends. We, in the scientific committee for this meeting, have selected to discuss topics such as sampling, analytical methods, data handling and finally specific groups of pollutants. The concepts of sampling methodology will be subject for thorough discussion during this meeting. Concerning analytical methods we will discuss of course, the well known techniques capillary gas chromatography and high performance liquid chromatography. In addition we will look specifically into the combination of HPLC and mass spectrometry, and we will look to mass spectrometry as a technique by itself, the socalled MS-MS-techniques. Particularly interesting subject in this area is also the computerized data collection handling and interpretation. Organic pollutants in water comprise a large number of compounds, or groups of compounds. Among the groups that we have selected as particularly important in the current state, are the organic halogenated compounds and phenolic compounds, but other compounds will also be discussed. I believe that this program comprises a combination of focusing on technical problems as well as the application of highly advanced techniques.

SESSION I - SAMPLING AND SAMPLE TREATMENT

Chairman: F. BRINKMANN, National Institute for
Water Supply, The Netherlands

Review paper :

- Recent concepts in sampling methodology

Short papers :

- A new method for the quantitative analysis of organochlorine
  pesticides and polychlorinated biphenyls

- Concentration and identification of the main organic micropollutants classes in waters

Poster papers :

- Recovery of organic micropollutants

- Influence of humus with time on organic pollutants and
  comparison of two analytical methods for analysing organic
  pollutants in humus water

- Improved accumulation of organophosphates from aqueous media
  by formation of ion-associates with tetraphenylarsonium
  cation

- The use of ECD and FID fingerprint techniques for the
  evaluation of river water purification contaminated with
  organic pollutants

- Lekkerkerk

RECENT CONCEPTS IN SAMPLING METHODOLOGY

B. Josefsson
Department of Analytical and Marine Chemistry
Chalmers University of Technology, University of Göteborg
S-412 96 Göteborg Sweden

Summary

Sampling of different kind of waters is discussed. Since the sampling is an integral part of the whole analytical procedure, different method designs are treated with such aspects as sensitivity, selectivity and convenience. Various small-volume sampling methods are exemplified for predetermined compounds. Results from comparative studies of different standard methods are shortly reviewed. Continuous automatic sampling by solvent extraction and by preconcentration on sorbents, to obtain representative samples over a period of time, are compared. Particulate matter in the water may cause systematic error in the sampling stage.

1. INTRODUCTION

Water sampling is not simple. Sampling theory cannot replace experience and good judgement. The reliability of any analytical measurement is directly coupled to the uncertainties of the sampling process, or to sample storage, preservation, or pretreatment prior to analysis. The analytical process involves a chain of operations, where a significant error may be introduced at any stage. No chain is stronger than its weakest link, and this axiom applies also to the analysis of water for its composition. If the sampling is not correctly performed, the most refined techniques and sensitive instrumentation will not automatically lead to reliable and accurate results. The sampling is the first step and probably the most essential step in the analysis. The classical statistical theory of sampling cannot be applied to water, since its environment is influenced by chemical physical, and biological processes, resulting in a pronounced variability of the momentary distribution. The large variety of different processes which influence the distribution of the compounds in question requires a good knowledge of water chemistry.

In this paper the sampling strategy will primarily be discussed. Since in most cases the sampling is an integral part of the whole analytical procedure, the method design will also be treated.

1.1 Sampling strategy for different waters

Knowledge about the nature of the water to be sampled is important. The distribution and transportation of organic micropollutants in natural water systems are influenced by the interaction between the dissolved and the particulate components. When liquid-liquid extraction procedures are used for the preconcentration of organics, the partition coefficient determines the yield. The partition coefficient in turn varies with the concentration of suspended solids. The nature and characteristics of the adsorbing solids are significant factors, e.g. the greater capacity of organic matter to adsorb in contrast to the sands. The volatilization of organics from water bodies is to a great extent dependent on oxygen content, temperature, hydrodynamics, etc. Thus, organic micropollutants may behave very different in river waters compared to sea water.

Sea water and ground water are very different in composition. Sea water contains besides inorganics a lot of particulate organic matter, e.g. organisms, cells, colloidals, and in addition dissolved organics. Ground water is protected from atmospheric influence and filtrated through e.g. sand, which means that the water is low in particulates and easily retains volatiles. Of course, these entirely different characteristics must be considered in the selection of sampling methods.

The dispersion pattern of organic substances in the sea is quite different from that in rivers. Meteorological circumstances may influence in such a way that sampling can only be carried out during controlled and stable conditions, which requires fast collection of samples.

The concentration of organics in urban run off water depends on the precipitation. Pollutants may have been accumulated during a long period of time with no rainfall. The first rainfall will then wash away high concentrations of particulate solids. The majority of the extractable organics, is carried by the solids.

Tap water is relatively convenient to collect, which often may be done already at the laboratory. The drinking water composition varies very little compared to other types of water. Therefore, sampling methods for drinking water may be easier to standardize.

Occasional discharges into recipients are very difficult to trace and especially quantitate. Grab sampling or integrated sampling techniques have to be considered according to convenience and the stated goal. The frequency of grab samples may be very high to balance integrated sampling performed in a continuous system. In many cases it is impossible to solve a pollution problem with grab sample collection. When monitoring a river system for occasional discharges, a continuous-collection strategy is a requirement for tracing suddenly appearing pollutants from unknown sources.

Continuous sampling may be performed in discrete steps, as with the spoon sampler. This water collector is not reliable for volatile compounds, e.g. oil constituents, which will partly evaporate during the sampling time. Continuous sampling involves very frequently a preconcentration step. Liquid-liquid extraction and adsorption on adsorbents are then the common techniques. The former technique leads to the extracted compounds being preserved in the organic phase. Grab sampling requires different preservation methods during storage. Ideally, full analysis should be conducted on the spot, immediately after the sample is taken.

The collection of sea water samples for oil analysis is connected with great contamination risks. The water surface around the ship is covered with a thin film of oil emanating from the ship. Furthermore, the atmosphere around the ship is polluted with diesel smoke, which necessitates an avoidance of atmospheric exposure of the sample. The sampler has to penetrate the oil film without being contaminated. This can be carried out with a closed sampling bottle, which opens at the desired depth. The bottle is then used for the solvent extraction; thereby manupulative steps, as pouring, are omitted, which otherwise cause exposure to the air. Closed sampling systems are always desirable in trace analysis.

2. SAMPLING AS AN INTEGRAL PART OF THE ANALYTICAL METHODS

An analytical method to determine trace organics in water often consists of a series of operations. It is desirable to use as few operations as possible, as every extra step means greater risk of error introduction. There are very few methods where the water sample is directly accessible for measurements. Fluorescence spectroscopy has been used to determine lignin sulphonates directly in sea water samples (1). Direct aqueous injection into a gas chromatograph equipped with an electron capture detector has been applied for haloform determinations (2,3). The method yields higher concentrations of some haloforms than the values obtained with other techniques. Haloforms are generated in the injection port (4).

In most cases a method begins with a preconcentration or isolation step followed by separation and concludes with measurements. These steps are more or less an integral part of the whole method. The different operations are connected to each other, e.g. the sampling volume is directly proportional to the detection capability. The sample, collected in a preconcentration step, will in some cases be accessible for direct measurement, thereby omitting the separation step. In this respect, different designs of fluorimetric methods have already been discussed (5). The sensitivity and selectivity of fluorescence detection make it possible to miniaturize the whole system with advantages such as low exposure to surfaces, smaller sample volumes, fewer manupulative steps.

In the following, different methods will be discussed from a small-volume sampling point of view. The organics to be measured are predetermined, socalled target compounds; thereby optimal conditions are maintained for the substance or the substance class in question.

2.1 Sampling of halocarbons by pentane extraction

The most common and simplest extraction of organics from water is performed by shaking with a water immiscible solvent. The extraction of halocarbons in water with 5 mL n-pentane in a 100-mL volumetric flask with a narrow neck is an uncomplicated procedure. The extraction efficiency is in the range of 80-90% for the common halocarbons. The combination of glass capillary columns and electron capture detection GC yields very high sensitivity as well as a certain selectivity for halocarbons (2). With the injection of large volume (10 μL pentane) extract directly on-column still higher sensitivity is obtained (range 0.1 - 1 ng/L in sea water). The analytical method allows halocarbons to be rapidly determined in the sea with great convenience (6) (Figure 1).

2.2 Sampling by extractive alkylation

Extractive alkylation is also called liquid-liquid phase-transfer catalysis. The technique is a combination of extraction of solute as an ion-pair into an aprotic solvent, where alkylation takes place. As the anion present is highly reactive in the poorly solvents, a nucleophilic displacement reaction occurs with high yield. After the displacement the phase-transfer catalyst may initiate a repetition of the entire sequence. At properly chosen pH, ion-pair regents and alkylation reagents, the extraction is very efficient. The method has been used to determine carboxylic acids and phenols simultaneously in different waters with GC^2 (7). Then pentafluorobenzylbromide was used as the alkylation agent, which results in highly electron capture sensitive derivatives. Extractive alkylation is a convenient method, since the procedure can be carried out in one step directly on a small volume water sample. Furthermore, there is no need for clean up or sample pretreatment.

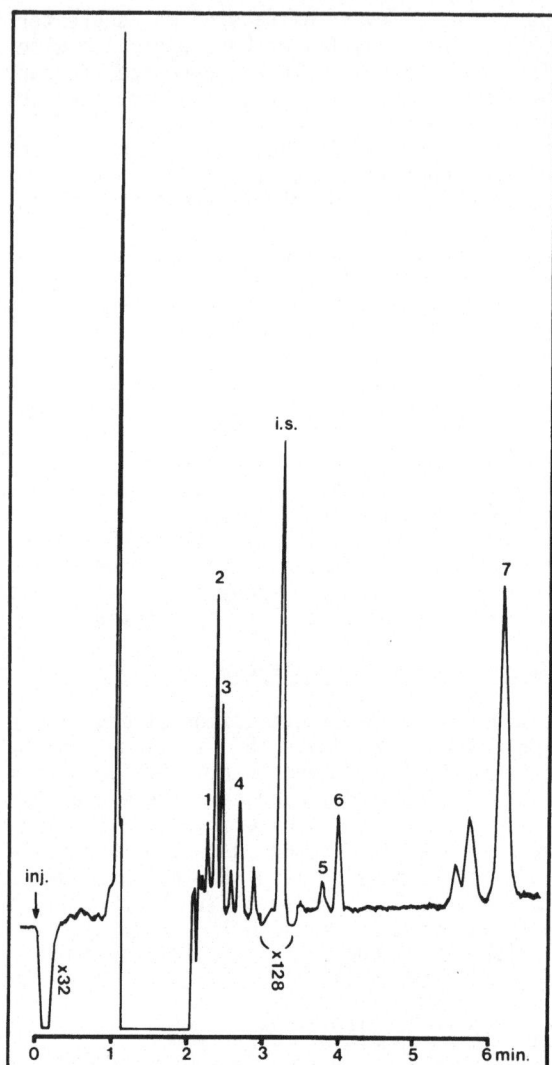

Fig. 1. Glass capillary gas chromatogram of a pentane extracted 100-mL sea water sample from the Norwegian Sea. Ten µL is injected on-column. The concentrations in the sea water were:

Peak: 1, $CHCl_3$ (<2 ng/L); 2, CH_3CCl_3 (1.8 ng/L)

3, CCl_4 (0.51 ng/L); 4, $CHCl=CCl_2$ (2.2 ng/L);

5, $CHBr_2Cl$ (0.9 ng/L); 6, $CCl_2=CCl_2$ (0.42 ng/L);

7, $CHBr_3$ (54 ng/L); i.s.: $CBrCl_3$

Reaction scheme:

$$R'COOH \longleftrightarrow R'COO^- \ H^+$$

$$NR_4^+ \ X^- \ \longleftrightarrow \ R'COO^-NR_4^+ + HX$$

$$NR_4^+Br^- \qquad\qquad R'COO^-NR_4^+$$

F F
F⟨◯⟩CH$_2$OOCR' ← F⟨◯⟩CH$_2$Br
F F   F F

2.3 Direct analysis of water samples by HPLC

Water samples may be directly injected onto a reversed-phase HPLC column. A reversed-phase column is compatible with large volume water samples, which in this case means up to several hundred millilitres. Lipophilic organic substances in the water sample will be concentrated in a narrow zone at the top of the column (8). A water-acetonitrile gradient is used to separate the solutes. The method has been used to determine 32 priority pollutants in water in the lower ppb-range using a UV detector at 202 nm (9). Tetrachloroethylene was determined in water when 10-mL samples were injected (10). The disadvantage of this sampling technique is that a complex mixture is concentrated, which creates a need for selective detectors.

2.4 Sampling column coupled on-line to a chromatographic system

Automatic sample handling techniques lead to less risk of contamination and lower blank values. The use of a pre-column for trace enrichment of organic pollutants in water may be practical even with a very complex matrix. The pre-column should be very small with, e.g. 2-mm thick sorbent layers to keep the band broadening small when the retained substances are to be eluted with the mobile phase solvent mixture employed (11). For the same reason a small particle size or the same size as in the analytical column is preferable. According to Van Vliet et al. (12), the loading capacity of a 2 x 4.6 mm I.D. pre-column is at least 10 µg. This means that 0.1 - 1 L water samples with, e.g., PCB and pentachlorophenol could be handled. The reported recoveries were over 90%. Between the pre-column and the analytical column is a loop injection valve, which can be switched off line when sampling and switched on-line when analyzing.

Many different types of adsorbents and solvent combinations can be used in the precolumn. The technique has been used to determine pentachlorophenols in river and lake water samples (12). The method has also been applied to determine the aqueous solubilities of PAH constituents (13).

The pre-column technique has also been used with gas chromatography for the determination of halocarbons. The water is first retained on a special stationary phase; thereafter it is vented out (14). Water can also been removed by diffusion across a permaselective membrane. A tube of perfluorosulphonic acid polymeric material is then used as an on-line pre--column (15).

2.5 Derivatization to increase lipophility and detectability

Polar compounds such as amines and phenols can be derivatized directly in the water sample. Thus, the products will be better amenable to extraction and detection. Coutts et al. (16) proposed forming acetate esters of phenols directly in the water sample with acetic anhydride as the derivatizing reagent. The products are extracted quantitatively in trace amounts using a small volume of extraction solvent. The stable derivatives are then separated by GC. The method has been used to determine chlorophenols in pulp mill effluents (17). Fluorescence labeling of pesticides has been performed directly in water, thereby obtaining products which are better suited for chromatographic determination (18).

3. COMPARATIVE STUDIES OF SOME STANDARD METHODS

The methods that will be discussed in general are Grob's closed loop stripping method (CLSA), batch solvent extraction, purge and trap methods, XAD-2 adsorption, and head space analysis. The first four methods were recently comprehensively investigated by Melton et al. (19) of the Environmental Protection Agency in USA. The methods were applied to the same waters, thereby making a careful comparison possible. All the methods were based on grab sample collection.

Grob et al. (20) compared the CLSA method with n-pentane microextraction and found better recovery with the CLSA method for more volatile substances, while extraction is suited for heavier materials. The CLSA method gave better quantitative reproducibility and is easier to standardize. Turbid water samples are unsuitable for solvent extraction since emulsification occurs. Solvent extracts which contain high molecular weight compounds are stressing the GC columns upon routine use. Our own findings are that the CLSA method is not suitable for quantitative haloform determinations (21). Melton et al. (19) found in their studies that the CLSA method was very sensitive and good mass spectra were obtained in the 1-10 ng/L range. The blanks were extremely clean. The main disadvantage of the method is that moderate and highly polar or ionizable organic species are poorly or not at all recovered.

Methods based on XAD and Tenax adsorbents are suffering from contamination problems. The plastic materials seem to deteriorate. The purity of the solvents is far easier to control.

Headspace sampling was cross-checked with solvent extraction for the determination of trace amounts of hydrocarbons in water (22). The solvent extraction method was then found to be more accurate. As concluding remarks, the CLSA method and solvent extraction procedure are preferable and they complement each other.

4. SAMPLING BY CONTINUOUS EXTRACTION OF A LARGE VOLUME WATER

The minimum detectable concentration of the organics in water is determined by e.g., the volume of water treated. However, this aspect will not be discussed here. Continuous automatic sampling by a preconcentration step is useful to get representative samples over a period of time in rivers, the sea, etc. Two main methods are used: liquid-liquid extraction and adsorption on sorbents placed in columns. The latter technique is the most common because the equipment is simple and straightforward to handle.

4.1 Adsorption methods

Recently a comprehensive review was published by Dressler (23) regar-

ding different porous organic polymers used as sorbents for organic trace
analysis in water. The sorbents have various polarities and pore sizes,
which makes them amenable to extraction of different organic classes. The
recoveries are very discriminative regarding type of sorbent used. There
is no sorbent for general analysis. However, different sorbents may be
mixed or coupled in series.

Musty and Nickless (24) carried out comparable studies with Amberlite
XAD-4, porous polyurethane foam, and solvent extraction applied to the same
water samples. They found that PCB concentrations in some waters were very
much higher with solvent extraction sampling compare to sorbent extraction
sampling.

Adsorption theory is very complicated, especially when water matrix
effects are involved. Sorption-desorption recoveries have generally been
studied with spiked water solutions. Our opinion is that this is not adequate since the water characteristics may vary widely. The water content
of particulate matter and the way the organics are stuck to the particles
are important. The nature of the active sites determines also the ability
of the organics to be extracted.

When the water is first filtrated colloidals and small particles will
still pass the filter and bring part of the organics more or less unaffected through the extraction column. When the water is not filtrated the adsorbent columns are easily clogged, resulting in chanelling. Even pores
may sometimes be blocked. Prefiltration is also not a good solution for
large volume sampling. Actually, there is a simple method for the determination of PCB and DDT in water, based on adsorption on membrane filters
(25).

Derenbach et al. (26) used a type of reversed-phase sorbent column to
study the recovery of some ^{14}C-labelled substances spiked into sea water
samples. They found that liquid extraction yielded somewhat higher recovery compared to the sorbent. During the Baltic Intercalibration Workshop
held in Kiel (1977), their sorbent method was compared with our continuous
liquid-liquid extraction system (see below). The sorbent system yielded
3.8 µg/L "oil equivalents", while the solvent extraction in series yielded
5.3 µg/L applied to the same sea water sample. Note, the reverse-phase
sorbent may be considered as a mixture of adsorbent and solvent extractant.

4.2 Continuous liquid-liquid extraction

An automatic continuous liquid-liquid extraction apparatus has been
constructed for use in the field (27,28). The extraction unit is based on
the mixer-settler principle and can be used with solvents lighter and heavier than water. The apparatus consists of two extractors in series for
serial extraction and can be applied at different depth in e.g., the sea.
The units are loaded with about 200 mL solvent and have a capacity to extract some hundred litres of water depending on the solvent used (Fig. 2).

The theory of solvent extraction is well founded. The extraction efficiency was found to be different for PCB in river water, 50% in a single
extraction, while 80% in sea water (29,30). The discrepancy depends on the
different particulate matter present in the waters. The apparatus has been
used to determine pesticides (31), oil components, chlorinated compounds,
(32), etc.

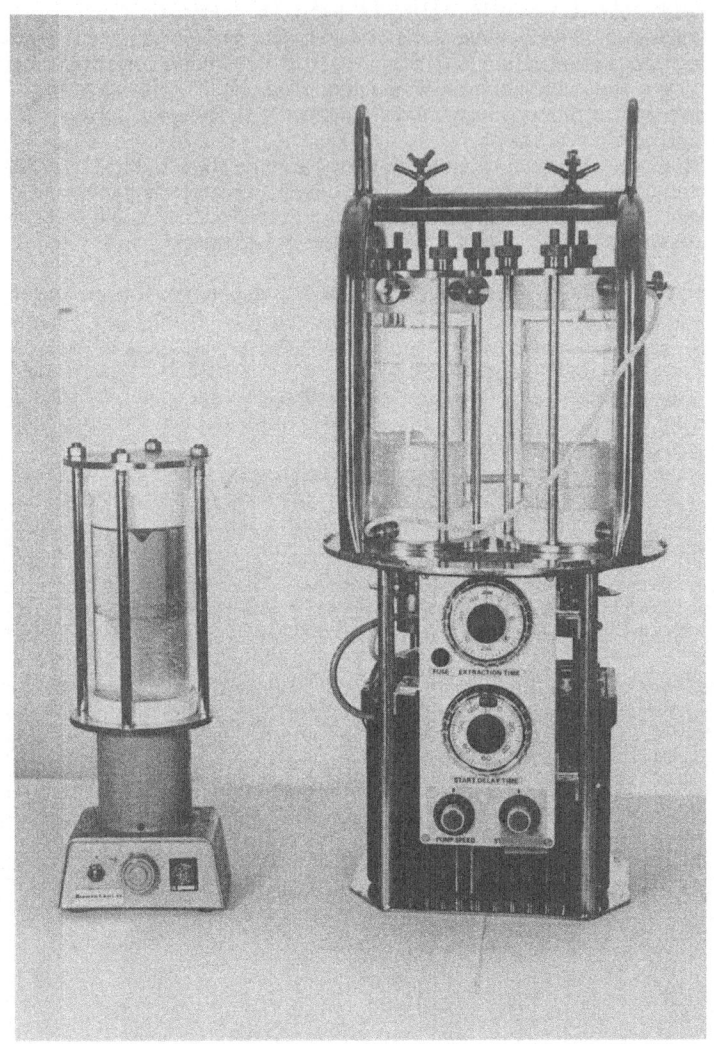

Fig. 2. Continuous in situ solvent extraction apparatus with two
extractor units in series. The electrical part is
uncovered. To the left an extraction unit for solvent
heavier than water.

REFERENCES

(1) T. Almgren, B. Josefsson and G. Nyquist, Anal.Chim.Acta, 78(1975)411
(2) G. Eklund, B. Josefsson and C. Roos,
     J.High Resol.Chromatogr.Chromatogr.Comm., (1)(1978)34
(3) A.A. Nicholson and O. Meresz, Bull.Environ.Contam.Toxicol.,
     14(1975)453
(4) A.A. Nicholson, O. Meresz and B. Lemyk, Anal.Chem., 49(1977)814
(5) S. Folestad, L. Johnson, B. Josefsson, S. Lagerkvist, P. Lindroth,
     G. Nyquist, Proceedings, European Symposium on Micropollutant in
     Water, Berlin FRG 1979
(6) E. Fogelqvist, B. Josefsson and C. Roos,
     Environ.Sci.Technol., submitted
(7) E. Fogelqvist, B. Josefsson and C. Roos,
     J.High Resol.Chromatogr,Chromatogr.Comm., 3(1980)568
(8) P. Schauwecker, R.W. Frei and F. Erni, J.Chromatogr., 136(1977)63
(9) K.A. Pinkerton, J.High Resol.Chromatogr.Chromatogr.Comm., 4(1981)33
(10) R. Kummert, E. Molnar-Kubica and W. Giger, Anal.Chem., 50(1978)1637
(11) R.W. Frei and U.A.Th. Brinkman, Trends in Anal.Chem., 1(1981)45
(12) H.P.M. Van Vliet, Th.C. Bootsman, R.W. Frei and U.A.Th. Brinkman,
     J.Chromatogr., 185(1979)483
(13) V.E. May, S.P. Vasik and D.H. Freeman, Anal.Chem., 50(1978)997
(14) T. Fuji, J.Chromatogr., 139(1977)297
(15) P.G. Simmonds and E. Kerns, J.Chromatogr., (1979)863
(16) R.T. Coutts, E.E. Hargesheimer and F.M. Pasutto,
     J.Chromatogr., 179(1979)291
(17) R.H. Voss, J.T. Wearing and A. Wong, "Advances in the Identification
     and Analysis of Organic Pollutants in Water II"
     L.H. Keith ed. Ann Arbor Sci.Pub.,Ann Arbor, 1981
(18) R.W. Frei and J.F. Lawrence, J.Chromatogr., 83(1973)321
(19) R.G. Melton et al., "Advances in the Identification and Analysis of
     Organic Pollutants in Water II"
     L.H. Keith ed. Ann Arbor Sci.Pub.,Ann Arbor, 1981
(20) K. Grob, K. Grob Jr. and G. Grob, J.Chromatogr., 106(1975)299
(21) G. Eklund, B. Josefsson and C. Roos, Vatten 3(1978)195
(22) W.J. Khazal, J. Vejrosta and J. Novak, J.Chromatogr., 157(1978)125
(23) M. Dressler, J. Chromatogr., 165(1979)167
(24) P.R. Musty and G. Nickless, J.Chromatogr., 120(1976)369
(25) D.A. Kurtz, Bull.Environ.Contam.Toxicol., 17(1977)391
(26) J.B. Derenbach, M. Ehrhardt, C. Osterroth and G. Petrick,
     Mar.Chem., 6(1978)351
(27) M. Ahnoff and B. Josefsson, Anal.Chem., 46(1974)658
(28) M. Ahnoff and B. Josefsson, Anal.Chem., 48(1976)1268
(29) M. Ahnoff and B. Josefsson, Ambio 4(1975)172
(30) M. Ahnoff, G. Eklund and B. Josefsson,
     Acta Hydrochim.Hydrobiol., 7(1979)171
(31) B. Stachel et al., Anal.Chem., 53(1981)1469
(32) M. Ahnoff, B. Josefsson, G. Lunde and G. Andersson,
     Water Res., 13(1979)1233

A NEW METHOD FOR THE QUANTITATIVE ANALYSIS OF

ORGANOCHLORINE PESTICIDES AND POLYCHLORINATED BIPHENYLS

M. GODEFROOT, M. STECHELE, P. SANDRA and M. VERZELE
Laboratory for Organic Chemistry, State University of Ghent,
Krijgslaan, 271 (S.4), B-9000 GENT (Belgium)

Summary

A simple method is described for the quantitative determination of organochlorine pesticides and polychlorinated
biphenyls (PCB's) in water at the sub ppb level. A micro
gas-phase extractor advantageously replaces other preconcentration and purification techniques. The extract is
analysed by capillary gas chromatography without further
enrichment. The recovery at the ppb level was nearly 100 %
for organochlorine pesticides and more than 80 % for PCB's.
The total procedure including sample preparation, steam
distillation-extraction and capillary gas chromatographic
analysis is carried out in less than four hours.

INTRODUCTION

Several methods are used for the isolation of organic compounds and pollutants in water e.g. solvent-solvent extraction
(1-7), adsorption on XAD-resin (8,9) or polyurethane foams (10),
direct head space analysis, gas stripping followed by concentration in a cold trap or on a porous polymeric adsorbent such
as Tenax GC (2,11,12) or on a charcoal trap (13). For exhaustive enrichment, solvent-solvent extraction is mostly used. This
technique is laborious and needs concentration of the solvent
by evaporation. This is time-consuming and can cause severe
losses (6,9,14). Even if the purification of the extraction
solvent is properly done and the concentration step is carefully performed, solvent impurities disturb to a large extent
the analysis (9,14). Micro extraction procedures (3,6) eliminate these problems but are selective and hence not quantitative. By solvent-solvent extraction non volatile compounds are
extracted as well. During gas chromatographic analysis they
contaminate the inlet part of the column. This phenomenon is
detrimental to the separation efficiency of a capillary column.
Moreover, decomposition of the non-volatile material causes
ghost peaks disturbing the analysis. The other methods mentioned are difficult to standardize due to incomplete and selective enrichment, which moreover strongly depends on several parameters. Steam distillation, one of the oldest methods for
enrichment, is not very often used in environmental research.
The extraction efficiency, especially for more polar and less
volatile compounds, of this technique which in fact is an enrichment by evaporation, is however much better compared to
conventional head space or gas stripping procedures. A disadvantage of steam distillation is that the distillate has to be
extracted with a suitable solvent and that evaporation of the
extraction solvent is required. Recently we transformed the

continuous gas phase extraction apparatus of Likens and Nickerson (15,16) into a microversion for essential oil extraction
with methylene chloride (17). By some modifications the micro
device can also be used for pentane gas phase extraction of low
concentration volatiles.
In the present contribution we show the possibilities of this
micro apparatus for organochlorine pesticide and polychlorinated biphenyls residue analysis in the sub ppb level.
      The micro gas-phase extractor replaces advantageously
techniques using chromatographic preconcentration and purification. The continuously generated steam distillate of the material (10 to 100 ml water; 1-20 g solid material (soil, cattle
feed etc.) blended with water) is continuously extracted with
n-pentane, which is an efficient extraction solvent for organochlorine pesticides and polychlorinated biphenyls and does not
disturb ECD analysis. The construction of the apparatus is
such that all the volatile material is collected in 1 ml n-pentane. The main advantage of the system is the high recovery in
a very short time even if only small amounts of contaminated material are available. This is due to the very effective enrichment in the small amount of extracting solvent (1 ml), requiring
no further enrichment by evaporation. The n-pentane solution is
then used as such for capillary gas chromatography in combination with ECD-detection.

EXPERIMENTAL

Continuous steam distillation (SD) - continuous liquid-liquid
extraction (E) micro apparatus (Fig. 1)

The water sample is placed in a 50 to 250 ml flask (A).
For the analysis of solid material, water is added so that the
material is sufficiently wet. A 1 ml volume of n-pentane containing a suitable internal standard is introduced in B, having
a content of 2 ml. Cleaned boiling chips are added to A and B.
Before starting the procedure approximately 1.5 ml of water and
1.5 ml of n-pentane are introduced into C with a syringe.
This is the amount needed to fill the return arms (D and E)
and to install (settle) a demixing equilibrium between the two
solvent layers in C. The cold finger in which ice water is
circulating is introduced and the pentane reflux is started by
placing B in a water bath at 80°C. After 5 min steam is generated by applying heat to flask A with an oil bath at 140°C.
The vapour channels F and G are isolated. The starting point
of the procedure is the moment when steam enters the central
part of the apparatus. During the whole procedure the temperature of 140°C is maintained. The vapours are condensed in C
by the cold finger. The construction of the apparatus is such
that the low density layer (n-pentane) returns through arm D
to B; the high density layer (water) returns through arm E to
A.
      After the enrichment is completed - normally 1 h is suffi-
cient - the steam generation is stopped while the solvent extraction is continued for 20 min more. In this way all the
steam distillable material is collected in ca 1 ml of n-pentane.
During the extraction some n-pentane from the demixing part
may get into flask B or eventually a small amount of the sol-

vent may be lost through evaporation. This does not matter
with the internal standard used as described. Subsequently 1
to 5 µl of the n-pentane solution are analysed as such by capillary gas chromatography.

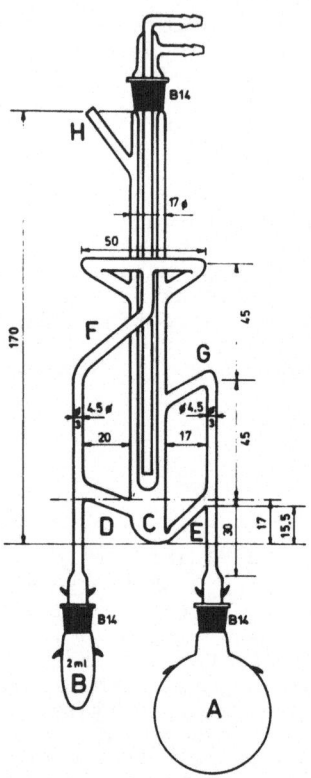

Fig. 1 : Scale drawing in millimeters of the micro steam dis-
         tillation-extraction (SD-E) apparatus (available from
         Alltech Europe, Eke, Belgium).

Capillary gas chromatography with electron capture detection

A Varian 3700 gas chromatograph equipped with a [63]Ni ECD was
used in this study. The samples were analyzed on a 30 m x
0.3 mm I.D. HTS-OV 1 glass capillary column (18). The organochlorine pesticide mixtures were introduced in the column with
splitting (1:10) in a home-built all-glass split device filled
with deactivated glass wool. The hot needle injection technique was applied (17,19). For the PCB mixtures a moving needle
injector equipped with water cooling was used (20,21). Peak

areas were measured with a Varian CDS 111 electronic integrator. The recovery was determined with the formula :

$$\% \ R = \frac{\left[\dfrac{\text{Peak area compound x}}{\text{Peak area internal standard}}\right]_{SD-E}}{\left[\dfrac{\text{Peak area compound x}}{\text{Peak area internal standard}}\right]_{reference}} \times 100$$

RESULTS AND DISCUSSION

Recovery for organochlorine pesticides in the ppb range

The quantitative aspect of the recovery was checked with a mixture of organochlorine pesticides added to 50 ml water. The composition and concentrations are given in table 1. 25 ng heptachlor was added as internal standard to bottle B. Steam distillation - solvent extraction was carried out during 1 h and the result is shown in fig. 2b. The same amount of pesticides and internal standard was dissolved in 1 ml n-pentane. The analysis of this reference mixture is shown in fig. 2a.

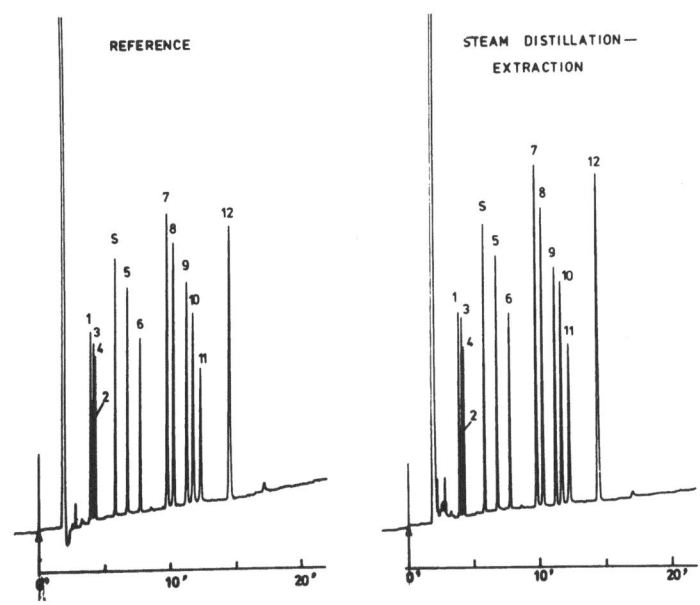

Fig. 2a and b : 2 µl pentane solution - split injection (1:10) with hot needle technique.
Column : 230°C, injector : 270°C, detector : 360°C. Carrier : Ar:CH$_4$ (90:10) : 1 ml min^{-1}.
Make-up gas : Ar:CH$_4$ (90:10) : 30 ml min^{-1}.

The total recovery and the recovery of the individual compounds for two experiments are given in table 1. A high recovery is obtained in a short time (1 h) and there is no advantage in proceeding longer. The experiments were carried out with O.4 to 4 ppb concentrations of pesticides and split injection (1:10) was used. By using on column injection or the moving needle injection technique 10-100 ppt can easily be detected. If higher sensitivity is desired concentration of the n-pentane layer to about 200 µl can be achieved in a Kuderna-Danish concentrator - micro Snyder column device. In this way 2 to 20 ppt can be detected.

Recovery for polychlorinated biphenyls (PCB's)

As a lower amount of HCB compared to the other organochlorine compounds was recovered, polychlorinated biphenyls were subjected to the same procedure. Water spiked with Arochlor 1260 in a concentration of 10 ppb was extracted with SD-E. 10 ng lindane was added as internal standard to bottle B (fig. 1). The chromatograms are given in fig. 3; the recovery data in table 2. Only the numbered peaks are Arochlor compounds. As can be deduced from table 2 the recoveries for polychlorinated aromatics are very high.

As can be deduced from the results obtained, the technique allows the analysis of pesticides in the sub ppb level. The total procedure, including sample preparation, steam distillation-extraction and capillary gas chromatographic analysis, is carried out in less than four hours.
One of the main advantages for pesticide analysis is that clean-up procedures before gas chromatographic analysis are avoided. The method also is especially suitable when low amounts of material are available. The technique was successfully applied in environmental studies. Lindane was determined in waste water in the 100 ppt to 30 ppb level, from which only 25 ml water samples were available.
Pesticide analysis in water was selected to show the possibilities of the technique. Pesticides in non aqueous samples such as soil, cattle feed, vegetables, fish tissues etc. can be analyzed as well in the same way.
The technique is very versatile and at present the possibilities to enrich halo-ethers, haloforms, hydrocarbons, polycyclic aromatic hydrocarbons, etc., are investigated.

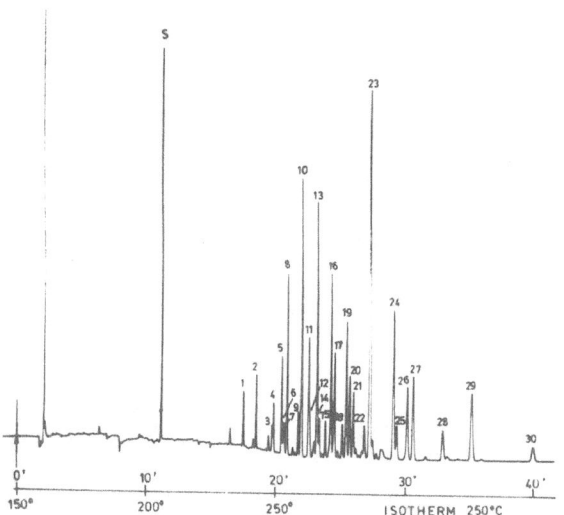

STEAM DISTILLATION — EXTRACTION

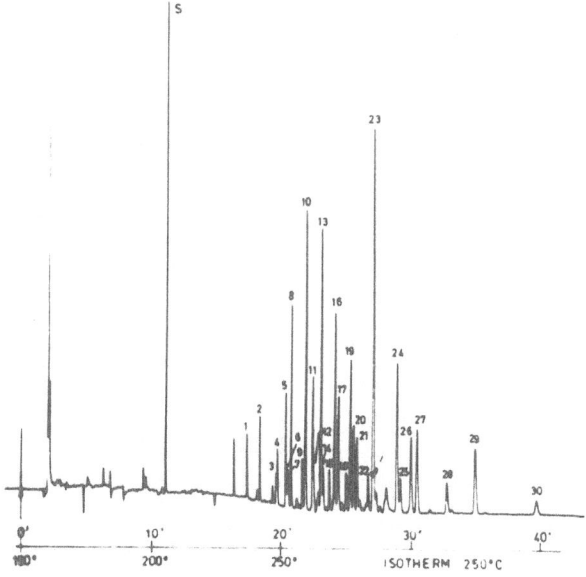

Fig. 3 : 2.5 µl pentane solution – Arochlor 1260 – with moving
needle injector at 270°C; detector : 300°C;
column : 150°C $\xrightarrow{5°/'}$ 250°C

Table 1 : Recovery of the method for the test compounds

peak nr	compound	amount in 50 ml water (ng)	conc. (ppb)	recoveries (%)		
				exp. 1	exp. 2	mean
1	HCH-isomer	20	0,4	100,8	101,7	101,3
2	HCB	20	0,4	80,4	85,2	82,8
3	HCH-isomer	20	0,4	104,9	107,6	106,3
4	lindane	25	0,5	95,6	110,4	103,0
5	aldrin	25	0,5	103,2	104,5	103,9
6	heptachlor-epoxide	50	1	102,0	98,3	100,2
7	p,p'-DDE	50	1	101,4	95,2	98,3
8	dieldrin	100	2	98,9	97,5	98,2
9	endrin	100	2	103,1	99,8	101,5
10	o,p'-DDT	100	2	98,4	96,9	97,7
11	DDD	100	2	98,5	94,8	96,7
12	p,p'-DDT	200	4	98,4	105,1	101,8
	Total	810	16,2	99,9	99,4	99,7

Table 2 : Recovery for PCB's

peak nr	recoveries			peak nr	recoveries		
	exp. 1	exp. 2	mean		exp. 1	exp. 2	mean
1	96,6	88,9	92,8	16	90,6	96,6	93,6
2	90,8	100,4	95,6	17	92,5	95,1	93,8
3	103,0	98,2	100,6	18	95,5	99,9	97,7
4	98,5	101,5	100,0	19	90,3	95,9	93,1
5	96,5	102,6	99,5	20	89,1	95,9	92,5
6	95,8	111,4	103,6	21	98,8	96,2	97,5
7	110,3	94,6	102,5	22	87,6	94,0	90,8
8	94,5	99,7	97,1	23	88,4	92,4	90,4
9	95,5	103,2	99,4	24	85,8	90,5	88,2
10	93,7	98,8	96,3	25	84,8	93,8	89,3
11 12	90,1	101,1	95,6	26	86,1	87,1	86,6
				27	86,5	91,2	88,9
13	94,0	97,6	95,8	28	80,7	87,2	83,9
14	91,1	99,7	95,4	29	80,2	86,1	83,1
15	97,9	103,6	100,8	30	79,1	82,7	80,9
				Total	94,7	95,1	94,9

REFERENCES

1. F. Drawert, A. Rapp, Chromatographia, 1968, 1, 446-457.
2. J.P. Mieure and M.W. Dietrich, J. Chromatogr. Sci., 1973, 11, 559-570.
3. K. Grob, K. Grob Jr. and G. Grob, J. Chromatogr., 1975, 106, 299-315.
4. G. Eklund, B. Josefsson and C. Roos, JHRC & CC, 1978, 1, 34-40.
5. W.A. Hoffman Jr., Anal. Chem., 1978, 50 (14), 2158-2159.
6. D.A.J. Murray, J. Chromatogr., 1979, 177, 135-140.
7. A. Rapp, W. Knipser, Chromatographia, 1980, 13 (11), 698-702.
8. A.K. Burnham, G.V. Calder, J.S. Fritz, G.A. Junk, H.J. Svec and R. Willis, Anal. Chem., 1972, 44, 139-142.
9. G.A. Junk, J.J. Richard, M.D. Grieser, D. Witiak, J.L. Witiak, M.D. Arguello, R. Vick, H.J. Svec, J.S. Fritz and G.V. Calder, J. Chromatogr., 1974, 99, 745-762.
10. H.D. Gesser, A.B. Sparling, A. Chow and C.W. Turner, J. Amer. Waterworks Assoc., 1973, 65, 220.
11. A. Zlatkis, H.A. Lichtenstein, A. Tishbee, Chromatographia, 1973, 6 (2), 67-70.
12. M. Novotny, M.L. Lee, K.D. Bartle, Chromatographia, 1974, 7 (7), 333-338.
13. K. Grob, J. Chromatogr., 1973, 84, 255-273.
14. J.A. Burke, P.A. Mills and D.C. Bostwick, J.A.O.A.C., 1966, 49 (5), 999-1003.
15. S.T. Likens and G.B. Nickerson, Proc. Am. Soc. Brew. Chem., 1964, 5-13.
16. G.B. Nickerson and S.T. Likens, J. Chromatogr., 1966, 21, 1-5.
17. M. Godefroot, P. Sandra and M. Verzele, J. Chromatogr., 1981, 203, 325-335.
18. M. Godefroot, M. Van Roelenbosch, M. Verstappe, P. Sandra and M. Verzele, JHRC & CC, 1980, 3, 337-343.
19. K. Grob Jr. and H.P. Neukom, JHRC & CC, 1979, 2, 15-21.
20. M. Godefroot and P. Sandra, unpublished results.
21. M. Verzele, G. Redant, S. Qureshi and P. Sandra, J. Chromatogr., 1980, 199, 105-112.

CONCENTRATION AND IDENTIFICATION OF THE MAIN
ORGANIC MICRO-POLLUTANTS CLASSES IN WATERS

A. SDIKA - R. CABRIDENC - C. HENNEQUIN

IRCHA - 91710 Vert-le-Petit- France

SUMMARY

During a previous symposium (Berlin Décembre 1979) a "total extractive technique" has been described, giving the majority of the organics (except volatile compounds).

After choosing a list of organic compounds and chemical classes which acute or chronic toxicity is well known, an analytical scheme is presented to analyse 13 classes.

After a first clean up, the total extract is separated in three fractions (acid, base and neutral). Each one is tested by TLC arrangements, versus known standards. The biggest spots are scratched off for confirmation (bi-dimensional TLC).

The semi-quantitative evaluation allows an assesment of the potential toxicological risk (Quality index).

By this way, we can compare different waters and treatment efficienties.

L'étude de la qualité des eaux et notamment la détermination de leur teneur en micropolluants organiques constituent, depuis de nombreuses années, une des préoccupations majeures des hygiénistes de tous les pays. L'identification de ces micropolluants présente, en outre, un intérêt indéniable dans de multiples buts :

. identification des origines des micropolluants.

. orientation vers la mise en oeuvre de filières de traitement spécialement adaptées à l'élimination de certaines classes de micropolluants organiques.

. mise en évidence de corrélation entre les effets biologiques (effets mutagènes, effets cytotoxiques, effets toxiques) de concentrats ou d'extraits obtenus à partir d'eaux et les substances chimiques présentes dans ces concentrats ou extraits.

Lors du 1er Symposium européen sur l'analyse des micropolluants organiques des eaux, nous avons eu l'occasion de présenter une technique d'extraction des micropolluants organiques applicable aux eaux. Cette technique est, rappelons le, basée sur la séquence fixe suivante :

. extraction par le chloroforme en milieu neutre.

. extraction par le chloroforme en milieu alcalin.

. extraction par le chloroforme en milieu acide.

. adsorption sur résine Amberlite IRA 68 suivie d'élution par solutions et réextraction par solvants.

. adsorption sur résine Amberlite XAD^2 suivie d'élution par solvants.

. adsorption sur charbon actif suivie d'élution par solvants.

Il est possible d'obtenir un extrait total, utilisé pour la réalisation parallèle d'études biologiques comparées et analytiques.

En fait, compte tenu des processus mis en jeu, au cours de l'extraction, les composés les plus volatils (haloformes, alcools légers) sont éliminés au cours des évaporations. Toutefois, pour la commodité, on retient le terme "extrait total" par opposition à des fractionnements ultérieurs.

Dans le but d'approcher la composition chimique de ces extraits, il nous a paru souhaitable d'envisager des recherches ayant pour objectifs, la mise en évidence et la quantification sommaire d'une liste de classes de micropolluants organiques considérés comme prioritaires du fait de leurs propriétés bioaccumulatives et toxiques à court ou à long terme.

. Amines aromatiques

. Azocycles

. Bioamines

. Chlorophénols

. Hydrocarbures chlorés

. Nitrosamines

. Oestrophénols

. Dérivés organomercuriels

. Phénols semi-polaires

. Phosphates organiques

. Phtalates

. Hydrocarbures polycycliques

. Uréines et Carbamates

. Dérivés organiques nitrés neutres.

I - PROTOCOLE EXPERIMENTAUX

I - 1. Principes généraux

Depuis quelques années, le développement de méthodes modernes (CG/SM -
HPLC...) a permis d'identifier de nombreux constituants présentant des
constitutions et structures chimiques diversifiées. Environ 400-500 constituants ont été détectés dans des eaux de provenances différentes. On
s'accorde à reconnaître que celà ne constitue qu'une fraction des matières
organiques présentes (de l'ordre de 10 %).

Il nous est apparu souhaitable de disposer d'un système analytique unitaire
plus simple et dont l'application systématique pouvait être envisagée paral-
lèlement à des analyses multi-techniques.

Le but visé étant d'établir des comparaisons entre des échantillons différents.

La chromatographie sur couche mince opérée sur des fractions de l'extrait
total a semblé la technique la plus polyvalente. Les méthodes retenues ont
été sélectionnées à la suite d'essais multiples de laboratoire réalisés en
tenant compte des informations scientifiques publiées dans ce domaine.

Le schéma analytique général est décrit en figure I.

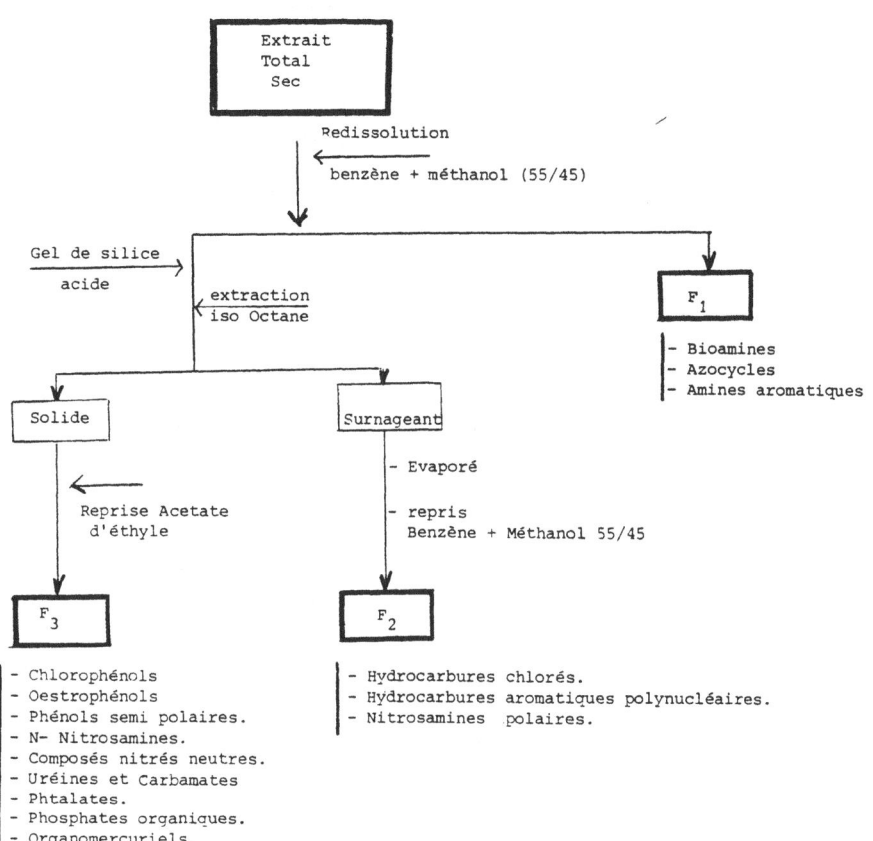

FIGURE I

SCHEMA ANALYTIQUE UNITAIRE

Ces pré-fractionnements ont pour but d'améliorer la sélectivité analytique en éliminant des "faux-positifs". L'analyse CCM est réalisée en deux temps sur chaque fraction F_1, F_2, et F_3.

- Les mises en évidence des classes chimiques sont réalisées dans des systèmes de développement et révélation d'étroite spécificité. L'estimation semi-quantitative est réalisée par référence à des témoins fixes en quantités connues en faisant appel à un système codé d'appréciation visuelle des aires relatives.

Les limites de détection, fonction des systèmes révélateurs et des composés peuvent atteindre 0,025 à 5μg par dépôt.

- Les résultats "positifs" sont ensuite retestés dans un système chromatographique différent pour confirmation.

- <u>Disposition systématique sur plaque</u>

- 8 à 9 échantillons d'extraits à tester déposés à raison de 500 μg/ 5μl.
- 1 Témoin "contaminants" (Essais à blanc de toute la chaîne extractive) à raison de 125 μg/5 μl.
- 1 Témoin de résolution chromatographique et de révélation comportant 4 à 7 composés/5μl
- 1 Standard quantitatif faisant partie des précédents et figurant 2[*] (équivalent à limite de détection x 2)
- 1 Standard quantitatif figurant 3[*]

Les détails du code d'évaluation pondéral seront développés plus loin.

I - 2 <u>Obtention de F_1, F_2, et F_3</u>.

L'extrait total sec est repris par le mélange benzène/méthanol (55/45) à raison de 10 % (500μg/5μl). Il faut 7,5 mg à 15 mg par échantillon dissous dans 150μl de solvant.

- une aliquote constitue directement F_1.

Une adsorption sur gel de silice acide activée ($\leqslant$ 80 μ) permet l'accès à F_2 et F_3.

Après trituration avec la silice et évaporation sous courant d'azote sec, le support est extrait par 2 volumes d'iso-octane ($\sim$ 2 ml).

Après centrifugation, la fraction iso-octane, évaporée, reprise par le mélange benzène-méthanol, constitue la fraction F_2.

Le gel de Silice repris par l'acétate d'éthyle conduit à F_3.

I - 3 <u>Analyses chromatographiques en couche mince</u>.

Le principe général repose sur l'application d'une première analyse (dite système général).

Dans le cas de résultats positifs, on pratique une confirmation avec un autre système, ce qui permet d'augmenter la fiabilité globale.

Les mélanges d'élution et de révélation font essentiellement appel aux formules décrites par : Egon Stahl. (TLC chromatography 2ème édition).

Les plaques de silice sont préparées au laboratoire à partir de silice Merck granulométrie 15μ .

Les autres supports proviennent de Macherey Nagel (plaques prêtes à l'emploi).

EXPLICATION DES TABLEAUX DE CLASSES :

Témoin :   composés destinés au contrôle du système chromatographique.

                le produit souligné sert à l'estimation quantitative.

Quantité déposée : Le volume déposé est fixé à $5 \mu l$.

                Les composés phénolés (F_3) necessitent deux chromatogrammes identiques quant aux phases fixes et mobiles mais à révélation double.

SYSTEME GENERAL — CONFIRMATION

	Témoin	Quantité µg/5µl	H.Rf ± 5%	Phase Fixe	Phase mobile	Révélateur	CCM	Révélateur
BIO AMINES	Tyramine,Hcl	1	60	Cellulose MN	Solvant Wolwood H_2O-125 Butanol 125 Acide Acétique 30	- Ninhydrine 1% - Acétate Cd 1%	Même système CCM	Réactif d'Ehrlich.
	Ethylamine,Hcl	"	45					
	Methylamine,Hcl	"	40					
	1,6 Hexane diamine	"	30					
	1,5 Pentane diamine	"	25					
	Histamine, Hcl.	"	20					
AMINES AROMATIQUES	α.Naphtylamine	2	60	Silice G	Hexane 70 Acétone 30	- Ninhydrine 0,5% dans alcool butylique + Acide acétique	Silice G - Benzêne 95 - Méthanol 5	Réactif d'Ehrlich.
	p.Chloraniline	"	50					
	P.Toluidine	"	45					
	P.Anisidine	"	35					
	Benzidine	"	20					
AZOALCALO CYCLES	Acridine	1	80	Alumine neutre Merck	Cyclohexane 75 $CHCl_3$ 20 Méthanol 4 Diéthylamine 1	Iodo Platinate de potassium Fixation à H_2SO_4	Benzêne 95 Méthanol 5 Diéthylamine 1	Tétraphényl borate de Sodium et Quercetine
	Quinoléine	"	75					
	Strychnine	"	40					
	Brucine	"	35					
	Cinchonidine	"	25					
	Atropine	"	20					
	Piperidine	2	15					
AZOCYCLES	Oxine	2	60	Silice G	Benzêne 90 Acetone 10	- UV 254 nm - Dragendorf	Silice G Benzêne 95 Méthanol 5	Même révélateur que Système Général.
	Prometryne	"	55					
	Propazine	"	50					
	Atrazine	"	40					
	Simazine	"	35					
	α. picoline	"	20					

AZO CYCLES

FRACTION F 2

SYSTEME GENERAL — CONFIRMATION

	Témoin	Quantité µg/5µl	H-Rf ±5%	Phase Fixe	Phase Mobile	Révélateur		Révélateur
HYDROCARBURES CHLORES	Hexachlorobenzène	2	90	Silice G	Ether Pétrole 94 dioxane 5 Huile de paraffine 1	Réactif de Mitchell + UV 254 nm	Silice G	Diphénylamine 1 % dans Ethanol + UV 254 nm.
	Aldrine	"	70					
	Heptachlor	"	65					
	pp' DDT	"	50				n. hexane	
	HCH	"	20					
	Dieldrine	"	10					
HYDROCARBURES AROMATIQUES POLYNUCLEAIRES	Fluorantène	0,05	60	Acétate de Cellulose		Fluorescence UV 254 nm	Acétate de Cellulose Solvant de Schaad	Fluorescence UV 254 nm.
	3 Methyl-Cholanthène	"	55					
	3-4 Benzo-Fluoranthène	"	40					
	Indeno pyrène	"	35		Solvant de Wieland			
	3-4 Benzo Pyrène	"	30					

FRACTION F3

	Témoin	SYSTEME GENERAL					CONFIRMATION	
		Quantité μg/5μl	H.Rf ±5%	Phase Fixe	Eluant	Révélateur		
CHLORO-PHENOLS	Chloranil	2	70	Silice G	i.octane 75 Acétate d'éthyle 25 Acide acétique 2	Révélation symétrique Folin !Tolidine +vap !Ethanol NH3 !amine !1% Me OH !+ UV 254nm	Silice G Chloroforme 4 Cyclohexane 25 Acide acétique 1	Révélation symétrique FeCl3 !Réactif de Ferry !Mitchell Cyanure K !+ UV
	Fentachlorophénol	"	50					
	2.4.5.Trichlorophénol	"	45					
	p.Chlorophénol	"	40					
	Hexachlorophène	"	35					
	2.4 Dichlorophénoxy-acétique							
OESTROPHENOLS	Hexoestrol	2	75	Silice G	Cyclohexane 3 Acetate ethyle 1	Révélation symétrique Folin !Vanilinne + !1% NH3 !H3 PO4 50% vapeur	id Système général	Naphtoquinone Sulfonate 1% H2 O +3 H PO4 - 4.N chauffage 110°. 20min.
	Oestrone	"	65	G				
	Diethylstilboestrol	"	60	acide				
	Oestradiol	"	50					
PHENOLS SEMI-POLAIRES	B - H - T	1	80	Silice G	i.octane 3 acétate ethyle 1	Réactif de Folin + vapeur NH3	id Système Général	Fe Cl3 1% dans H Cl 2 N + Ferricyanure K 2%
	O. cresol	"	55	G				
	Phényl-Phénol	"	50	Acide				
	Résorcinol	"	10					
NITROSAMINES	Diphényl nitrosamine	2	75	Silice G	Hexane 66 Ether 20	UV 254- 30 min. puis Acide sulfonilique à 0,8% α Naphtylamine	id Système Général	Pd Cl2 Diphénylamine UV 254nm. 20 minutes.
	Dibutyl	"	60	G				
	Dipropyl	"	50		CH2 CL2 14			
	Nitrosopiperidine	"	36					
	Nitroso-methyl nitro guanidine	"	18					

FRACTION F₃ (Suite)

	Témoin	SYSTEME GENERAL Quantité µg/5 µl	H.Rf ± 5%	Phase fixe	Eluant	Révélateur	CONFIRMATION	Révélateur
UREINES ET CARBAMATES	Propham	5	70	Silice	n. Hexane 70	p. dimethylamino benzaldehyde	Alumine	id
	Promecarb	"	65		Acetone 30	0,5 % Méthanol	CH_2 Cl_2	Système Général
	Carbaryl	"	45	G		H_2 SO_4 4 N		
	Chlortoluron	"	30	Acide				
	Phenuron		20					
PHTALATES	Dioctyl phtalate	10	60	Silice	n. Hexane 3	Thymol à 20 % dans Ethanol	Silice G	id
	Dibutyl phtalate	"	40		Acétate d'ethyle 1	+ H_2 SO_4 4 N	CH_2 Cl_2	Systeme General
	Diethyl phtalate	"	20	G		+ chauffage		
	di methyl phtalate	"	10			120° - 30 min.		
PHOSPHATES	Tricresyl phosphate	5	35	Silice	n. Hexane 3	- UV 254 15 min.	Silice G	- UV 254 30 min.
	Triphenyl phosphate	"	30	G	Acétate d'éthyle 1	réactif Hanes I vol.	Cyclohexane 3	- reactif Hanes
	Tributyl phosphate	"	20			+ cristal violet 0,1 % 3 - 3 vol.	Acetate Ethyle 2	110 ° 15 min.
ORGANO MERCURIELS	Acetate phenyl Hg	2	70	Silice	i.Octane 3	Dithizone 0,5%		
	Chlorure d'éthyl Hg	"	40	G	Acetate Ethyle 1	CH_2 Cl_2		
	Hydroxyde Methyl Hg	"	40					
COMPOSES NITRES NEUTRES	Pentachloronitrobenzène	5	88	Silice -Gr	Acetate Ethyle 20	Fluorescence Na 0,2% + diethylène triamine 1%	Silice G acide	id
	2.4.5-Trichloronitro-benzène	"	78				Cyclohexane 95	Système -Général
	2.4. Dinitro.chloro-benzène	"	42	acide	Cyclohexane 80		Acetate Ethyle 5	
	m. dinitro benzène	"	35					
	9. nitroanthracène		57					

II - CONTROLE DE L'EFFICACITE - EXPRESSION DES RESULTATS

Lors du précédent symposium, le contrôle d'efficacité (dopage d'un échantillon extraction totale et analyse) avait porté sur une cinquantaine de composés appartenant aux diverses classes examinées. Ces essais ont été repris sur 140 constituants à concentration aqueuse égale à celle des étalons 2 + et ont confirmé l'applicabilité de la technique (90 % des produits récupérés).

Un des objectifs de ce schéma analytique était de servir de base à des comparaisons entre eaux, efficacité de filières de traitement, études épidémiologiques. Il était nécessaire de regrouper les estimations semi-quantitatives des spots révélés en un paramètre unitaire caractéristique de l'extrait de micro polluants.

Cet"indice de richesse", ou Indice global de micropollution I G M, est défini comme suit :

- Evaluation semi quantitative codée :

  On définit :

  2^* = réponse de l'étalon de chaque classe (Témoin 1)

  3^* = réponse de 50 fois la quantité d'étalon (Témoin 2)

Par comparaison avec ces étalons, on attribue à chaque tache détectée les valeurs suivantes :

  0^* = non détecté.

  1^* = limite de détection ; tache inférieure à celle du Témoin 1.

  2^* = tache sensiblement égale au Témoin 1.

  3^* = tache   "   "   Témoin 2.

Pour chaque dépôt, on totalise le nombre de croix. Cette valeur, déduction faite de l'indice dû aux contaminants, constitue pour un extrait de micropolluants son indice de richesse.

- Dans le cas où les dépôts correspondent à un même poids d'extrait, on peut directement comparer les indices de richesse et donc la composition intrinsèque des extraits.

On peut également rapporter ces indices au litre d'eau en tenant compte de la teneur de l'eau en extrait étudié.

- Dans le cas où les dépots correspondent à un volume d'eau, on peut également comparer les indices de richesse et donc la qualité des eaux. Cependant, dans le cas des eaux pour lesquelles les teneurs en micropolluants organiques extractibles sont très faibles, certaines classes ne peuvent être mises en évidence du fait des limites de détection des techniques utilisées. On peut également , tenant compte de la teneur de l'eau en micropolluants organiques rapporter les indices à une même masse d'extrait.

EXEMPLE :
_____

Classe "Chlorophénols"

Elution de la plaque CCM "Chlorophénols".

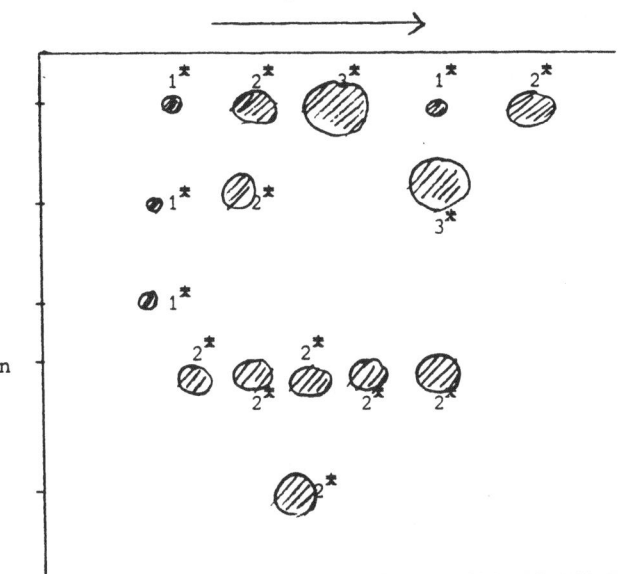

500 µg F$_3$ d'extrait (E$_1$)

    d'une eau (I)

500 µg F$_3$ d'extrait (E$_2$)

    d'une eau (II)

"contaminants"

Témoin résolution/révélation
(2µg de chaque constituant
type)

Standard quantitatif
2 µg pentachlorophénol.

Témoin   : 2^*
Contaminants : 1^*
E1   : $9^* - 1^* = 8^*$
E2   : $6^* - 1^* = 5^*$

Si, dans le cas de l'eau I, l'extrait E1 est de 3 mg/l l'indice de richesse est :

$$\frac{8 \times 3}{0,5} = 48^*$$

De même si la teneur de l'eau II en extrait E$_2$ est de 0,5 mg/l on obtient :

$$\frac{5 \times 0,5}{0,5} = 5^*$$

D'autres mises en évidence de tendances, par une exploitation différente des
résultats sont possibles.

III - CONCLUSION

Le schéma présenté et son exploitation volontairement simplifiée constituent
un outil utilisé dans nos laboratoires depuis plusieurs années. Des amélio-
rations successives apportées, permettent maintenant de mettre en évidence des
familles chimiques dont le caractère gênant ou toxique est largement reconnu.

Ce système conventionnel a été appliqué dans divers domaines, notamment :

- en vue de l'optimisation des filières de traitement d'eaux de surface
  pour l'obtention d'eaux destinées à la consommation humaine. On a
  pu mettre en évidence une augmentation, dans les extraits, du nombre
  de dérivés organo-chlorés du fait des traitements de préchloration.

- étude épidémiologique (effectuée sur contrat C.E.E.) dans le but
  de mettre en évidence des corrélations entre la qualité chimique
  des eaux (classes principales de micropolluants) et les effets sur
  la santé des consommateurs.

Ces techniques simples et peu coûteuses (2 jours pour la réalisation du schéma
analytique) donnent des résultats dans des délais courts. Toutefois, du fait
d'une sensibilité relativement limitée, elles ne peuvent prétendre se substituer
à des techniques plus élaborées, telles que celles basées sur des méthodes plus so-
phistiquées (CG/SM - H.P.L.C....).

BIBLIOGRAPHIE

Articles généraux:

- Thin layer chromatography
  Egon Stahl, Springer Verlag
- Handbook of analytical Toxicology

Articles spécifiques:

Composés azotés

- Booth - Boyland
  Biochem. J. 91 1964-364

- Smyth.Mc. Korand
  J. Chrom. 16 1964-454

- Gillio Toss
  J. Chrom. 13 1964-571

- Munier
  Bull. Soc. Chim. France 19 1952-852

- Parishar
  J. Chrom. 24 1966-443

- Preussmann
  Nature 201 1964-502

Composés phénolés

- Mitchell
  J. Off. Anal. Chemists 41 1958-781

- Keith
  J. Chrom. 1 1958-534

- Richter
  J. Chrom. 18 1965-164

- Metz
  Naturwissenschaften 48 1961-569

Composés hydrocarbonés

- Abbott
  J. Chrom. 16 1964-481

- Beynon - Elgar
  Analyst 91 1966-161

- Schaad
  J. Chrom. 41 1969-120

RECOVERY OF ORGANIC MICROPOLLUTANTS

A. SDIKA - C. HENNEQUIN - R. CABRIDENC

IRCHA - 91710 Vert-le-Petit (France)

SUMMARY

To carry out comparative identifications and bioassays on water micropollutants, we try different extractive-concentration procedures on the
same water samples.

The efficiency of these technics are tested under strict conditions
according to:

- quantity
- purity of the extract
- presence of the 13 organics classes (quality index)
- repetability

A critical review of the experimental procedures is presented and gives
some indications on the reliability and cost.

Organic micropollutants are found in water in a large range :
- drinking water : 0,15 - 1,5 mg/l
- surface water : 0,6 - 3 mg/l
- sewages : 5 - 50 mg/l sometimes more

To perform ⇓	We need ⇓	
- Analytical index (TLC)	10	mg
- Cytotoxicity	20	mg
- Ames test	20	mg
- GC/MS - HPLC	40	mg
- Cancerogenesis (Mice)	100	mg
- Sarcomatogenesis (Rats)	1200	mg
- Chronic toxicity	100000	mg

On the same water (River), before and after treatment, we compare the following procedures. ⟹
Our assesment depends on :
- Recovery
- Distribution of the analytical classes
- Evaluation of the cost

Total extraction procedure (IRCHA) Liquid/Liquid extraction CHC13-CH2Cl2-Cyclohexane Freeze drying-solvent extraction.
Reverse Osmosis-solvent extraction
Solid adsorption - Xad2
Xad2 improved/Activated Carbon + Ion exchange Résin

Extract Purity is a criteria of the contaminants (Apparatus + reagents blank

values) = 1 - $\dfrac{\text{Contaminants}}{\text{contaminants + micropollutants}}$

Solvent efficiency on raw water

	Liquid/liquid extraction					Liquid/solid F.D.		
	IRCHA ⩽ T	CHCl3 pH 7	CH2Cl2 ⩽ T	CH2Cl2 pH 7	Cyclo Hexan	CHCl3	CH2Cl2	Cyclo Hexan
Quantity mg/l	0,97	0,31	0,54	0,20	0,15	1,23	1,0	0,67
Compounds detected	17	17	17	12	7	12	13	10
Extract purity	97	96	93	90	93	n o n t e s t e d		

Raw water { -Total CHCl3 (IRCHA) gives:;;;........ 2,85mg/l
{ -XAD2 improved .... 4,1 mg/l
{ -Freeze drying (Polar + non Polar).. 7,4 mg/l

• Pump contaminants ?
• What is the effect of mineral matrix ?

- FD is the best way to extract polar compounds, but it needs actually complementary studies

- As a solvent : CHCl3 is good, but it must be used with another extractive system.
- We have only a few comparative results, but it seems that the total ⩽T IRCHA method is a good compromise.
- XAD2 improved is certainly a good and cheap way to make a routine survey but we have to find, why we detect only 5 compounds, versus 14 in the ⩽ T CHCl3.

	∿ T. IRCHA CHCl3		CHCl3 pH. 7		XAD2 alone		XAD2 +Ac.Carbon +IRA		Reserve Osmosis +XAD2 imp		Freeze Drying Non Polar		F.D. polar compounds	
	1	2	1	2	1	2	1	2	1	2	1	2	1	2
Quantity mg/l	2,85	0,58	0,29	0,17	0,04	0,03	4,1	0,97	non tested	1,04	1,08	0,45	6,4	non tested
Extract Purity %	91	95	99	94	85	90	#99	#99	non tested		#98	#96		
Compounds detected (TLC)	14	14					5	5		7	3	6	2	5
≤ 25 l	20 KF		20 KF				20 KF		50 KF		120 KF			
25 - 150 l	100 KF				20 - 50 KF				50 KF		750 KF			
150-1000 l	1000 KF		1000 KF						200 KF		1000 KF			

(1) ⟹ before treatment

(2) ⟹ after treatment

Evaluation of the cost of materials and reagents

INFLUENCE OF HUMUS WITH TIME ON ORGANIC POLLUTANTS AND COMPARISON OF TWO ANALYTICAL METHODS FOR ANALYSING ORGANIC POLLUTANTS IN HUMUS WATER

G.E. CARLBERG and K. MARTINSEN
Central Institute for Industrial Research
P.B. 350 Blindern, Oslo 3, Norway

Summary

In a laboratory study the influence of aquatic humus with time on low concentrations of ftalates, PAH and chlorinated hydrocarbons have been investigated. The feasibility of a liquid/liquid extraction method and a resin (XAD-2) adsorption method for analysing these pollutants in humus water has also been investigated. The influence of the aquatic humus on the organic pollutants was found to vary considerably between different classes of compounds as well as between compounds of the same class. None of the two analytical methods investigated were found feasible for the determination of the total amount of organic pollutants present in humus water.

1. INTRODUCTION

Humus, present in all surface waters, consists of very complex organic compounds of terrestrial origin. This organic matter effects the quality of the water in a number of ways. Humus has been found to act as a chelating agent for inorganic and organic compounds thereby affecting the state of these chemicals. Humus also affects the biological conditions in water. The presence of humus generally lowers toxicity but it may enhance the bioaccumulation of chemical compounds. It has been shown that humus in surface waters combines with PAH and chlorinated organic compounds (1,2). Furthermore dialkylftalates and normal plus branched/cyclic alkanes have been found in water soluble soil fulvic acid (3,4).

From the results of previous studies it is probable that the interaction between humus and organic pollutants will influence the determination of these pollutants in humus water. The aim of this laboratory study has been to investigate the influence of humus with time on low concentrations of selected organic pollutants, and to compare to widely used analytical methods for analysing organic water pollutants. This has been done by adding small amounts of ftalates, PAH and chlorinated aromatics to humus water and analysing the mixture after various storage times. Replicate samples were analysed by a liquid/liquid extraction method and a resin adsorption method to evaluate the feasibility of these methods for analysing organic pollutants in humus water.

2. EXPERIMENTAL

Filtered brook water with a colour of 90 mg Pt/l was used throughout these experiments. The water was added to amber glass bottles and an acetone standard mixture (10 µl) was injected into the water using a syringe. The mixture contained: diisobutylftalate, dibutylftalate, fluoranthene, phenanthrene, triphenylene, 3-bromobenzene, pp'-DDD and 3-chlorobiphenyl. The concentration was 0.02 ppb for the chlorinated compounds and 2.5 ppb for the other compounds. The samples were agitated and kept in the dark at 4 °C until analysis. Replicate sample bottles were analysed after a storage time of 4, 11, 29 and 61 days by solvent extraction (cyclohexane) and resin adsorption (XAD-2) (5). The amount of the organic pollutants that adhered to the glass bottle walls were monitored throughout the experiment. 1-chlorodecane and 1-chlorotetradecane were used as internal standards and the recoveries were determined by gas chromatography with electron capture and flame ionization detector.

3. RESULTS AND DISCUSSION

The recoveries of the added pollutants as a function of storage time is shown in figure 1 for dibutylftalate, 3-chlorobiphenyl and fluoranthene. The recovery trends as shown in the figure were typical for each class of compounds.

We found that the reaction rates between the humus and the PAH were quite large, and that the equilibria were reached quickly. The reaction rates between humus and the ftalates were also found to be large, but no equilibria were reached. The reaction rates between humus and the chlorinated aromatics were found to be lower than for the other compounds. No equilibria were reached for the chlorinated aromatics either.

Even though the recovery trends shown in figure 1 were typical for each class of compounds, variations were observed between compounds of the same class (5).

We have only investigated a few compounds of each class and the behaviour of these model compounds are not necessarily representative for all compounds within that class. Investigations with a larger number of compounds must therefore be performed to obtain further knowledge about the behaviour of these compounds towards humic materials.

The amount of the organic pollutants that adhered to the glass bottle walls were monitored throughout the experiment. As shown in table I, only small amounts of the compounds adhered to the glass even after sixtyone days of storage. In a similar experiment where distilled water was used instead of humus water, a large part of the organic pollutants had adhered to the glass walls after a similar storage period (table I). These experiments indicate that the humic material has a large influence on the behaviour of organic pollutants in the aquatic environment.

TABLE I

Amount of pollutants adhered to the glass bottle walls in humus and distilled water after sixtyone days storage (percentage of added amount, average values for each class of compounds)

	Ftalates	PAH	Chlorinated organics
Humus water	−	−	2
Distilled water	20	36	54

− not detected

It was found throughout this experiment that the liquid/liquid extraction method gave higher recoveries than the resin adsorption method. One explanation for this might be that the presence of humus influences both the adsorption and desorption of the compounds to the resin. Colour measurements indicated that ninety per cent of the humic material went through the column, while ten per cent was adsorbed on the column. It will be a competition between the resin and the humus for adsorption/complexation of the pollutants. Part of the pollutants which is complexed to the humic material might therefore pass through the column. Small amounts of the pollutants were actually found in the water that had passed through the column, thereby showing that the humus interferes with the adsorption process.

When distilled water was used instead of humus water, the recoveries of organic pollutants were very good for both methods. This shows that humus in surface waters will influence the extraction efficiencies of the two methods. The concentration of humus will probably have a significant influence on the recoveries of the pollutants.

The results obtained in this work indicate that the levels of pollutants which are present in humus water are greater than those determined by these two widely used analytical methods. Of the two methods, the liquid/liquid extraction method seems better for analysing organic pollutants in humus water than the XAD-2 adsorption method.

4. ACKNOWLEDGEMENT

This work has been financially supported by the Royal Norwegian Council for Scientific and Industrial Research.

5. REFERENCES

1. E.T. Gjessing and L. Berglind; Arch. Hydrobiol., in press (1981).

2. D.E. Weidhass, M.C. Bowman and C.H. Schmidt; J. Econ. Entomol., 53 (1961) 175.

3. G. Ogner and M. Schnitzer; Geochim et Cosmochim, Acta, (1970) 921.

4. G. Ogner and M. Schnitzer; Science 170 (1970) 317.

5. G.E. Carlberg and K. Martinsen (in preparation).

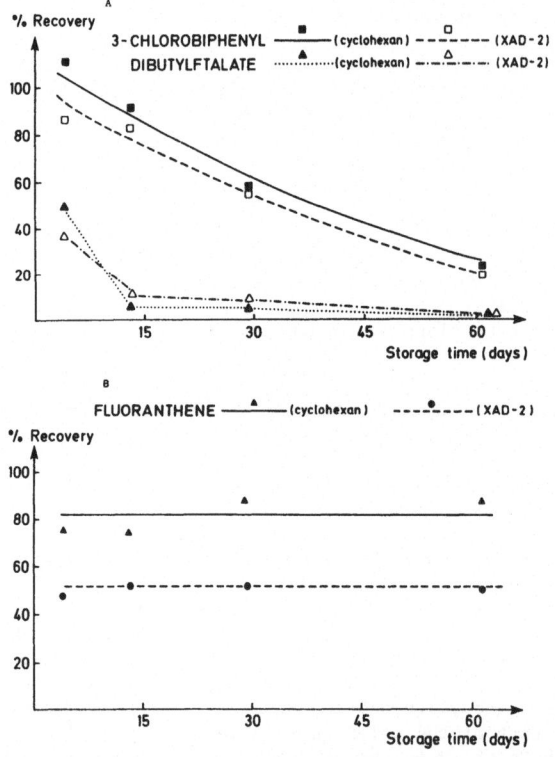

Fig. 1 Per cent recovery of dibutylftalate and 3-chlorobiphenyl (a) and fluoranthene (b) as function of storage time

IMPROVED ACCUMULATION OF ORGANOPHOSPHATES FROM AQUEOUS MEDIA BY FORMATION OF ION-ASSOCIATES WITH TETRAPHENYLARSONIUM CATION

V. DREVENKAR, Z. FRÖBE, B. ŠTENGL and B. TKALČEVIĆ
Institute for Medical Research and Occupational Health, Zagreb, Yugoslavia

Summary

The recovery efficiencies of dialkyl phosphates, thiophosphates and dithiophosphates achieved by extraction of ion-associates with tetraphenylarsonium cation from water and by elution of these species adsorbed on activated carbon are compared with values obtained in absence of ion-pairing effect.
The addition of tetraphenylarsonium chloride promotes the extraction of dimethyl and diethyl dithiophosphates from aqueous solutions while the elution of ion-associates from activated carbon seems to be promising for the accumulation of all organophosphates investigated.

1. INTRODUCTION

The control of aquatic environment polluted with organophosphorus pesticides comprises not only the determination of parent compounds but of the degradation products as well. Therefore adequate procedures for the accumulation of dialkyl phosphates, thiophosphates and dithiophosphates from water are essential for a reliable monitoring of organophosphorus pesticides.

The accumulation of phosphorus containing degradation products by adsorption on Amberlite XAD-4 (1) did not afford satisfactory results in our experience because of an incomplete and irreproducible desorption of thio- and dithiophosphates. On the other side by the extraction with pure organic solvents-benzene, diethyl ether, chloroform and methylene chloride-only a negligible recovery (< 1%) of dimethyl dithiophosphate was obtained.

An improvement of the accumulation procedure has been attempted by the extraction with organic solvents of the ion-associates of organophosphates with tetraphenylarsonium cation $(Ph_4As)^+$ as well as by the elution of these species adsorbed on activated carbon.

2. EXPERIMENTAL

The 50 ml deionized water sample (7.5<pH<9) containing organophosphorus salt and $(Ph_4As)^+Cl^-$ in the ratio not lower than 1:2 was treated by following procedures:

a) extraction with 2 x 5 ml methylene chloride and evaporation of the extract to 2 ml under a stream of nitrogen
b) percolation through an activated carbon microcolumn (5 x 6 mm) and elution either with 10 ml (Table I) or 5 ml (Fig. 1) methylene chloride, the 5 ml eluate being evaporated under a stream of nitrogen to 1 ml.

The extracts and eluates were analysed by gas chromatography after methylation of an aliquot of the sample with diazomethane.

Gas chromatographic analysis: a Varian 1400 gas chromatograph equipped with an Alkali Flame Ionization Detector (Rb_2SO_4) was used.

The 1.5 m x 2 mm i.d. glass column was packed with 80-100 mesh Chromosorb W-AW/DMCS coated with 20% w/w Triton X-305 liquid phase.

3. RESULTS

The addition of $(Ph_4As)^+Cl^-$ to the aqueous phase in the solute to extractant ratio not lower than 1:2 offers a possibility to extract dimethyl and diethyl dithiophosphates from deionized water with 70% recovery under optimum conditions of $7.5 < pH < 9$ and the concentration of the solute ≥ 1 µg/ml (Table I).

Table I. Extraction of dialkyl phosphates, thiophosphates and dithiophosphates as ion-associates with $(Ph_4As)^+$

	Concentration, µg/ml	Recovery, %
Dimethyl dithiophosphate	1	70
Diethyl dithiophosphate	1	70
Dimethyl thiophosphate	1	6
Diethyl thiophosphate	1	30
Diethyl phosphate	1	not detected
Inorganic phosphates	3	<1

The average percentage recoveries tend to decrease with decreasing concentration in water. On the detection limit of 3-4 ng/ml both dithiophosphates are extracted with only 3-6 % recovery. The extraction of corresponding thiophosphates is found to be less efficient while diethyl phosphate and inorganic phosphates are not extracted at all.

The results obtained for the extraction of dithiophosphates encouraged the attempt to improve the elution by adsorption on activated carbon microcolumn of organophosphates as ion-associates with $(Ph_4As)^+$.

Compared with extraction the elution performed with methylene chloride gives an inverse sequence of the recoveries for dithiophosphates, thiophosphates and phosphates (Table II).

Table II. Elution of dialkyl phosphates, thiophosphates and dithiophosphates adsorbed on activated carbon microcolumn (A) without and (B) with $(Ph_4As)^+$ added to water sample

	Recovery of elution with methylene chloride, %	
	A	B
Dimethyl dithiophosphate	8	40
Diethyl dithiophosphate	13	40
Dimethyl thiophosphate	8	84
Diethyl thiophosphate	7	51
Dimethyl phosphate	1	61
Diethyl phosphate	not detected	77
Inorganic phosphates	<1	14

With exception of 15 % of originally added inorganic phosphates no other compound investigated was detected in the water passed through the activated carbon.

The adsorption and elution efficiency was determined as a function of the initial concentration of the aqueous solution of organophosphates.

4. CONCLUSION

The recovery efficiencies of organophosphates from deionized water sample using either extraction or elution of the species adsorbed on activated carbon are significantly increased by formation of ion-associates with tetraphenylarsonium cation.

The addition of $(Ph_4As)^+Cl^-$ to water sample promotes the extraction of dimethyl and diethyl dithiophosphates while the elution from activated carbon seems to be more suitable for accumulation of all phosphorus containing degradation products of organophosphorus pesticides.

The greater part of inorganic phosphates present in water, which interfere in the determination of dimethyl phosphate or of total amount of organophosphates (1) can be eliminated by adsorption on activated carbon.

Testing of the applicability of the proposed procedure for the accumulation of organophosphates from different natural waters is currently in progress.

Reference

1. V. Drevenkar, Ž. Vasilić, B. Štengl, B. Tkalčević and B. Stilinović: Organophosphorus pollutants in surface waters, European Symposium "Analysis of Organic Micropollutants in Water", Berlin 1979, Preprint of the Proceedings, Volume I, p. 138.

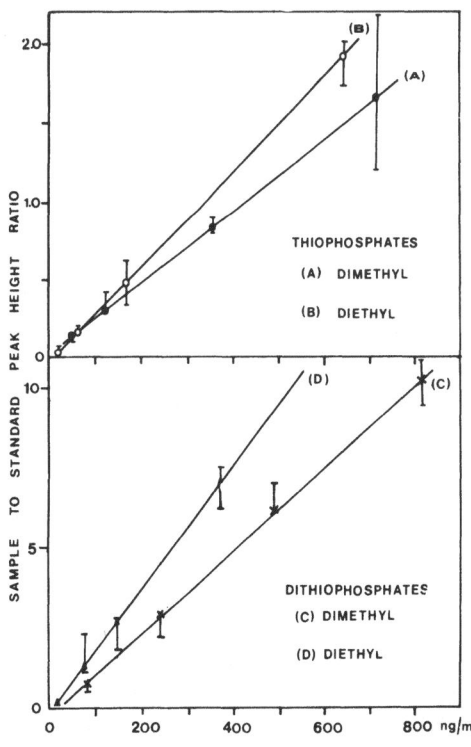

Fig. 1

a—d show calibration diagrams, linearity ranges and reproducibilities obtained for the quantitative determination of dimethyl and diethyl thio- and dithiophosphates when adsorption on activated carbon was applied

THE USE OF ECD AND FID FINGERPRINT TECHNIQUES FOR THE EVALUATION OF RIVER WATER PURIFICATION CONTAMINATED WITH ORGANIC POLLUTANTS

M. PICER

Center for Marine Research, "Rudjer Bošković" Institute, Zagreb, Yugoslavia

SUMMARY

For obtaining some more informations **about** organic pollution of water and for a rough estimation of total volatile organic matter in water samples it is suggested to use FID and ECD fingerprint methods. These methods are relatively simple and consist of concentrating the organic matter either with solvents or with XAD-2 resin. After evaporating the solvent to a small volume, the sample is analysed by means of gas chromatography. By using several solvents of various polarity for the elution of organic matter from the XAD-2 resin, it is possible to obtain some additional information about the nature of eluted organic matter in water samples. For obtaining more volatile compounds from higher polar phases, organic matter is derivatized by acylation or silylation process.

INTRODUCTION

Besides already classical parameters for the estimation of organic pollution of water (for example BOD, COD, TOC, DOC) certain new "group parameters" such as EOCl, EOHal., SAS (surface active supstances) etc. have recently appeared. The main advantage of these group parameters is their relative analytical simplicity.

As a "screening" method for a rough estimation of total volatile organic matter in water samples, we have used FID and ECD fingerprint methods. We have evaluated these methods during the investigation of the levels of organic pollutions in river waters and in ground waters by a polluted river.

METHODOLOGY

The ECD and FID fingerprint methods consisted of concentrating an organic matter from the water sample either with solvents (for example pentane) or with macroporous resin or some other adsorbent. After evaporating the solvent (or the eluate, in case of adsorbent concentration) to a small volume, the sample is analysed by means of gas chromatography. By using

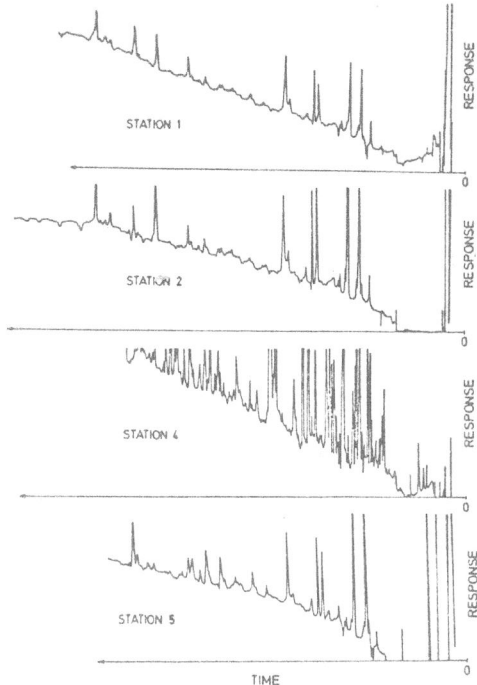

Fig. 1

ECD chromatograms of river
water samples obtained within
two hours at 4 stations.
Distances between stations
are seen in Fig. 2

the EC detector the peaks on the chromatogram are compared with
the DDT standard and the values obtained represent the ECD matter in the equivalents of DDT. The same concentrate is analysed
also by using the FI detector. The peaks on the chromatogram
are compared with the n-heneicosane standard and the values
obtained present FID matter in the equivalents of n-heneicosane.

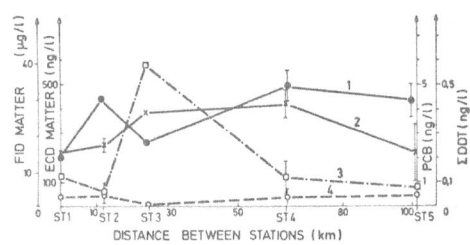

Fig. 2

Comparison of the concen-
trations of ECD and FID matter
with ⋜ DDT and PCBs in river
water obtained within two
hours at stations presented.
(1) FID matter; (2) ECD
matter; (3) PCBs; (4) ⋜ DDT

RESULTS AND DISCUSSION

As can be seen in Fig. 1, there are significant differences in the structure of ECD chromatograms of water samples
collected at presented stations. These differences are especially significant between Stations 2 and 4. It must be pointed
out that several kilometers upstream from Station 4 there is an
inflow of waste waters from a big town.

From the values for ECD matter, $\leq$ DDT and PCBs presented
in Fig. 2, it is evident that $\leq$ DDT and PCBs present only a
small fraction of the total ECD matter (between 1-5% in equivalents of DDT).

There are three levels of concentrations of ECD matter in
river water, samples from piezometers (positioned at different
distances from the river bank) and the well as can be seen in
Fig. 3. Variations of ECD matter concentrations in piezometer
samples are relatively high and it is not possible to see the
influence of the pumping time on their values.

CONCLUSION

As seen from these preliminary results obtained using the
ECD and FID fingerprint methods in the evaluation of water
samples polluted with volatile organic matter, it is not possible to identify a single organic compound, or even a group of
compounds. However, by using a relatively simple method and
instrument, it is possible to screen many samples in the monitoring program for the identification of only a limited number
of the most important samples to be further analysed applying
a very tedious method and expensive GC-MS or GC-IR systems.

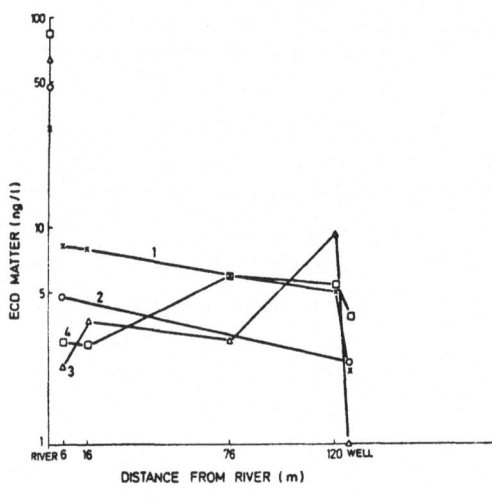

Fig. 3

Concentrations of ECD matter
obtained from the river water,
piezometers and the well.
(1) after 1 day pumping period;
(2) 15 day pumping period;
(3) 25 day pumping period;
(4) 32 day pumping period.

LEKKERKERK

Dr. F.J.J. BRINKMANN
National Institute for Water Supply, Voorburg, The Netherlands

SITUATION, HISTORY, STRATEGY, ORGANIZATION

In the last few years several cases of severe soil pollution have been discovered in The Netherlands. The best known one is the 'Lekkerkerk' case.

Lekkerkerk is located at about 20 kilometers east of Rotterdam on a branch of the river Rhine. A quarter of the village has been built on a former pasture, its area is about 600 x 150 m^2. There are 271 houses with nearly 1000 inhabitants.

In 1970 and 1971 before constructing the houses the ditches (depth between 2 and $3\frac{1}{2}$ m) were filled with 'household and demolition' waste.

The entire area, pasture as well as filled ditches had been covered with a layer of sand of approximately 70 cm. As has been shown later the rubble had been mixed with chemical waste originating from paint industries, industries of building materials, printing offices and others.
The area is located just near the river. Its surface is below the level of the river and therefore, the area experiences upward flowing groundwater. This 'reversed' groundwater flow prevented the diffusion of the pollution.

In 1978 the first signs of severe pollution appeared:
deterioration of plastic drinking water distribution pipes; polluted groundwater; dirty clothes and limbs of children playing in the sand box; poorly growing vegetation in some gardens and later also penetration of bad odours and inflammable gasses in a number of houses.
A sensory survey on odour, a survey with a mine-detector and also analyses of samples of soil and groundwater clearly indicated a coincidence of pollution and ditch pattern.

It was concluded to carry out a preliminary excavation. Below a surface of 10 x 5 m^2 the presence of 42 filled and empty drums at a depth of 1,5 to 3 meters has been shown in a filled ditch.

At nearly the same time, April 1980, two cases of drinking water pollution occurred due to percolation of the waste through the polyethylene piping system.
Drinking water had to be distributed in bottles.

It had been decided by the authorities to evacuate the population and to dig out the contaminated content of the ditches to a depth of $2\frac{1}{2}$ till $3\frac{1}{4}$ meter depending on the original situation.

This operation also covered the ditches below the houses. Because of the possible occurrence of scattered rubble the upper layer of 70 centimeters had to be removed over the entire area.

On June 1, the population of the area had moved to their temporary living quarters.

After the preparatory activities the clean-up operation which was managed by the Department of Public Works of the Province of South Holland, started the 4th of August 1980.

As decided earlier the strategy was to excavate the ditches completely and to remove the upper sand layer over the entire area. Waste, polluted soil and uncovered drums had to be transported by ship to an incineration plant.

The cleaned areas had to be filled up with non-polluted peat as well as 'flugsand'. To prevent pollution of cleaned areas the excess of ground-water had to be pumped. Before its discharge to the river the groundwater had to be purified and for this purpose a treatment plant was installed.

The operation was escorted by an extended analysis programme. This programme as well as the environmental control was assigned to the National Institute for Water Supply. A team of environmental hygienists sampled soil, water and waste, made photographical records and finally reported and listed their observations.
The chemical and toxicological analyses concerned polluted ground, polluted water, waste, coagulation sludge, clean peat, flugsand, purified water and also the badges worn by the workers in the field.
The clean-up operation finished on the 16th of January 1981.

RESULTS

In total 86.700 m^3 polluted soil was transported to the incineration plant 30 km downstreams in 117 shiploads.

1652 drums were found, some were still filled with the original che-micals, some were filled with mixtures of chemicals or mixtures of chemi-cals, soil and water and many drums were empty.

In the entire area large quantities of xylene, toluene, ethylbenzene and other aromatic compounds as well as alkanes were found. The concentra-tion in the ground ranged up to 1 g/kg ground for toluene and 2 - 3 g/kg ground for total lower aromatic compounds. The highest concentrations of organic solvents in removed soil were measured in a composed sample with the following composition.

Benzene	0.3 mg/kg		$\Sigma\ C_9H_{12}$	300 mg/kg
Toluene	1000 mg/kg		C_{10}-Alkane	30 mg/kg
Ethylbenzene	30 mg/kg		C_{10}-Alkane	30 mg/kg
m/p-Xylene	300 mg/kg		C_{10}-Alkane	30 mg/kg
o-Xylene	100 mg/kg		C_{10}-Alkane	300 mg/kg
C_9-Alkane	100 mg/kg		Cyclohexene	30 mg/kg
Carene	30 mg/kg		C_{11}-Alkane	30 mg/kg
C_{10}-Alkane	100 mg/kg		$\Sigma\ C_{10}H_{14}$	300 mg/kg

Organic compounds other than lower aromates and alkanes were found only very rarely and on specific locations. These compounds were isophoron, tetrachloromethane, polycyclic aromatic hydrocarbons, phenols, butanol, ketones, cyclic hydrocarbons, butylacetate and carene. PCB's were observed only at one place (0.4 mg/kg ground).
High concentrations of the heavy metals lead, cadmium and antimony were measured in many composed samples of removed soil. In a number of samples other trace elements were found also. The maximum and minimum contents of trace elements in composed samples of removed soil were as follows:

Antimony	<0.2 – 230 mg/kg ground	Copper	3.6 – 490	mg/kg ground
Arsenic	<0.2 – 9 mg/kg ground	Mercury	<0.05 – 8.2	mg/kg ground
Cadmium	1.1 – 97 mg/kg ground	Lead	8 – 740	mg/kg ground
Chromium	<0.5 – 140 mg/kg ground	Zinc	37 – 1670	mg/kg ground

250 of the 1652 uncovered drums were selected for the analysis of their content.

Alkydresins on basis of orthophthalic acid and epoxy resins on basis of epichlorohydrine and bisphenol-A were found regularly. In addition aliphatic hydrocarbons, fatty acids, ketones, carbonic acids, esters, polyurethanes and polyvinychloride were found also in some or more drums. In only one charge chloroparaffines were observed.

As for the inorganic constituents especially lead, cadmium and also barium and selenium were found frequently.

At the moment the inhabitants have returned to the homes in these parts of the quarter that have been filled up with peat or flugsand.

SESSION II - GAS-CHROMATOGRAPHY

Chairman: R. SCHWARZENBACH, EAWAG, Switzerland

Review papers :

- Sampling techniques for capillary GC

- Recent developments of selective detectors in GC

- Progress in column technology

Short papers :

- Recent developments in capillary column preparation

- Techniques for quantitation and identification of organic
  micropollutants by high resolution gas chromatography and
  element specific emission spectroscopy

- Applications in gas chromatography

- The use of fused-silica capillary columns in gas chromatography and gas chromatography/mass spectrometry

Poster paper :

- The determination of linear PTGC retention indices for use
  in environmental organics analysis

SAMPLING TECHNIQUES FOR CAPILLARY GC

K.Grob Jr.
Kantonales Labor, P.O. Box, CH-8030 Zürich
Switzerland

The sampling technique is seldom regarded as a problem in packed column GC. In capillary GC, however, there is a considerable variety of injection techniques, and each of them has its range of applications. On one hand this means greater flexibility allowing one a better chance of finding a technique which produces more accurate results than by packed column GC. On the other hand one needs more knowledge and experience to choose the appropriate sampling method for a particular application. Furthermore, one must know the working rules and the limitations of these injection techniques.

This paper reviews three important sampling methods and the special techniques involved in their proper use.

1. SPLIT INJECTION

The split injection is, upon first impression, the most simple method of introducing a sample into a capillary column. Nearly any solvent may be injected at any column temperature without risking band broadening effects. The sample size is usually small enough to avoid that the sample influences its own chromatography. Band broadening of the chromatography is rarely a problem and the reproducibility of the retention is relatively high.

However, quantitative work based on split injections becomes more complex. Standard deviations are often high and, even more serious, there is a considerable danger of becoming a victim of systematic errors. Systematic errors are usually difficult to recognize. The common method to check for repeatability does not reveal them.

The processes involved in a split injection are complex. Furthermore, they are higly dependent upon each other. This means that if one injection parameter is changed to improve one aspect of the injection, several other aspects also are influenced, often in the undesirable fashion.

The problems start when introducing the sample into the injector. The composition of the sample reaching the injector may be changed by a selective elution out of the syringe needle. Furthermore, this partial elution is so poorly reproducible that it contributes a major proportion of the standard deviation (2).

The evaporation inside the injector is normally incomplete. It is even impossible to obtain a complete evaporation if a sample contains non-volatile constituents (as do most of the environmental samples). In this "real" sample, the sample components of interest become partitioned between the droplets or particles of the "dirt" and the gas phase. Splitting of a mixture of droplets and gases, however, is seldom linear, and therefore, the sample components are not split by the same ratio (1).

Finally the split ratio fluctuates during the splitting process. Thus in a non-homogenous vapour the proportion of sample entering the column depends on the location of the compound within the vapour cloud.

The user of a split injector should realize these problems in order to minimize the errors and to keep deviations constant allowing their correction by calibrated factors. There are few simple working rules because there are too many parameters to optimise (sample size, solvent, column temperature, injector temperature, syringe needle handling technique and injector design). Each type of sample ends up being a particular case, calling for its own set of compromises.

Quantitation based on the external standard method is difficult because of possible problems reproducing the split ratio. The split ratio is heavily influenced by the recondensation of the sample in the column inlet (3), by the pressure wave in the injector created by the rapid evaporation of the sample (4) and by the uncomplete evaporation of the sample. Reproduction of the split ratio requires reproduction of the deviations of the true from the pre-set split ratio. This reproduction becomes particularly difficult if the deviation reaches a factor of 10 to 30.

Quantitation with internal standards is independent of split ratio. Problems may arise from discrimination of one portion of the sample if the mixture contains components with a wide range of physical characteristics (e.g. volatility and solubility in the injector). Again, calibration may solve the problems if a good reproducibility is achieved. The reproduction of the results, i.e. the reproduction of the discrimination, requires primarily a constant sample volume, a constant column temperature, an approximately constant composition of the sample (and the calibration mixture!) and the use of the same syringe.

2. SPLITLESS INJECTION

The splitless injection has become the most popular sampling method in trace analysis. It achieves a higher sensitivity than the split injection, and is unaffected by the processes which creates errors during splitting. However, one of the major sources of errors in split injection, the selective elution out of the syringe needle, remains the same in splitless sampling.

The movement to more narrow capillaries has drastically increased the problem of the sample transfer in splitless injection. If the sample is not recondensed in the column inlet, a 95% transfer of the sample vapours from the injector to the column requires a minimum carrier gas flow rate of about 2.5 ml/min. (5). This value was determined for an injector with an I.D. of 3.6 mm and an internal volume of 1 ml. The minimum carrier gas flow rate rapidly increases for larger injectors due to the increased dilution with the carrier gas. A flow rate of 2.5 ml/min. is appropriate for capillary diameters down to about 0.25 mm if hydrogen is used as carrier gas. The column should have an I.D. of about 0.30 mm if helium is used as carrier gas. The use of nitrogen as carrier gas cannot be recommended.

The sample transfer is accelerated by the recondensation of the solvent in the column inlet. The acceleration is particularly efficient in the first few seconds of the sample transfer, when the vapours are relatively concentrated. If hydrogen is used as carrier gas, the minimum flow rate is lowered to 1.5 to 1.8 ml/min. when using an injector volumn of 1 ml.

The sample transfer during 40 to 100 seconds causes intolerably broad bands of the sample components in the column inlet. We call this "band broadening in time" as all bands have the same width in terms of the (isothermal) migration time (6). Band broadening in time may be eliminated by creating an increased retention in the column inlet during a period of time which must be at least equivalent to the duration of the sample transfer. This "retention hill" is used to retain the first sample material until the last of the transferred material has reached the column inlet. Thus the sample is reconcentrated in the head of the column. This reconcentration may be achieved by cooling the column during the sampling period (cold trapping) or by recondensing the solvent in the column inlet (solvent trapping effect) (7).

The thick layer of recondensed solvent used as a retention barrier for for the solvent trapping effect is not stable. The liquid flows further into the column, driven by the flow of carrier gas. The sample components dissolved in this liquid are spread out over the whole length of the flooded column section. As the components end up to have the same broadness measurable in terms of length inside the column, we call the phenomenon "band broadening in space" (6). Peaks are typically not only broadened but also distorted or even split.

In splitless injection the band broadening in space is further complicated by the fact that only part of a sample component reaches the recondensed solvent. The rest of it is trapped in the temperature gradient between the hot injector and the cool column, thus in the column section located in the area of the fitting to the injector.

The bands broadened in space may be reconcentrated by the used of a retention gap (8). The column inlet which is expected to be flooded by the recondensed solvent (40 to 80 cm) is kept free of stationary phase. The sample components spread in this zone migrate more rapidly or at lower temperature (in temperature programmed runs) than in the regularly coated column and become reconcentrated at the beginning of the regular film of the stationary phase.

The retention gap may be prepared by extracting the stationary phase from the column inlet, by the use of deactivated uncoated pre-columns or by leaving the inlet section uncoated during the column preparation procedure (9).

The introduction of large amounts of solvent or of a greatly dominating sample component may cause undesired solvent effects which result in peak broadening or distortion. The "reverse solvent effect" was described as the peak broadening due to a later eluting large peak (10). Distorted peaks may also be seen when the retention of the recondensed solvent is not sufficient for a full trapping of a volatile sample component ("partial solvent trapping", (10)). This is often observed if very polar solvents (e.g. water or alcohols) are used. But medium to polar sample components in apolar solvents may produce similar peaks.

Cold on-column Sampling

The vaporisation of the sample inside the syringe needle and in the injector, as well as the splitting and transfer of the vapours (or aerosols) from the injector to the column accounts for the majority of the problems in quantitative GC. Polar materials may be adsorbed, high boiling substances condensed and labile components degraded, e.g. on the hot metallic needle surface. On the other hand, there is often little benefit from the vaporisation in the injector. The great majority of the important sample components are recondensed in the column inlet. Thus, unless the sample requires splitting, there is little motivation to evaporate a sample merely for its introduction into the column.

Cold on-column sampling bypasses all processes occurring inside the injector since it allows deposition of the entire sample directly into the capillary column. Accordingly, the results are highly accurate. As there is no discrimination, there is no need for correction factors (other than response factors which need to be determined only once). Thus, the analysis by cold on-column injection saves most calibration runs. Furthermore, the results are reliable because systematic errors are practically excluded.

Cold on-column sampling requires consideration of the following four rules:

1.   The injection must be carried out by a rapid push of the plunger to ensure a complete mechanical transfer of the sample from the tip of the needle to the column (12). Liquid which remains on the sampling device, is fractionated by partial evaporation and causes discrimination of the untransferred sample materials.

2.   The column temperature during the injection must be below the boiling point of the sample solvent to avoid an exceedingly rapid evaporation of the liquid introduced, which would result in a pressure increase in the column head and a backflow of solutes (12).

3.   The sample size should be no smaller than 0.25 µl to assure a complete mechanical transfer and no greater than 3 µl to avoid plugging of the column by the liquid (12).

4.   The introduction of 0.5 to 3 µl of liquid creates the problem of the band broadening in space. In contrast to the splitless injection technique, the whole sample is spread out in the flooded section of the column. This renders peak distortion or splitting more obvious. Columns used for cold on-column sampling should therefore contain a retention gap.

One of the major problems during the analysis of crude extracts is caused by the non-volatile byproducts of the sample. During a splitless injection a considerable fraction of this material enters the column in the form of an aerosol. With the cold on-column sampling, all of the "dirt" is deposited in the column inlet. These byproducts may a) degrade the stationary phase by their chemical activity, b) form small droplets or c) disrupt the film of the stationary phase. As these products are not volatile, they remain in the flooded column section.

If this section is kept free of stationary phase (retention gap), two of the three effects of the "dirt" are eliminated. Thus larger amounts of "dirt" may be tolerated until the column inlet needs to be washed or removed.

References

(1)   K. Grob Jr., Proceedings of the Fourth International Symposium in Capillary Chromatography, Hindelang 1981, Hüthig, Heidelberg, Basel, New York, 1981, p. 185; references cited there.

(2)   K. Grob Jr. and H.P. Neukom, J. High Res. Chromatogr. Chromatogr. Commun. 2 (1979) 15; K. Grob Jr. and S. Rennhard, ibid. 3 (1980) 627.

(3)   K. Grob Jr. and H.P. Neukom, J. Chromatogr. (in press).

(4)   K. Grob Jr. and H.P. Neukom, J. High Res. Chromatogr. Chromatogr. Commun. 2 (1979) 563.

(5)   K. Grob Jr. and A. Romann, J. Chromatogr. 214 (1981) 118.

(6)   K. Grob Jr., J. Chromatogr. 213 (1981) 3.

(7)   K. Grob and K. Grob Jr., J. High Res. Chromatogr. Chromatogr. Commun. 1 (1978) 57.

(8)   K. Grob Jr., J. Chromatogr. (in press).

(9)   K. Grob Jr. and R. Müller, (in preparation).

(10) R.J. Miller and W. Jennings, J. High Res. Chromatogr. Chromatogr. Commun. 2 (1979) 72.

(11) K. Grob Jr. (in preparation).

(12) K. Grob Jr. and H.P. Neukom, J. Chromatogr. 189 (1980) 109.

RECENT DEVELOPMENTS OF SELECTIVE DETECTORS IN GC

E. MANTICA
Politecnico - Istituto di Chimica Industriale "G.Natta"
Piazza Leonardo da Vinci, 32 - 20133 Milano (Italy)

Summary

The problem of identifying the substances eluted from the chromatographic column has always been of great importance in gas chromatography. Many types of detectors, capable of converting the physical and/or chemical properties of eluted compounds into useful, generally electrical signals, have been proposed to monitor the effluent from a gas chromatograph over about thirty years of gas chromatographic work. Some devices are general purpose universal detectors, very useful because they respond to anything (or almost anything) present in the carrier gas. Other devices are selective detectors which can give information on some elements or chemical groups present in the molecules. In complex mixtures of natural, environmental or biological origin, selective detectors provides chromatographic profiles complementary to the response of universal detectors and thus more useful information on the chemical functionalities required for structural elucidation.
A very critical application domain for gas chromatographic detectors concerns capillary columns. High sensitivities and small detection cell volumes are mandatory in view of the severe needs of high resolution gas chromatography. Selective detectors have played an important role in capillary GC; many compounds have been traced, which otherwise would have remained unnoticed with less sensitive universal detectors.
The versatility of GC systems when selective detectors are used can be increased by using different configurations of multiple column and/or detectors. It is now technically feasible to place two or more detectors in parallel or in series to obtain dual or multiple detection systems which can simultaneously detect compounds belonging to different chemical classes and produce very useful selective chromatograms. Another very important application of selective detectors is the analysis of particular compounds which are unstable and/or not volatile. Selective derivatization of the samples might be carried out to permit their GC analysis and detection. This derivatization enhances the usefulness of selective detectors since it increases the sensitivity of the system, converts the substances to be analyzed in more stable and volatile compounds, and reduces the sample manipulations and possibility of mistake.

1. INTRODUCTION

Great importance has always been attached to the problem of the identification of the compounds eluted after their fractionation in a chromatographic column, since, from the solution of this problem depends the qualitative knowledge of the composition of the analyzed sample. Doubtless, the gas chromatography is one of the best separation techniques at our disposal, but, unlike other analytical techniques considered specific, it has serious deficiencies under the point of view of the possibility of identifying the compounds it is capable of separating. The chromatogram contains little information on the identity of the substances which produce the different peaks. Such information can be derived from the gas chromatographic retention parameters (retention times or volumes, relative retentions, retention indices) from which, preferably by comparison with the same parameters of known compounds, it is possible to go back to the identity of the analyzed substances. Furthermore, it is necessary to bear in mind the difficulties which arise with this procedure, due to the fact that, occasionally, two or more different compounds can have undistinguishable gas chromatographic retention parameters in the same working conditions, in spite of the high resolving power attainable in many chromatographic systems. When very complex mixtures of natural, environmental or biological origin are analyzed, identification mistakes which affect the reliability of the analytical work done can be easily made. In these cases, it is recommended to use detectors which can simultaneously give qualitative analytical information on the detected compounds.

The device of the gas chromatograph which monitors the different compounds present in the carrier gas, is the detector which acts as a transducer, converting any variation of a physical and/or chemical property of the gaseous stream flowing from the column, into a signal, generally electrical, which can be subsequently processed and presented in several ways. The functions of the detector are both qualitative and quantitative: it must convert, as faithfully as possible, the concentration profiles of the matters eluted from the gas chromatographic column during the analysis into a profile of voltage, current or frequency (chromatogram) having a similar trend. To meet these requirements, an ideal detector should linearly and equally respond to all components present in the analyzed mixture. In practice, none of the available detectors is capable of doing this, but remarkable response differences are noticed (up to 10^6 and more).

From these preliminary considerations come out the reasons why a close attention was paid, in about thirty years of gas chromatographic work, to the problem of the detection of the chromatographic peaks, with the introduction and development of a few dozen types of detectors based on very different physical and/or chemical principles. No other analytical technique saw the introduction of such a great number of detectors. Even though not all detectors had a wide diffusion or became marketable, their great number stands in witness of the efforts made by many researchers aiming at solving this problem in the most satisfactory way.

The many types of gas chromatographic detectors which have been proposed since 1952, the year of the first paper of James and Martin on gas-liquid chromatography, can be grouped according to different criteria of classification. The more interesting one, concerning the present communication, is that based on the versatility of the detection which makes us classify the detectors as universal or semiuniversal and selective.

Universal or semi-universal detectors are those devices which have a general usefulness in that they respond to all (or almost all) the components of the analyzed sample. The selective detectors, instead, have a response limited to groups of substances containing in their molecules certain chemical elements, such as halogens, sulphur, nitrogen, phosphorus, silicium, tin, etc., or some functional groups which impart to the compounds special properties, such as acidity, basicity, electronic affinity, radiation absorption, fluorescence, etc., which may be used for detection. In the universal detectors, the response depends on concentration (concentration sensitive detectors) or on mass flow-rate (detectors sensitive to the mass of substance fed in the time unit) rather than on the chemical nature of the detected substances: the response factors vary within narrow limits from one compound to the other. In the selective detectors, instead, the response depends on the nature, structure and qualitative composition of the solutes; the response factors may vary within wide limits, from one component to the other, with considerable difficulties for the quantitative measurements.

The choice of the type of detector (universal or selective) depends on the desired response and the analytical problem to solve. A universal detec tor can be chosen to obtain more exhaustive information on the number and, if necessary, the quantity of the components present in the analyzed sample, whilst a selective detector can be chosen when the desired analytical re- sponses regard only one or few classes of compounds. The selective detec- tors are generally more sensitive than the universal ones and this is one of the points in their favour, chiefly regarding the analysis of traces which has become more and more important in recent years. Moreover, when no inter ference occurs, they can detect the presence of the compounds to which they are sensitive, even if these compounds are not chromatographically separa- ted in a complete way from much larger quantities of non-detectable substan ces or from compounds which can be detected with very small response factors.

The combined use, in parallel or in series, of two or more universal or selective detectors, having different response characteristics, may turn out to be particularly useful. This is a point which will be dealt with again later on: now, we would like to point out that from the comparison of the greatly different responses of the two detector types (universal and se lective), it is possible to obtain the necessary qualitative information which integrate the often insufficient data obtainable by means of the com- moner detectors (TCD, FID and similar).

Owing to the large number of proposed detectors and papers published to describe their operating principles, the fabrication procedure, the im- provements introduced and the numerous applications, it is impossible to give a complete picture, in the short time at our disposal, of the recent situation of the said subject. We have, therefore, been obliged to make choices which are typically individual and not such as to be approved by everybody. Among the selective detectors listed in Table I, we chose, for successive considerations, only a few types of detectors commonly conside- red "gas chromatographic" (electron capture detectors - ECD - flame photo- metric detectors - FPD - thermionic detectors - TID - electrochemical de- tectors) excluding, also on request of the organizers of this meeting, the spectroscopic and radiochemical detectors which can be a subject for di- scussion in future meetings. We can illustrate the special requirements that the selective detectors, used in combination with the glass or fused silica capillary columns which have been so successful in recent years, have to meet. We shall also deal with the advantages offered by the com- bined utilization in series or in parallel of two or more detectors of dif ferent types, capable of producing reciprocally complementary responses, and with the extension of the use of selective detectors to many classes of compounds, commonly considered non-detectable, which was made possible by the derivatization reactions.

Before closing this introduction, we would like to recommend a series of publications of the last ten years, which have a general character and can be useful to those who are interested in detectors. They are books (1- -7) or general reviews on gas chromatography (8-13) published every two years by Analytical Chemistry: each review gives information about the si- tuation and the developments in the field of detectors, as well as a com- prehensive bibliography of the papers published in the two years to which the review refers. Specific reviews (14-39, 140) have been published by

TABLE I - SELECTIVE DETECTORS FOR GAS CHROMATOGRAPHY

IONIZATION DETECTORS	Hydrogen atmosphere flame ionization detectors (HAFID) Photoionization detectors (PID) - UV one-photon - Laser excited resonance enhanced 2-photon Argon ionization detectors (ArD)- iodine-doped Electron capture detectors (ECD) Thermionic (TID) or alkali flame ionization detectors (AFID) Aerosol ionization detectors (AID)	
ELECTRO-CHEMICAL DETECTORS	Coulometric detectors (CD) Electrolytic conductivity detectors (ElCD) Ion selective electrode detectors	
SPECTROSCOPIC DETECTORS	Emission	- Flame photometric detectors (FPD) - Graphite-microcuvette emission detectors - Glow discharge emission detectors - Inductively-coupled plasma atomic emission spectrometry detectors (ICP-AESD) - Microwave-excited plasma detectors (MEPD) - Vacuum ultraviolet plasma atomic emission detectors - Fluorescence detectors - Chemiluminescence detectors (e.g. TEA)

	Absorption	- Ultraviolet-visible spectrometric detectors (UV-VIS)
		- Infrared spectrometric detectors
		    - Dispersive IR
		    - FTIR
		- Atomic absorption spectrometric detectors (AAS)

	Mass spectrometric detectors (MS)	
RADIOCHE-MICAL DETECTORS	Radiocarbon detectors Neutron activation analysis detection systems	

many authors, dealing with detectors in general, their characteristics and their applications in the different fields of analysis. Some of them are dealing specially with selective detectors (15, 17, 18, 21-27, 29-35, 38, 39, 140).

2. SELECTIVE DETECTORS AND CAPILLARY COLUMNS

The selective detectors have become more and more important in recent years, in high resolution gas chromatography and analysis of traces, in combination with the capillary column. As previously said, these detectors can supply analytical data which are complementary to those obtainable with universal detectors and, in many application, they ensure the high sensitivities and selectivities requested by the most complex analytical problems. However, their connection at the outlet of the capillary column poses many problems which are more serious than those which must be solved when packed columns are used. The requirements put forward to the detector when working with capillary columns, are summarized in table II. They are difficult to satisfy for many reasons and mainly for the low flow-rate of carrier gas (a few cm^3/min.) due to the geometric characteristic of the column.

TABLE II - CAPILLARY GC REQUIREMENTS FOR THE SELECTIVE DETECTORS

1. The highest sensitivity

2. The highest selectivity for detected elements or chemical groups

3. Good base line stability also in programmed temperature analyses

4. The lowest distortion of the peaks (broadening, skewness, etc.)
   and consequently the highest system efficiency

5. The lowest noise to obtain a better signal to noise ratio

In these conditions, difficulties are met with, owing to the dead volumes between the end of the column and detector inlet, inner volume of the detector and the appearance of adsorption or high temperature reactivity phenomena due to the materials used. If the high efficiency obtainable with a capillary column must be preserved, it is necessary to avoid broadening or skewness of peaks due to the abovementioned reasons. The problems which arise are different, according to the type of detector considered. Up to today, the "chromatographic" selective detectors used in combination with the capillary columns are the electron capture detectors (ECD), the thermionic detectors (TID) and the flame photometric detectors (FPD). Many papers were published in the last ten years about the use of selective detectors at the outlet of the capillary columns (40-88). Some of them cover the electron capture detector (40-42, 44-46,48,49,51-52,54-61,66,67,69,72-74, 76, 79, 80, 82-84, 87, 88), others the thermionic detectors (47, 56, 62-65, 68, 75, 77, 78, 81, 85, 86) and the flame photometric detectors (43, 50, 53, 70, 71, 73). Most of these papers deal with examples of applications of capillary columns combined with the selective detectors; only a small part of them (40, 42, 46, 58, 72, 73) examines the problems actually posed by this combination. In the paper published by Yang and Cram (73) in July 1979, it was still stated that: "very little attention has been given to-date to a systematic and rigorous study of the operating characteristics and performance of GC detectors for high resolution glass capillary systems", which emphasizes the still unsatisfactory knowledge of these problems.

Even after the overcoming of the difficulties depending on the adsorption phenomena on the materials used and the reactivity of the compounds to be detected on the surfaces maintained at high temperatures, there are, at the low flow-rate of carrier gas used in capillary column gas chromatography, residence times in the column-detector connecting region and in the detector inner volume, which are not compatible with the times of the peak output from the column. This is quite evident with electron capture and flame photometric detectors which have such effective volumes as to adversely affect the results of the separation in the column, specially regards to the detected compounds having small values of k partition ratio. There are two different approaches to obviate these difficulties:
1. To minimize the inner volume of the detector. This operation can be made only within certain limits which are fixed by the operating principles of the different systems, as it will be specified in detail hereinafter, when the different types of detectors will be taken into consideration.
2. To mix with the chromatographic column effluent, more or less large quantities of make-up gas, properly selected for each detector in such a way as to shorten the residence times within admissible limits.

The latter approach may be realized in a comparatively easy way by inserting between the end of the capillary column and the detector, a Tee coupling for the admittance of the make-up gas, and, furthermore, it offers the advantage of bringing back to normal values the operating conditions of the detector. This solution does not cause serious inconvenience with mass flow sensitive detectors, such as the flame ionization detector (FID), the thermionic detector (TID) and the flame photometric detector (FPD) which are not affected by the dilution made in the gas stream entering the detector. The situation changes remarkably with the concentration sensitive detectors, such as the thermal conductivity detector (TCD) and the electron capture detector (ECD). Special attention has been paid to the latter detector, for both its importance and diffusion in many application fields and the greater complications which set in when its performances are improved in combination with a capillary column. In this case, two contrasting requirements should be fulfilled: on one side the necessity of preserving the high efficiency inherent in the capillary column, which would dictate the use of high make-up gas flow-rates to reduce the residence times of the compounds to be detected in the cell of a comparatively large volume (in the order of cm^3), on the other side the necessity of maintaining the high sensitivity of the system which would be adversely affected by the excessive dilution with the high make-up gas flow-rates. The reduction of the inner volume of the detector which represents, as previously said, the alternative approach to the solution of the problem, cannot be extended beyond certain limits in relation to the geometry of the cell and the nature of the radioactive source used for its construction.

Now, let us consider in detail the practical experience relating to the capillary column/selective detector connection, for the three detectors which have been used so far in common practice.

A. THERMIONIC DETECTORS (TID) OR ALKALI FLAME IONIZATION DETECTORS (AFID)

They are selective detectors particularly useful in the analysis of compounds containing nitrogen or phosphorus, but less interesting, owing to their lower sensitivity, for compounds containing halogens, sulphur, arsenic, antimony, tin, lead. So far, no detailed information is available on the differences in behaviour of these detectors when their connection is changed from a packed column to a capillary column. It can be assumed that the differences are the same as those noticed in the flame ionization detectors for which Yang and Cram (73) reported that, with capillary columns, in the absence of make-up gas, there is a loss of efficiency in the system which may be as high as 15% and the appearance of peak distortion or tails. When make-up gas flow-rates are higher than 20 ml/min the column efficiency, expressed as plates per meter of column, improves and then keeps constant up to approx. 60 ml/min. Simultaneously, it is noticed that peak tails are reduced until they are no longer visible: then, the trend is towards a constant value when flow-rates exceeds 50 ml/min, but, beyond this value, disturbances may occur in flame conditions. With the capillary columns and the mass flow-rate dependent detectors, the peak heights are higher than those which are obtained with a packed column operating in the same conditions of temperature and with the same elution time, and the lower detection limit (LDL) is reduced, thus favouring the detection of compounds present in traces.

B. FLAME PHOTOMETRIC DETECTORS

These are selective detectors widely used for the analysis of compounds containing sulphur or phosphorus, which they are very sensitive to, but they can also be used for substances containing halogens, CN groups or heavy metals. The detectors available on the market, produced by different manufacturers, have in general rather large inner volumes which involve severe losses of system efficiency and remarkable deterioration of peak shapes. In this case too, to shorten the residence times of the compounds to be detected, it is possible to resort to the addition of make-up gas (nitrogen at 60 ml/min (50)), after modifying the position of the detector in such a way as to make it possible to insert the capillary column directly in the burner jet, or of flame gases (air + oxygen added to column effluent (71) or air and hydrogen successively in the dual flame system of Varian (73). When operating in these conditions, the contributions of the factors which adversely affect the system performances become negligible: peaks of a very satisfactory shape are obtained as well as a very stable base line also during the programmed temperature operations. The lower detection limit (LDL) for capillary columns is lower than that which can be obtained with a packed columns and this is most useful in the analysis of traces.

C. ELECTRON CAPTURE DETECTORS (ECD)

These are selective detectors which, owing to their high sensitivity and selectivity, are widely used in the analysis of substances containing halogens or nitro-groups (insecticides, pesticides, chlorinated hydrocarbons, drugs and their metabolites, etc.) or groups of atoms having high electronic affinities (conjugated carbonyl groups, sulphonamides, and so on). Their connection to the capillary column is very critical since generally the detectors used have rather large inner volumes which cannot be greatly reduced for reasons depending on detector construction and operating principles. In the three different types of geometry used (a- plane parallel b- coaxial cylinder, c- displaced coaxial cylinder (fig.1) (89)), it is necessary to have sufficient space for particles β emitted from the radioactive source to deactivate thereby producing by interactions with the carrier gas molecules the thermal electrons necessary to the operation of the detector.

The distance between the source and the anode shall therefore be greater than the β particle range which is in the order of a few millimeters, in order to obtain the complete deactivation of these particles and the highest production of thermal electrons, avoiding the direct collisions between high energy β particles and the anode with consequent surface corrosion and higher adsorption. The energy of β particles emitted by the source depends on the type of β particle emitter used as electron source. The more common β particle emitters are ^{63}Ni and ^{3}H which have the properties shown in table III (89). As it can be noticed from the quoted values, ^{63}Ni

TABLE III - SOME PROPERTIES OF ELECTRON SOURCES FOR THE ECD. (89)

P R O P E R T Y	S O U R C E		
	^{63}Ni foil	^{3}H Titanium foil	^{3}H Scandium foil
β particle energy (KeV)	66	18	18
β particle range (mm)	10	2,5	2,5
Maximum activity (mCi cm^{-2})	10	170	
Upper temperature limit (°C)	350	220	325
Maximum current (pA)	9	30	
Rate of electron production, R ν (sec^{-1})	6.10^{10}	2.10^{11}	
Noise level^{+} (pA)	1,5	3	

+ Measured at ambient pressure in nitrogen at 21°C

sources have a good thermal stability (350°C) and therefore they can be used also for analytical applications relating to little volatile compounds which are frequent in biological and medical analyses, a high energy of the β particles emitted (66 KeV) which limits the possibilities of reduction of the detector effective volume (for example, 0.38 ml in ECD detector produced by Varian (73)), a low specific activity which involves the use of foils having a wider surface to obtain a sufficiently high standing current. Tritium sources have a smaller energy of β particles (18 KeV) which facilitates their use in the fabrication of small volume detectors and a higher specific activity which allows the reduction of the foil dimensions. The thermal stability of the titanium tritide (Ti^{3}H) source is not satisfactory, since 220°C cannot be exceeded without losses of tritium which quickly reduce the activity of the source. This low value of the maximum temperature of use limits very much the use of this detector in many analyses; it represented for many years a severe handicap for ECD detectors. With these operating conditions, serious contaminations of the foil surface may occur reducing the sensitivity of the detector, the linearity of its response, the operating stability and increasing the background noise. The use of scandium tri tide (Sc³H) detectors greatly improved the operating conditions (maximum temperature 325°C, better specific activity with a higher standing current, smaller inner volume of the cell (down to 0.14 ml (58)). With tritium sources the necessity still exists of avoiding the use of hydrogen as carrier gas, owing to its possible exchanges with tritium and the consequent quick deterioration of the source.

In consideration of the difficulties arising from the reduction of the detector volume for the reasons previously explained, it is necessary to use, in the capillary column/ECD systems, make-up gas (nitrogen, argon/methane, of the highest purity) in such a way as to restore a flow compatible with the detector optimal operating conditions. On fig. 2 (73) are represented the effects of the variation of the total gas flow-rate passing through the detector on the efficiency of the system and on the shape of the peaks. It is evident that the increase of the flow-rate improves the efficiency of the system, mainly regarding the peaks for which the k partition ratio value is high; however, this takes place to the detriment of the sensitivity which lowers as the flow-rate rises, as shown on fig. 3. The deformations of the peaks are reduced with flow-rates exceeding 30 ml/min. The lower detection limit for a capillary column/EDC system is remarkably lower than that which can be obtained with a packed column and it can attain 1 fentogram in optimal operating conditions. Finally, in fig. 4, are shown the results obtained with a sample of Aroclor 1260 on a OV-101 (25 m x 0.25 mm i.d.) capillary column and on a packed column containing the same stationary phase in the same temperature conditions (isotherm at 210°C). It is evident that it is extremely interesting to have at our disposal high resolution and high sensitivity systems suitably designed to better exploit the advantages offered by the capillary column/ECD assembly.

3. SELECTIVE DETECTORS AND MULTIDETECTION

We previously mentioned the reasons why the selective detectors have been introduced and developed; now, we would like to examine the reasons for the multidetection, namely the use of two or more selective or universal detectors arranged in parallel or in series at the outlet of the chromatographic column (dual or multiple channel gas chromatography). In the complex mixtures which are analyzed to solve problems arising from environmental pollution studies, from clinical analyses, from composition investigation of substances of natural origin, the assignment of a peak to a given compound on the basis of its retention parameters on one chromatographic column and with one universal detector, is often unreliable on account of the possible overlapping of two or more components present in the analyzed mixture. To get a high confidence level of identification it is necessary to obtain further analytical qualitative information which may be collected in different ways. Among these, the most versatile one is doubtless the connection between the outlet of the chromatographic column and the ion source of a mass spectrometer supplying specific information of great usefulness. However, this solution may be limited by the low resolving power of the column used for the separation, by the insufficient sensitivity of the system at the level of traces, by the lenght of time requested for the analysis and the processing of the data, by the high cost and complexity of the apparatus. Another alternative approach of great interest, already experimented in the sixties and intensively developed in the last five years, is offered by the multidetection with the separation of the column effluent into many streams simultaneously feeding a universal detector and one (or more) selective detectors or two selective detectors. The first approach assigns to the uni-

versal detector the task of visualizing as many peaks as possible (fingerprint of the sample) and reserves for the more sensitive selective detectors
the identification of the peaks due to special classes of substances which,
also in traces, may have a special interest for certain properties they pos-
ses (toxicity to man or to animal or plant life, persistence in the environment, deterioration of living conditions, degrading of contamined materials,
non-suitability of a product to a given application and so on. The differences of response from the two types of detectors for the analyzed compounds
and the relations between the responses of the two detectors for the same
substance, supply further parameters which can be used in the qualitative
analysis together with the more conventional parameters.

In other cases, when the full knowledge of the composition of a given
sample is to necessary, it is better to resort to the combined use of two
or more selective detectors and to show, with each detector, the presence
of substances containing a given element (halogens, phosphorus, sulphur,
nitrogen and similar) or a given functional group (nitro groups, conjugated
carbonyls, sulphonamidic groups, etc.). This is a situation which is met
with frequently when residues of pesticides in water, foods, animal or
plant tissues are analyzed. With the combined use of many selective detectors it is possible to solve, with a considerable saving of time and of separation and purification work, the problem of quickly discriminating contaminated and non-contaminated substances.

The results which can be obtained by the multidetection systems may be
illustrated with some examples picked out from the great number reported in
the literature covering the application of different detectors. In fig. 5
(99) are shown two chromatograms obtained with FID (upper curve) and the
FPD S-mode (lower curve) on a sample of mouse urine; from the two curves
it is evident how the selective detector is able to quickly generate the
clear chromatographic profile of the sulphur compounds. In fig. 6 (49) are
presented the chromatograms simultaneously recorded with FID and ECD on
the non-polar fraction of the extract from sewage water; in this case too,
the profile of the halogenated compounds can be easily obtained from the response of the selective detector, whilst it is notably masked in the chromatogram obtained with the universal detector. In fig. 7 (30) are shown the results of an analysis made on a mixture of organophosphorus pesticides and
Aroclor 1232 with an ECD and a NPD detector; in the chromatogram recorded
with the former detector, many intense peaks of the halogenated compounds
hide the presence of some organophosphorus compounds which are, instead,
clearly visualized in the chromatogram obtained with the phosphorus selective detector. In fig. 8 (58) are reported the chromatograms obtained with
a FID and an ECD on a water sample analyzed at the EAWAG Laboratories of
Dübendorf, before and after a chlorination treatment; the different curves
show the formation of compounds hazardous to health during the disinfection
treatment. In fig. 9 (120) are presented the responses of a FID (upper curve) and NPD (lower curve) obtained for a 0.5 μl DDS lube-oil spill sample.
The FID chromatogram is typical of a lube-oil, whereas the NPD chromatogram
shows the peaks of nitrogen compounds, usually non present in an unused
oil, which have been formed during the operation of the engines. This information can be useful for locating the source of the pollution. In fig.10
(130) are shown the chromatograms recorded with FID, ECD and FPD for the
gases and vapours present in the head space of two samples of beer; the
qualitative and quantitative differences between the two drinks and, in
particular, the presence of considerable quantities of dimethylsulphide
in one of them and of halogenated hydrocarbons and alpha-diketones in different ratios in both liquids, are easily seen.

In recent years, the following different solutions to the problem of the multidetection were suggested:

Universal detector/selective detector: FID/ECD (30, 49, 58, 90, 91, 95, 111, 112, 115, 116, 119, 122, 125, 128); FID/FPD (99, 105, 107, 121, 123, 125, 129); FID/TID (30, 35, 38, 103, 104, 108, 113, 114, 118, 120, 125,126); FID/MPED (96); FID/EFID (109); FID/Radio (124); FID/HECD (127); FID/ECD/FPD (110); FID/TID/FPD/Nose (106).

Selective detector/selective detector: ECD/TID (30, 35, 38, 92, 93, 95, 125); ECD/FPD (30); ECD/FPD/CECD (101, 102).

To realize these solutions, it is necessary to solve a few problems which are associated with the type of column used (packed or capillary), with the particular configuration selected (in parallel or in series), with the operating characteristics and the special requirements of each one of the coupled detectors, with the distance between the column end and the inlets of the different detectors and, therefore, with the geometrical location of these detectors in the gas chromatograph. The difficulties which arise from the coupling with the capillary columns are always greater than those arising from the packed columns which have rather high carrier gas flow-rates and are less affected by the dead volumes let in by the splitters. Limiting our considerations to the systems which do not involve variations in the concentration of the compounds to be detected in the carrier gas during the coupling, two configurations (parallel coupling and series coupling) may be obtained which, combined in a different way, lead to the four solutions shown on fig. 11.

The parallel coupling of the detectors involves the use of spitters for the partition of the chromatographic column effluent into two or more streams of equal or different rate as a function of the different of sensitivities of the coupled detectors. The parallel coupling does not present particular problems in that it does not depend on the type of detector used and does not involve any exclusion. Therefore, this is the most commonly used configuration, as proved by the numerous papers dealing with this subject (30, 35, 38, 58, 90, 92-94, 103-105, 107, 109, 111-115, 118-121, 123--130). The series coupling of the detectors is more critical in respect to the preservation of the high system efficiency in that, in this coupling, an important role is played by the inner volumes of the detectors, which can be very different one from another, by the adsorptions on reactive surfaces, and by the transformations of the analyzed compounds inside the detector. A destructive detector (FID, FPD, TID and similar) cannot be generally used as the first detector of the series in that, inside of it, occur irreversible modifications of the substances to be detected which can not be transferred to the second detector. Therefore, the series couplings must use as a first step a non-destructive detector like ECD, in spite of the difficulties previously pointed out which depend on its comparatively large volume. Furthermore, the metal β emitting foils may produce adsorption effects or promote transformations of the detected substances owing to the high temperatures used. For these reasons, the series couplings are less used than parallel couplings, as it is proved by the limited number of papers dealing with this subject (35, 38, 91, 93, 94, 116, 122, 130).

The problem of splitting the effluent from the chromatographic column requires special attention both for the operating procedure and for the materials used. This applies in particular to the capillary columns which do not tolerate dead volumes at the low carrier gas flow-rates used, or peak deformations due to the adsorptions of the substances to be detected on the active surfaces of the materials used. One of the proposed solutions involves the use of splitters obtained with small glass-lined stainless steel tubes produced by SGE. However, the most practical method is that suggested by Neuner-Jehle, Etzweiler and Zarske (97, 98, 100) in 1973-74 and recently re-examined by Anderson and Bertsch (117). In this method, metal capillary tubes (platinum, platinum/iridium) and glass capillary tubes are combined to obtain couplings of different types which proved to be particularly inte-resting in the field of capillary column chromatography. The metal capillary tubes are available in different lengths and diameters and their welding to the glass tubes can be easily made heating the tubes kept in a fixed posi-tion by means of a natural gas (or butane) and oxygen microtorch, after a short trial period.

Thus, simple low dead volume Tees can be obtained for dividing the ga-seous stream into two others; more and more complex couplings, up to mani-fold line couplings for multi-detector systems requiring make-up gas, can be made. The modifications of the ratios between the flow-rates in the dif-ferent branches can be obtained by varying the resistance to the flow by means of small restrictions, by modifying the length of the capillary tubes in the different branches or by inserting a small diameter wire into the metal capillary tube. Constant partition ratios may be obtained only if the restrictions are thermostatted. The values of the partition ratios depend on the relative sensitivity of the coupled detectors: indicatively, values 1 : 1 have been reported for couplings ECD/TID, FID/TID, FID/FPD and 10 : 1 for FID/ECD.

4. SELECTIVE DETECTORS AND DERIVATIZATION (131, 132, 133, 134)

In the gas chromatographic separations, some classes of compounds may bring about difficulties which remarkably affect the reliability of the re-sults, owing to the insufficient volatility and the unsatisfactory thermal and/or chemical stability. In general, they are substances (acids, alcohols, phenols, mercaptans, amines, amides, oximes, carbonyl compounds in enolic form and similar) which, owing to their high content of strongly polar fun-ctional groups, give very asymmetric chromatographic peaks, are adsorbed by interaction with the column walls or with the non completely deactivated supports, may decompose, mainly at high temperatures, by contact with the metallic surfaces or with other possible centres of catalytic activity. The low volatility of a compound may depend sometimes on its high molecular weight: in these conditions, the formation of a derivative does not substan-tially modify its chromatographic properties, since the intermolecular in-teractions are mainly of dispersive type. In other cases, instead, the vola-tility depends on possible intermolecular associations among small size mo-lecules which, owing to the formation of hydrogen or ionic bonds, make the abovementioned inconveniences occur. These difficulties may be obviated by the formation of derivatives which eliminates the intermolecular associa-tions and increases the stability of the modified molecules. This result may be obtained on condition that it is possible to identify derivatizing agents capable of converting quickly and completely the compounds to be de-termined into others having more satisfactory chromatographic behaviour. In this direction were oriented the many reactions which, in the past, brought about the formation of esters, ethers, silyl derivatives, acyl de-rivatives, oximes and hydrazones, metallic chelates, etc., and solved quite satisfactorily many of the difficult analytical problems. Moreover, in re-cent years (see the review of Poole and Zlatkis (132) in the August 1980 issue of Analytical Chemistry) the role of derivatization was extended, to enable the analysts to profit from the high sensitivity and discriminatory power inherent in the available selective gas chromatographic detectors

through the introduction into the molecole to be derivatized of groups having a stabilizing and simultaneously a sensitizing detector oriented effect (ECD, FPD, TID, HECD, etc.). This modification of the compounds to be
analyzed has allowed to use selective detectors in new very important fields
and in the solution of many difficult problems of analytical chemistry. Lower limits of detection at the level of traces (pg or less) are obtained
for substances, which are otherwise not suitable for the common gas chromatographic techniques, with a high discriminatory degree in respect to other
chemical non-derivatizable species present in a complex matrix.

Among the selective detectors, special attention was paid to ECD in
conjunction with the formation of derivatives containing electrophores or
electron-capturing groups responsible for the absorption of the thermal e-
lectrons present in the detector reaction chamber. Any derivatizing agent
which can be used in ECD gas chromatography must have in its molecule a
part containing the atoms or groups sensitive to the selective detection,
and another part which, owing to its reactivity, represent the means of attaching to the molecule which must be modified. The first part must impart
to the newly-formed compound, in addition to the required sensitivity to
the selective detector, also the volatility and the thermal and chemical
stability necessary for a correct chromatographic separation. The other
part, instead, must control the selectivity of the reagent towards the functional groups which must be converted and the rate and completeness of the
conversion. Both size and shape of the organic chain of the reagent are
considerably important in that steric factors may affect the reaction rate
limiting the access of the reacting part of the derivatizing agent to the
functional groups of the substrate to be modified.

The groups which introduce electron capturing properties into the commonly used reagents, are mainly the halogens and the nitro groups. The former groups are very important and their order of response to ECD is:
I ⪢ Br ⪢ Cl ⪢ F, which is a reversed order of volatilities of the molecules in which they are contained. The halogens which give the highest sensitivity have the disadvantage of remarkably increasing the boiling points
and of reducing the thermal stability of the derivatives which are obtained
from them, so that, usually, recourse must be made to a compromise between
the requirements of sensitivity and those of volatility. The atoms of fluorine have demonstrated to be specially interesting in these applications,
since the effect of the increment of the molecular weight, they bring about
when they are introduced into a molecule, is balanced by the decrement of
the intermolecular forces, which is always found in fluorinated compounds.
This is the reason why it is possible to have more atoms of fluorine in the
molecule of the reagent and to obtain, without serious variations of its
volatility, a strong increase of its sensitivity to ECD, which is particularly evident for the derivatives in which the captured electrons are stabilized by delocalization. This is the case, for example, of heptafluorobu-
tyryl derivatives, in which the presence of the carbonyl group helps the
stabilization of the captured electrons, or of pentafluorophenyl derivatives in which the same function is assigned to the aromatic group.

The main classes of derivatizing agents used in combination with the
electron-capture detectors are the following:

A. <u>Halomethyldimethylsilyl or pentafluorophenyldimethylsilyl reagents</u>:
These reagents are similar to the silylating agents which are used for introducing into a molecule a trimethylsilyl group replacing an atom of hydrogen of a highly polar functional group. These reagents (trimethylchlo-
rosilane, hexamethyldisilazane, silylamines, silylamides, etc.) have been
known and utilized for many years for the derivatization of the classes

of compounds shown in fig. 12 (132); however, the derivatives obtained from
them do not give special responses to the electron-capture detector. The
sensitivity to ECD can be obtained in two different ways:
1. By inserting an atom of halogen (generally Cl, Br, I) into one of the
   three methyl groups in such a way as to obtain halomethyldimethylsilyl
   reagents which convert the substances to be derivatized into compounds
   detectable by the electron-capture detector, with a sensitivity more or
   less high as a function of the atom of halogen present.
2. By replacing one of the methyl groups with a pentafluorophenyl group and
   obtaining the reagent known as flophemesyl which derivatizes alcohols,
   phenols, acids, amines, steroids, drugs, etc., allowing their detection
   at the level of traces.
The derivatives obtained with these special silylating agents, when operating under closely controlled reaction conditions, are sufficiently volatile and stable; they are satisfactory for their chromatographic behaviour
and are sensitive to the selective detector. Flophemesyl, in particular,
represents a good compromise between the necessary volatility of the derivatives and the detection sensitivity and it is, therefore, widely used.

B. Haloacyl reagents: The haloacyl anhydrides, such as monofluoroacetic,
monochloroacetic, monobromoacetic, chlorodifluoroacetic, dichloroacetic,
trichloroacetic, trifluoroacetic, pentafluoropropionic, heptafluorobutyric,
but mainly the last three, are the most used acylating agents for the introduction of a halogenated acyl group replacing an atom of hydrogen of the
functional groups indicated in fig. 13 (132). When specially sensitive molecules are concerned, when undesirable side reactions, due to the haloanhy-
dride or to the haloacid formed from the former one, must be avoided, good
results are obtained from haloacylimidazole reagents which have a milder
action and give as a by-product imidazole, a weakly amphoteric compound,
instead of the haloacid formed from the anhydrides. The response to ECD detector is greater for the derivatives containing chlorine or bromine which
are, however, less volatile, less thermally stable, less satisfactory from
the chromatographic point of view. Therefore, the perfluorinated anhydrides
are used by preference, as they ensure that the former derivatives have a
satisfactory volatility, a good chromatographic behaviour, a sensitivity to
the ECD which increases with the length of the chain: among these derivatives, those of the heptafluorobutyric anhydride represent the best compromise between the two different contrasting requirements and are the most widely used.

C. Derivatizing reagents containing a pentafluorophenyl group: These are
a group of reagents having a different chemical composition as a function
of the type of functional group which must be derivatized, all containing
the pentafluorophenyl group as a sensitizing group detector-oriented (table IV) (132). The derivatives are easy to be prepared, and have a sufficient volatility, a satisfactory gas chromatographic behaviour and a high
response to ECD.

D. Reagents for the selective derivatization of bifunctional compounds:
In the gas chromatographic analysis the behaviour of the substances containing two or more high-polarity functional groups often gives rise to difficulties. Consequently, the proposal of using specific reagents, capable of
selectively derivatizing compounds of this type having two equal or different functional groups in close proximity, with formation of cyclic compounds like those shown in fig. 14 (132), found general agreement. In many
fields of analytical chemistry and, in particular, in certain clinical analyses, it is necessary to identify and quantitate some compounds which have

unsatisfactory reactive properties, but which, after derivatization, can be well analyzed at the level of traces, in that the derivatives obtained are volatile, have a correct chromatographic behaviour, are stable and present a good responde to ECD detector.

E. Other reagents: In recent times, other reagents for the derivatization of different types of compounds, besides those previously pointed out, have been proposed. The trichloroethylchloroformate was used by Hartvig et al. (135) for the analysis of the nicotine converted into carbamate and by Yamaguchi et al. (136) for the derivatization to trichloroethylcarbamate of the antianginal agent ANP-4364 in plasma. Doshi and Edwards (137) used 2,6-dinitro-4-trifluoromethylbenzenesulphonic acid as derivatizing agent for the primary amines, applying it to the analysis of catecholamines, histamines, etc. Hoshika and Muto (138) used bromine as derivatizing agent for the analysis of phenols; the obtained bromophenols are detectable with the ECD detector at levels of approximately 0.01 ng which are much lower than those obtainable with the FID detector for non-brominated phenols. Tanaka et al. (139) determined traces of nitrites in milk by making the nitrite react with 1-hydrazinophthalazine in acid solution and obtaining tetrazolophthalazine, a stable compound which can be detected with the ECD detector with a lower detection limit of 2 ng/ml.

TABLE IV - REAGENTS FOR THE INTRODUCTION OF THE PENTAFLUOROPHENYL GROUP INTO ORGANIC MOLECULES

REAGENT	FUNCTIONAL GROUP TYPE
Pentafluorobenzoyl chloride	Amines, phenols, alcohols
Pentafluorobenzyl bromide	Carboxylic acids, phenols, mercaptans, sulfonamides
Pentafluorobenzyl alcohol	Carboxylic acids
Pentafluorobenzaldehyde	Primary amines
Pentafluorobenzyl chloroformate	Tertiary amines
Pentafluorophenacetyl chloride	Alcohols, phenols, amines
Pentafluorophenoxyacetyl chloride	Alcohols, phenols, amines
Pentafluorophenylhydrazine	Ketones
Pentafluorobenzylhydroxylamine	Ketones

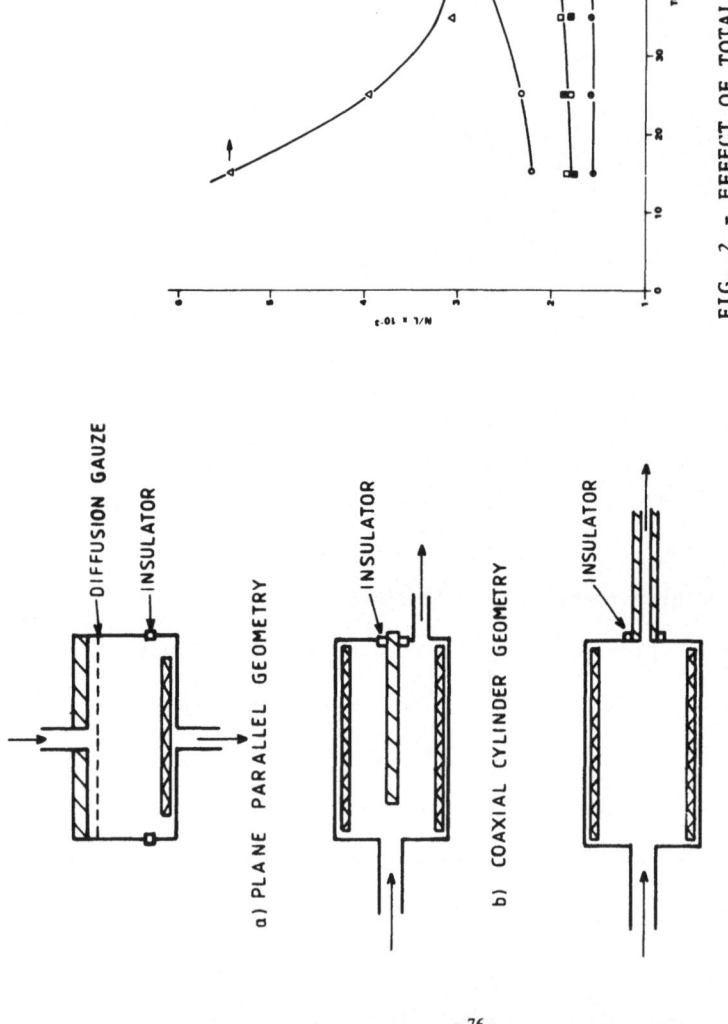

FIG. 2 – EFFECT OF TOTAL DETECTOR GAS FLOW-RATE ON THE COLUMN
EFFICIENCY AND PEAK SHAPE FOR CAPILLARY-ECD. COLUMN
WAS AN OV-101 (25 m x 0.25 mm i.d.) ISOTHERMAL AT
170°C (73)

a) PLANE PARALLEL GEOMETRY

b) COAXIAL CYLINDER GEOMETRY

c) DISPLACED COAXIAL CYLINDER GEOMETRY

→ GAS FLOW
▨ RADIOACTIVE SOURCE
▨ COLLECTOR ELECTRODE

FIG. 1 – SCHEMATIC DIAGRAM OF ALTERNATIVE
ECD GEOMETRIES (89)

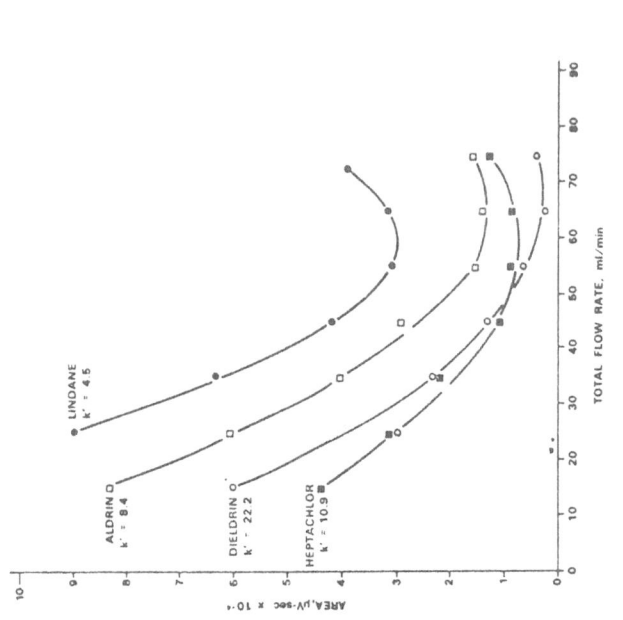

FIG. 3 – EFFECT OF TOTAL DETECTOR GAS FLOW-RATE ON ECD SENSI-
TIVITY. COLUMN WAS AN OV-101 (25 m x 0.25 mm i.d.),
ISOTHERMAL AT 170°C (73)

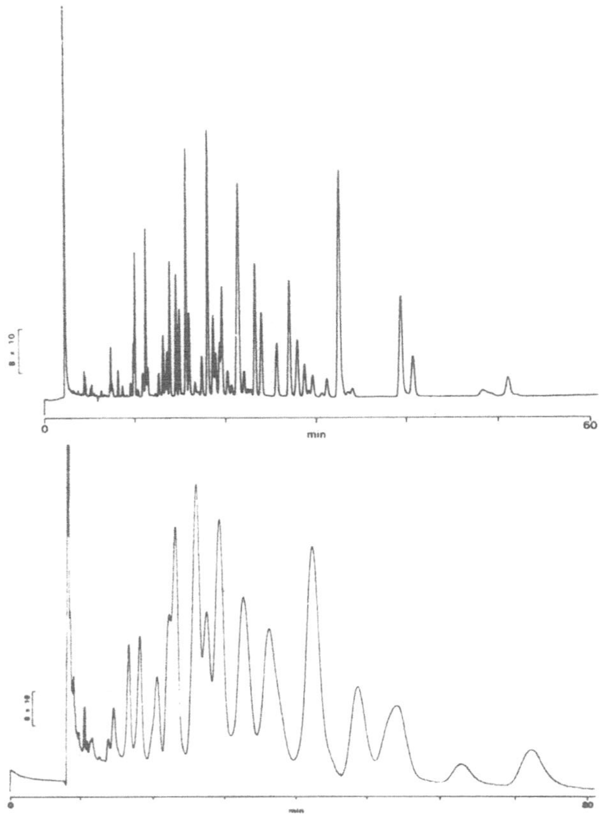

FIG. 4 – CHROMATOGRAMS OF AROCLOR 1260 ON OV-101 GLASS
CAPILLARY AND PACKED COLUMNS (73)

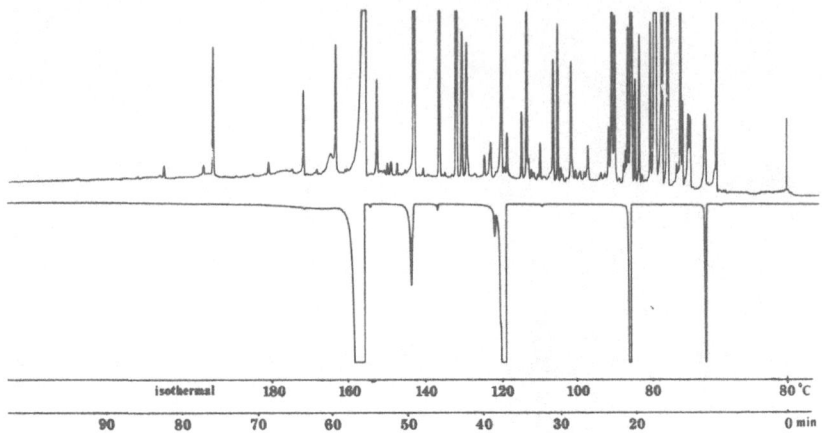

FIG. 5 - PROFILE OF SULFUR COMPOUNDS IN MOUSE URINE (99)
LOWER TRACE - SULFUR PROFILE WITH FPD;
UPPER TRACE - TOTAL CHROMATOGRAM WITH FID
SAMPLING CONDITIONS: 40 ml OF URINE + 10 g $(NH_4)_2SO_4$
COLUMN: 100 m x 0.5 mm I.D. NICKEL, COATED WITH
EMULPHOR ON-870
CARRIER GAS: HELIUM, 10 psig.
TEMPERATURE PROGRAM: 18 minutes at 80°C, 80-180 at 2°C/min.

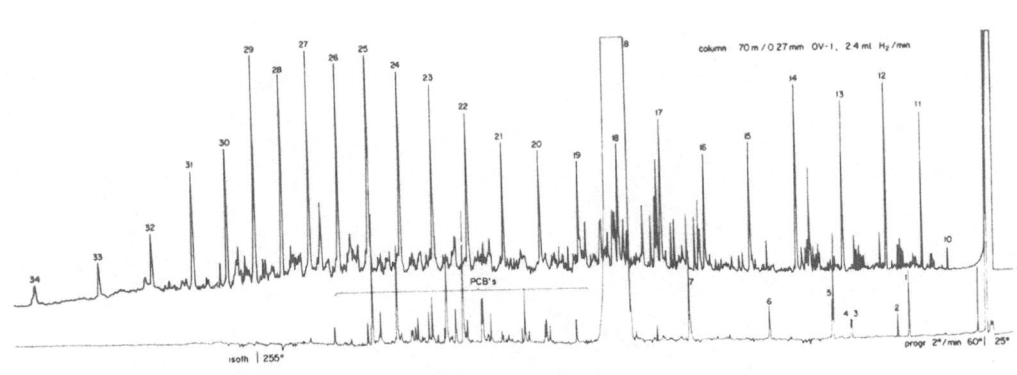

FIG. 6 - ANALYSIS OF NONPOLAR FRACTION OF EXTRACT FROM SEWAGE
WATER, SIMULTANEOUSLY RECORDED WITH FID AND ECD (49)

1. 1,2,4-TRICHLOROBENZENE
2. 1,2,3-TRICHLOROBENZENE
3. 1,2,3,5-TETRACHLOROBENZENE
4. 1,2,4,5-TETRACHLOROBENZENE
5. 1,2,3,4-TETRACHLOROBENZENE
6. PENTACHLOROBENZENE
PCB's MIXTURE OF POLYCHLORINATED BIPHENYLS
10-34 C-NUMBER OF N-ALKANES

- 78 -

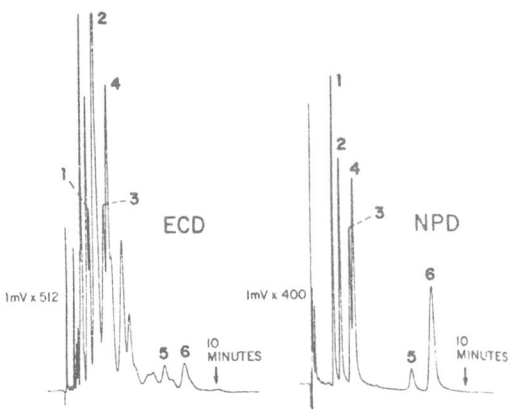

FIG. 7 - ANALYSIS OF THE PESTICIDE MIXTURE TO WHICH AROCLOR 1232
WAS ADDED, WITH THE ELECTRON CAPTURE AND NITROGEN - PHO
SPHORUS DETECTORS (30)

CONDITIONS: 6 ft x 0.16 mm ID (GLASS) PACKED COLUMN
CONTAINING 3% OV-101 METHYL SILICONE OIL ON GAS CHROM Q
80-100 mesh; COLUMN TEMPERATURE 190°C; SAMPLE VOLUME
1 mm^3. THE SOLUTION INJECTED CONTAINED 5 ng OF THE
EQUAL AMOUNT PESTICIDE MIXTURE AND 50 ng AROCLOR.
PEAKS: 1 DI-SYSTON, 2 METHYL PARATHION, 3 MALATHION
4 PARATHION, 5 METHYL TRITHION, 6 ETHION

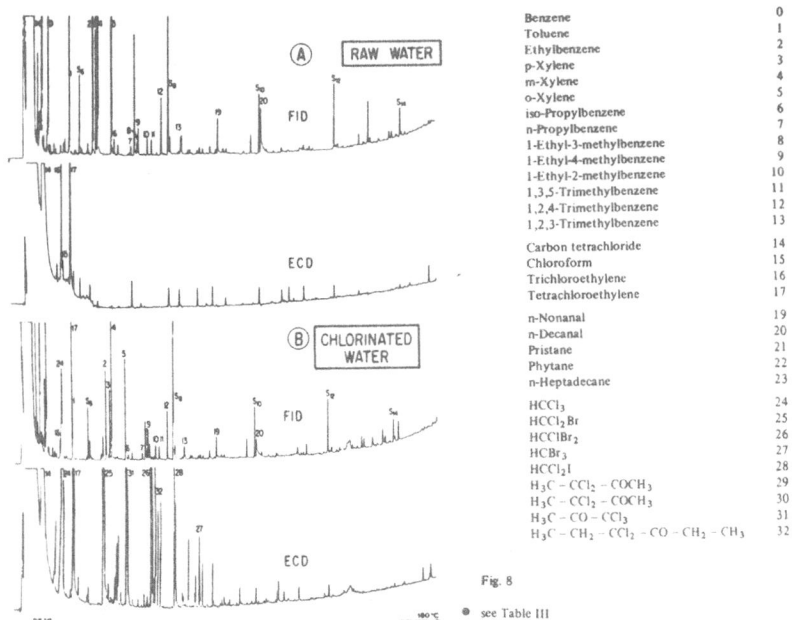

Benzene	0
Toluene	1
Ethylbenzene	2
p-Xylene	3
m-Xylene	4
o-Xylene	5
iso-Propylbenzene	6
n-Propylbenzene	7
1-Ethyl-3-methylbenzene	8
1-Ethyl-4-methylbenzene	9
1-Ethyl-2-methylbenzene	10
1,3,5-Trimethylbenzene	11
1,2,4-Trimethylbenzene	12
1,2,3-Trimethylbenzene	13
Carbon tetrachloride	14
Chloroform	15
Trichloroethylene	16
Tetrachloroethylene	17
n-Nonanal	19
n-Decanal	20
Pristane	21
Phytane	22
n-Heptadecane	23
$HCCl_3$	24
$HCCl_2Br$	25
$HCClBr_2$	26
$HCBr_3$	27
$HCCl_2I$	28
$H_3C - CCl_2 - COCH_3$	29
$H_3C - CCl_2 - COCH_3$	30
$H_3C - CO - CCl_3$	31
$H_3C - CH_2 - CCl_2 - CO - CH_2 - CH_3$	32

Fig. 8

● see Table III

FIG. 8 - ANALYSIS OF RAW AND CHLORINATED WATER WITH
FID AND ECD (58)

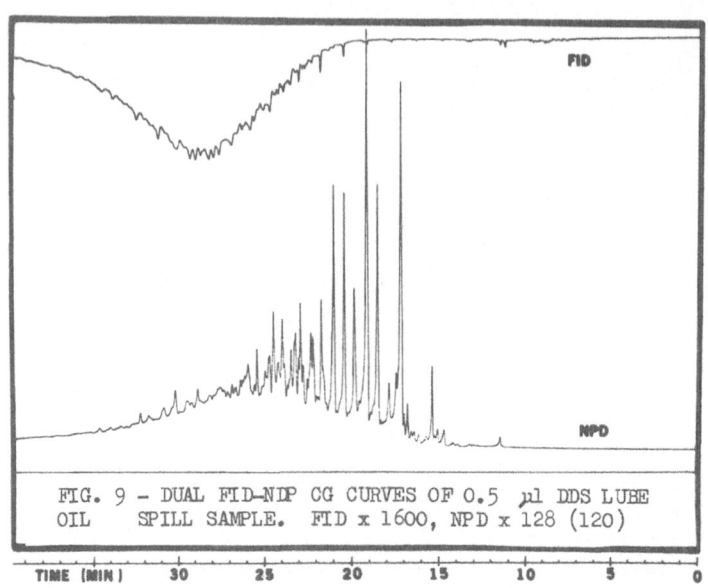

FIG. 9 – DUAL FID–NPD CG CURVES OF 0.5 μl DDS LUBE OIL SPILL SAMPLE. FID x 1600, NPD x 128 (120)

TIME (MIN)   30   25   20   15   10   5   0

Fractovap 2900 series with
HS SAMPLER Mod. 250

Samples: Beers
Detectors: FID – ECD – FPD
Column: Glass capillary WWCOT
    42 m x 0,6 mm CW 400
Carrier gas: Hydrogen 4,5 ml/min
Injector temp: 150°C
Detectors temp: 175°C
Sample turntable temp: 35°C
Sampling syringe temp: 80°C
Equilibration time: 90 min
Head space gas volume injected:
    2 ml (splitter mode 1:4)
Make up gas: Nitrogen

      Sample C

FIG 10 – ANALYSIS OF BEER
SAMPLES WITH FID, FPD, ECD (130)

- 80 -

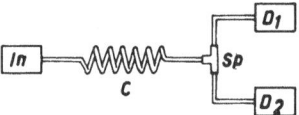

a – two detectors, parallel coupling

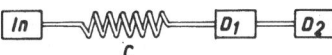

b – two detectors, series coupling

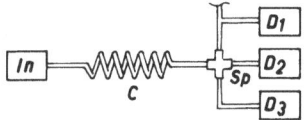

c – three(or more) detectors, parallel coupling

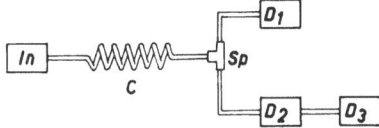

d – mixed system, parallel and series coupling

FIG. 11 – SOME SOLUTIONS OF THE MULTIDETECTION PROBLEM.

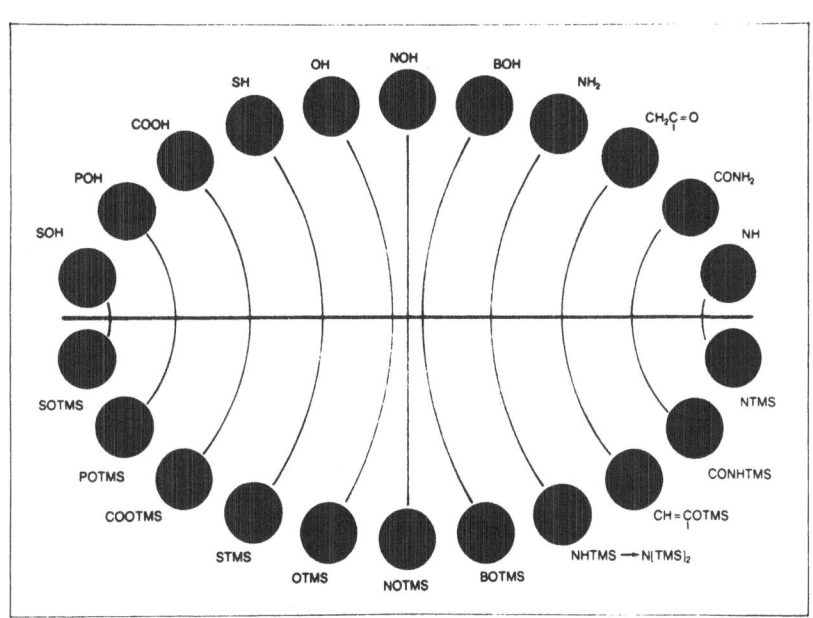

FIG. 12 - FUNCTIONAL GROUPS FORMING TRIMETHYLSILYL
DERIVATIVES (132)

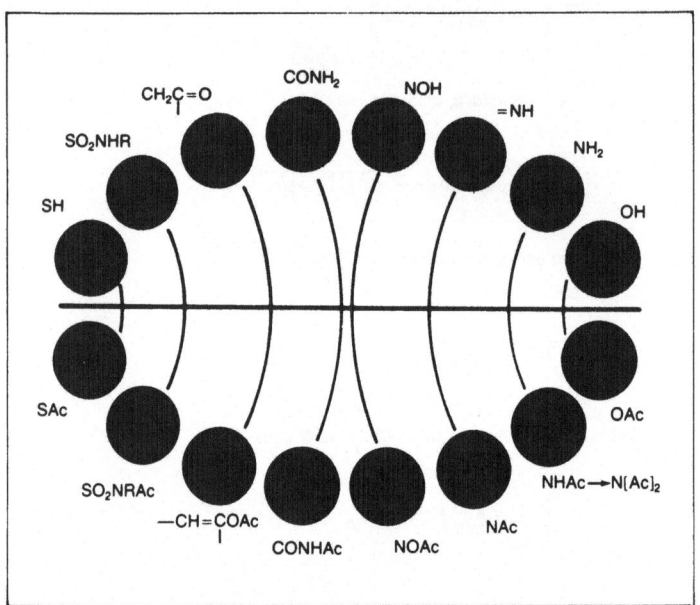

FIG. 13 - FUNCTIONAL GROUPS FORMING ACYLATED DERIVATIVES (132)

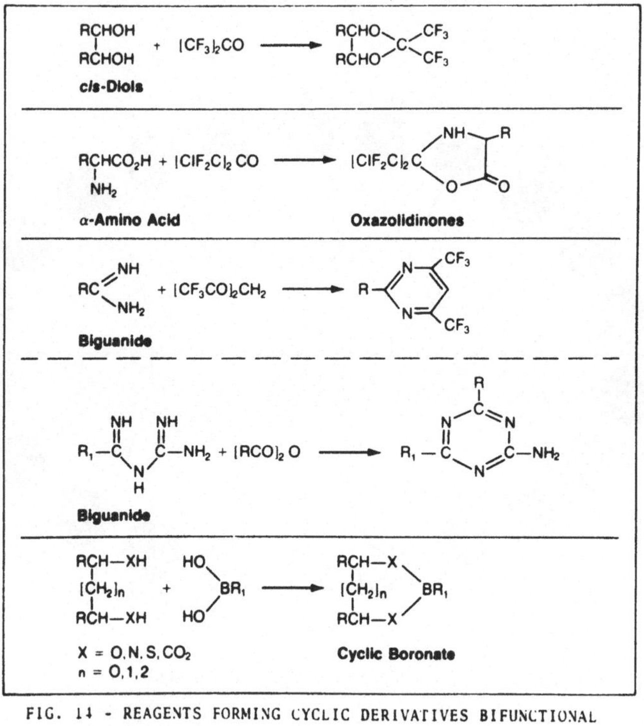

FIG. 14 - REAGENTS FORMING CYCLIC DERIVATIVES BIFUNCTIONAL
COMPOUNDS (132)

5. CONCLUSIONS

In conclusion, we can say that the recent developments in the field of the "gas chromatographic" selective detectors have fully confirmed the importance and the usefulness of these devices as an aid to identifying the compounds which give origin to the chromatographic peaks. In the last years, no development of special interest was reported in this field: only a limited number of new detectors was introduced, but none of them reached a reason able diffusion. However, the more complete investigation on the principles and detection mechanisms, the extended knowledge of the dependence of the detector signals upon the operating conditions, the modifications made to the detector design, have enhanced their performances as concerns sensitivity, selectivity, linearity of response, ease of operation. With the introduction and development of the low inner volume detectors, studied purposely for use with capillary columns, the elimination of the adsorption and dead volume effects, the improvement of the sample introduction techniques and the quality of the capillary columns, very highly efficient and sensitive systems, specially suitable for the analysis of compounds contained in traces in different matrices, were made available to the analysts. The performances of the capillary column/selective detector systems were enhanced by the introduction of the derivatizing reagents which impart to the molecules to be detected the capability of responding to a selective detector, introducing in the molecule certain hetero atoms or special groups. The selective derivatization allowed the simultaneous elimination or reduction of many manipulations aiming at isolating certain compounds from the matrices in which they are contained, reducing the risk of altering the representativity of the final sample to be introduced into the gas chromatograph. Also the introduction of the multidetection systems, capable of producing simultaneously and for the same very small sample injected many analytical qualitative responses, was very helpful.

It is difficult to forecast the future developments, mainly in relation to the possibility of introducing new, more sensitive detection systems. It can be affirmed that the tendency to further developing the systems now used will continue, as a result of a still more complete knowledge of the phenomena occurring in each detector and also of the future improvements of the systems of processing and presenting the detector responses. Very promising developments are expected to occur in the field of the spectroscopic selective detectors and an extended debate about this subject in the course of a future meeting is to be desired.

REFERENCES

1 . Jentzsch D., Otte E., "Detektoren in der Gas Chromatographie", Akademische Verlagsgesellschaft, Frankfurt a.M., 1970.

2 . David D.J., "Gas Chromatographic Detectors", Wiley-Interscience, New York, 1974.

3 . Brazhnikov V.V., "Differential Detectors for Gas Chromatography", Nauka, Moscow, 1974

4 . Rotin V.A., "Radioionization Detection in Gas Chromatography", Atomizdat, Moscow, 1974.

5 . Sevcik J., "Detectors in Gas Chromatography", Elsevier Scientific Publishing Company, Amsterdam, 1976.

6 . Krejci M., Pajurek J., "Classification of Detectors Respecting their Sensitivity and Selectivity in GC and LC" in "Trace Analysis in Column Chromatography", M. Krejci (editor), House of Techniques, Ostrava, 1977, pp. 11-36.

7 . Sullivan J.J., "Detectors" in "Modern Practice of Gas Chromatography", R.L.Grob (editor), J.Wiley & Sons, New York, 1977, pp. 213--288.

8 . Juvet R.S.Jr., Cram S.P., "Gas Chromatography", Anal. Chem. 42 (5), 1R - 22R (1970).

9 . Cram S.P., Juvet R.S.Jr., "Gas Chromatography", Anal. Chem. 44 (5), 213R - 241R (1972).

10 . Juvet R.S.Jr., Cram S.P., "Gas Chromatography", Anal. Chem. 46 (5), 101R - 124R (1974).

11 . Cram S.P., Juvet S.R.Jr., "Gas Chromatography", Anal. Chem. 48 (5), 411R - 442R (1976)

12 . Cram S.P., Risby T.H.,   "Gas Chromatography", Anal. Chem. 50 (5), 213R - 243R (1978).

13 . Cram S.P., Risby T.H., Field L.R., Yu W.-L., "Gas Chromatography", Anal. Chem. 52 (5), 324R - 360R (1980).

14 . Karmen A., "Ionization Detectors for Gas Chromatography", Advances in Chromatography 2 293-336 (1966), M.Dekker, New York.

15 . Coulson D.M., "Electrolytic Conductivity Detection in Gas Chromatography", Advances in Chromatography 3 197-214 (1966), M.Dekker, New York.

16 . Winefordner J.D., Glenn T.H., "Non-ionization Detectors and their Use in Gas Chromatography", Advances in Chromatography 5, 263-300 (1968), M.Dekker, New York.

17 . Brazhnikov V.V., Gur'ev M.V., Sakodynsky K.I., "Thermionic Detectors in Gas Chromatography", Chromatogr. Revs. 12, 1-41 (1970).

18 . Krejci M., Dressler M., "Selective Detectors in Gas Chromatography", Chromatogr. Revs. 13, 1-59 (1970).

19 . Bocek P., Janak J., "Flame Ionization Detection (Flame Ionization Phenomena)", Chromatogr. Revs. 15, 111-150 (1971).

20 . Hartman C.H., "Gas Chromatography Detectors", Anal. Chem. 43 (2), 113A - 125A (1971).

21 . Aue W.A., "Flame Detectors for Residue Analysis by Gas-Liquid Chromatography", Adv. Chem. Ser. 104 (Pesticide Identification of the Residue Level), 39-72 (1971).

22 . Selucky M.L., "Specific Gas Chromatography Detectors", Chromatographia 4, 425-434 (1971).

23 . Selucky M.L., "Specific Gas Chromatography Detectors. Part II: Electrolytic Conductivity Detector", Chromatographia 5, 359-366 (1972).

24 . Natusch D.F.S., Thorpe T.M., "Element selective Detectors in Gas Chromatography", Anal. Chem. 45, 1184A - 1194A (1973).

25 . Aue W.A., Kapila S., "The Electron Capture Detector - Controversies, Comments and Chromatograms", J.Chromatogr.Sci. 11, 255-263 (1973).

26 . Pellizzari E.D., "Electron Capture Detection in Gas Chromatography", J.Chromatogr. 98, 323-361 (1974).

27 . Aue W.A., "Detectors for Use in GC Analysis of Pesticides", J. Chromatogr.Sci. 13, 329-333 (1975).

28 . Adlard E.R., "Review of Detectors for Gas Chromatography. I. Universal Detectors", CRC Crit.Rev.Anal.Chem. 5, 1-11 (1975).

29 . Adlard E.R., "Review of Detectors for Gas Chromatography. II. Selective Detectors", CRC Crit.Rev.Anal.Chem. 5, 13-36 (1975).

30 . Pigliucci R., Averill W., Purcel J.E., Ettre L.S., "The Routine Use of Selective Gas Chromatographic Detectors", Chromatographia 8, 165-175 (1975).

31 . Farwell S.O., Rasmussen R.A., "Limitations of the FPD and ECD in Atmospheric Analysis: a Review", J.Chromatogr.Sci. 14, 224-234 (1976)

32 . Cochrane W.P., Greenhalg R., "Evaluation and Comparison of Selective Gas Chromatographic Detectors for the Analysis of Pesticide Residues", Chromatographia 9, 255-265 (1976).

33 . Taylor J.F., "Selective Detectors in Gas Chromatography", Proc. Anal. Div.Chem.Soc. 13 (6), 168-175 (1976).

34 . Poole C.F., "ECD in Organic Chemistry. Soliciting a Response from the Electron Capture Detector toward organic Molecules", Chem. Ind. (London) 1976, 479-482.

35 . Ettre L.S., "Selective Detection in Column Chromatography", J. Chromatogr.Sci. 16, 396-417 (1978).

36 . Knapman E.E.H. (editor), Developments in Chromatography - 1, Applied Science Publishers, London,1978.

37 . Sevcik J., Lips J.E., "Meaning of GC Detector Characteristics", Chromatographia 12, 693-703 (1979).

38 . Ettre L.S., "Selective Detection in Chromatographic Analysis", in "Trace Organic Analysis: A New Frontier in Analytical Chemistry", H.S.Hertz, S.N. Chesler (editors), NBS SP 519, National Bureau of Standards, Washington, 1979, pp. 547-585.

39 . Vessman J., "Quantitation of Non-halogenated Compounds by Electron Capture Gas Chromatography", Chromatogr.Revs. _24_, 313-324 (1980).

40 . Devaux P., Guiochon G., "Etude et Réalisation d'un Detecteur à Capture Electronique fonctionnant avec des Colonnes Capillaires", Chromatographia _2_, 151-157 (1969).

41 . Grob K., Grob G., "Trace Analysis on Capillary Columns. Selected Practical Applications: Insecticides in Raw Butter Extract; Aroma Head Space from Liquors; Auto Exhaust Gas", J.Chromatogr.Sci. _8_, 635-639 (1970).

42 . Fenimore D.C., Loy P.R., Zlatkis A., "High Temperature Tritium Source for Electron Capture Detector. Application to a low Volume Detector", Anal.Chem. _43_, 1972-1975 (1971).

43 . Goretti G., Possanzini M., "Coupling of the Flame Photometric Detector and the Free Fatty Acid Phase (FFAP) Capillary Column in the Trace Analysis for Sulphur Compounds", J.Chromatogr. _77_, 317-321 (1973).

44 . Franken J.J., Vader H.L., "Open Hole Tubular Columns in Pesticide Analysis", Chromatographia _6_, 22-27 (1973).

45 . Franken J.J., Trijbels M.M.F., "Preliminary Studies in the Analysis of Biological Amines by Means of Glass Capillary Columns. I. Model Compounds", J. Chromatogr. _91_, 425-431 (1974).

46 . Pellizzari E.D., "High Resolution Electron Capture Gas-Liquid Chromatography", J.Chromatogr. _92_, 299-308 (1974).

47 . Hartigan J.J., Purcell J.E., Novotny M., Mc Connell M.L., Lee M.L., "Analytical Performance of a Novel Nitrogen Sensitive Detector and its Applications with Glass Open Tubular Column", J.Chromatogr. _99_, 339-348 (1974).

48 . Schulte E., Acker L., "Gas-Chromatographie mit Glascapillaren bei Temperaturen bis zu 320°C und ihre Anwendung zur Trennung von Polychlorbiphenylen", Z.Anal.Chem. _268_, 260-267 (1974).

49 . Grob K., "The Glass Capillary Column in Gas Chromatography. A Tool and a Technique", Chromatographia _8_, 423-433 (1975).

50 . Krijgsman W., Van de Kamp C.G., "Analysis of Organophosphorus Pesticides by Capillary Gas Chromatography with Flame Photometric Detection", J.Chromatogr. _117_, 201-205 (1976).

51 . Nygren S., Mattsson P.E., "Flow Programming in Glass Capillary Column-Electron Capture Gas Chromatography by Using the Valve in the Splitter Line", J.Chromatogr. _123_, 101-108 (1976).

52 . Mattsson P.E., Nygren S., "Gas Chromatographic Determination of Polychlorinated Biphenyls and Some Chlorinated Pesticides in Sewage Sludge Using a Glass Capillary Column" , J.Chromatogr. _124_, 265-275 (1976).

53 . Blomberg L., "Gas Chromatographic Separation of Some Sulphur Compounds on Glass Capillary Columns Using Flame Photometric Detection", J.Chromatogr. _125_, 389-397 (1976).

54 . Buser H.-R., "High-Resolution Gas Chromatography of Polychlorinated Dibenzo-p-dioxins and Dibenzofurans", Anal.Chem. _48_, 1553-1557 (1976).

55 . Buser H.-R., "Preparation of Qualitative Standard Mixtures of Polychlorinated Dibenzo-p-dioxins and Dibenzofurans by Ultraviolet and gamma-Irradiation of the Octachloro Compounds", J.Chromatogr. 129, 303-307 (1976).

56 . Novotny M., Maskarinec M.P., Steverink A.T.G., Farlow R., "High-Resolution Gas Chromatography of Plasma Steroidal Hormones and Their Metabolites", Anal.Chem. 48, 468-472 (1976)

57 . Rejthar L., Tesarik K., "Performance Characteristics of a System Containing a Capillary Column and an Electron-Capture Detector", J.Chromatogr. 131, 404-407 (1977).

58 . Brechbühler B., Gay L., Jaeger H., "A Micro Electron Capture Detector for Temperature Programmed Analysis with Capillary Columns for a Wide Range of Applications", Chromatographia 10, 478-486 (1977)

59 . Franken J.J., De Nijs R.C.M., Schulting F.L., "Deactivation of Glass Open-Tubular Columns with PEG 20M via the Gas Phase", J. Chromatogr. 144, 253-256 (1977)

60 . Evrard E., Razzouk C., Roberfroid M., Mercier M., Bal M., "Isothermal Gas Chromatography with Wall-Coated Glass Capillary Columns, Electron-Capture Detection and a Solid Injector. I. Avoidance of Ghost Peaks", J. Chromatogr. 161, 97-102 (1978).

61 . Razzouk C., Evrard E., Lhoest G., Roberfroid M., Mercier M., "Isothermal Gas Chromatography with Wall-Coated Glass Capillary Columns, Electron-Capture Detection and a Solid Injector. II. Application to the Assay of 2-Fluorenylacetamide N-Hydroxylase Activity in a Rat-Liver Microsomal System", J.Chromatogr. 161, 103-109 (1978).

62 . Jacob K., Falkner C., Vogt W., "Derivatization Method for the High-Sensitive Determination of Amines and Amino Acids as Dimethylthiophosphinic Amides with the Alkali Flame Ionization Detector", J. Chromatogr. 167, 67-75 (1978).

63 . Hild J., Schulte E., Thier H.P., "Trennung von Organophosphor-Pestiziden und ihren Metaboliten auf Glaskapillarsäulen", Chromatographia 11, 397-399 (1978).

64 . De Leenheer A.P., Gelijkens C.F., "Quantification of 5-Fluorouridine in Human Urine by Capillary Gas-Liquid Chromatography with a Nitrogen-Selective Detector", J.Chromatogr.Sci. 16, 552-555 (1978).

65 . Matisova E., Krupcik J., Liska O., "Quantitative Analysis of s-Triazine Herbidides by Glass Capillary Columns Gas-Liquid Chromatography', J. Chromatogr. 173, 139-146 (1979).

66 . Brötell H., Ahnfelt N.O., Ehrsson H., Eksborg S., "Electron Capture Gas Chromatography with Splitless Injection on Isothermally Operated Wide-Bore Glass Capillary Columns", J.Chromatogr. 176, 19-24 (1979).

67 . Fitzpatrick F.A., Stringfellow D.A., Maclouf J., Rigaud M., "Glass Capillary Gas Chromatography with Electron-Capture Detection", J.Chromatogr. 177 51-60 (1979).

68 . Magin D.F., "Preparation and Gas Chromatographic Characterization of Benzyloximes and p-Nitrobenzyloximes of Short-Chain (C_1-C_7) Carbonyls", J.Chromatogr. _178_, 219-227 (1979).

69 . Boe B., Egaas E., "Qualitative and Quantitative Analyses of Polychlorinated Biphenyls by Gas-Liquid Chromatography", J.Chromatogr. _180_, 127-132 (1979).

70 . Wright B.W., Lee M.L., Booth G.M., "Determination of Triphenyltin Hydroxide Derivatives by Capillary GC and Tin-Selective FPD", HRC&CC _2_, 189-190 (1979).

71 . Farwell S.O., Gluck S.J., Bamesberger W.L., Schulte T.M., Adams D.F., "Determination of Sulfur-Containing Gases by a Deactivated Cryogenic Enrichment and Capillary Gas Chromatographic System", Anal. Chem. _51_, 609-615 (1979).

72 . Kern H., Brander B., "Precision of an Automated All-Glass Capillary Gas Chromatography System with an Electron Capture Detector for the Trace Analysis of Estrogens" HRC&CC _2_, 312-318 (1979).

73 . Yang F.J., Cram S.P., "Characteristics and Performance of Gas Chromatographic Detectors with Glass Capillary Columns", HRC&CC _2_, 487-496 (1979).

74 . DeLeon I.R., Maberry M.A., Overton E.B., Raschke C.K., Remele P.C., Steele C.F., Warren V.L., Laseter J.L., "Rapid Gas Chromatographic Method for the Determination of Volatile and Semivolatile Organochlorine Compounds in Soil and Chemical Waste Disposal Site Samples", J. Chromatogr. Sci. _18_, 85-88 (1980).

75 . Lieu Fong-Yi, Jennings W., "Separation of Histamine and its Metabolites Using Glass Capillary Gas Chromatography and a Nitrogen Detector", HRC&CC _3_, 89-90 (1980).

76 . Edlund P.O., "Determination of Clonidine in Human Plasma by Glass Capillary Gas Chromatography with Electron-Capture Detection", J.Chromatogr. _187_, 161-169 (1980).

77 . Pitts F.N.Jr., Yago L.S., Aniline O., Pitts A.F., "Capillary Gas Chromatography with a Nitrogen Detector for Measurement of Phencyclidine, Ketamine and Other Arylcycloalkylamines in the Picogram Range", J. Chromatogr. _193_, 157-159 (1980).

78 . Good B.W., Parrish M.E., Douglas D.R., "Volatile Phase Profiling of Mainstream Smoke by Glass Capillary Gas-Chromatographic Techniques", HRC&CC _3_, 447-451 (1980)

79 . Fogelqvist E., Josefsson B., Roos C., "Determination of Carboxylic Acids and Phenols in Water by Extractive Alkylation Using Pentafluorobenzylation, Glass Capillary GC and Electron Capture Detection", HRC&CC _3_, 568-574 (1980).

80 . Oliver B.G., Bothen K.D., "Determination of Chlorobenzenes in Water by Capillary Gas Chromatography", Anal.Chem. _52_, 2066-2069 (1980).

81 . Hsu F., Anderson J., Zlatkis A., "A Practical Approach for Optimization of a Selective Gas-Chromatographic Detector by a Sequential Simplex Method"; HRC&CC _3_, 648-650 (1980).

82 . Huibregtse-Minderhout I., Van Der Kerk-Van Hoof A.C., Wijkens P., Biessels H.W.A., Salemink C.A.,"Glass Capillary Column Gas Chroma

tographic Method for Simultaneous Quantitative Determination of Insect Juvenile Hormones at the Picogram Level: A Comparative Study of Various Halogenated Derivatives", J.Chromatogr. 196, 425-434 (1980).

83 . Guerret M., "Determination of Pindolol in Biological Fluids by an Electron-Capture Gas-Liquid Chromatographic Method on a Wall-Coated Open Tubular Column", J.Chromatogr. 221, 387-392 (1980).

84 . Edlund P.O., "Determination of Opiates in Biological Samples by Glass Capillary Gas Chromatography with Electron-Capture Detection", J.Chromatogr. 206, 109-116 (1981).

85 . Wolf M., Deleu R., Copin A., "Separation of Pesticides by Capillary Gas Chromatography. Part 2: Organophosphorus Insecticides", HRC&CC 4, 346-347 (1981).

86 . Wehner T.A., Seiber J.N., "Analysis of N-Methylcarbamate Insecticides and Related Compounds by Capillary Gas Chromatography", HRC&CC 4, 348-350 (1981).

87 . Douse J.M.F., "Trace Analysis of Explosives at the Low Picogram Level by Silica Capillary Column Gas-Liquid Chromatography with Electron-Capture Detection", J. Chromatogr. 208, 83-88 (1981).

88 . McKague A.B., "Phenolic Constituents in Pulp Mill Process Streams", J. Chromatogr. 208, 287-293 (1981).

89 . Connor J., "The Electron Capture Detector. II Design and Performance", J.Chromatogr. 210, 193-210 (1981).

90 . Oaks D.M., Hartmann H., Dimick K.P., "Analysis of Sulfur Compounds with Electron Capture/Hydrogen Flame Dual Channel Gas Chromatography", Anal. Chem. 36, 1560-1565 (1964).

91 . Williams I.H., "Gas Chromatographic Techniques for the Identification of Low Concentrations of Atmospheric Pollutants", Anal.Chem. 37, 1723-1732 (1965).

92 . Camoni I., Gandolfo N., Ramelli G., Sampaolo A., Binetti L., "Rapid Screening Method for the Simultaneous Determination of Chlorinated and Phosphorus-containing Pesticide Residues in Fruits and Vegetables", Boll.Lab.Chim.Prov. 18, 579-612 (1967).

93 . Wessel J.R., "Collaborative Study of Three Gas Chromatographic Dual Detection Systems for Analysis of Multiple Chlorinated and Organophosphorus Pesticides", J.Assoc.Off.Anal.Chem. 51, 666-675 (1968).

94 . Krejci M., Dressler M., "Selective Detectors in Gas Chromatography", Chromatogr. Revs. 13, 1-59 (1970).

95 . Brandenberger H., "Gas Chromatography in Toxicological Analyses. Improvement of Selectivity and Sensitivity by Inserting Multiple Detectors", Pharm.Acta Helv. 45, 394-413 (1970).

96 . McLean W.R., Stanton D.L., Penketh G.E., "Quantitative Tunable Element-Selective Detector for Gas Chromatography", Analyst (London) 98, 432-442 (1973).

97 . Neuner-Jehle N., Etzweiler F., Zarske G., "Platinkapillaren als Interface bei der Direktkopplung von Glaskapillarsäulen mit einem Massenspektrometer", Chromatographia 6, 211-216 (1973).

98 . Etzweiler F., Neuner-Jehle N., "Eine einfache Vorrichtung zur Strom-
      teilung am Ausgang von Glaskapillarsäulen hoher Trennleistung",
      Chromatographia 6, 503-507 (1973).

99 . Bertsch W., Shunbo F., Chang R.C., Zlatkis A., "Preparation of High
      Resolution Nickel Open Tubular Columns", Chromatographia 7,
      128-134 (1974).

100 . Neuner-Jehle N., Etzweiler F., Zarsche G., "Zur Frage der Chromato-
      gramm-Registrierung bei der Direktkopplung von Kapillarsäulen
      mit einem Massenspektrometer. Der Anschluss eines FID paral-
      lel zum Massenspektrometer", Chromatographia 7, 323-332
      (1974).

101 . McLeod H.A., Butterfield A.G., Lewis D., Phillips W.E.J., Coffin D.E.,
      "Gas-Liquid Chromatography System with Flame Ionization, Pho-
      sphorus, Sulfur, Nitrogen and Electron Capture Detectors
      Operating Simultaneously for Pesticide Residue Analysis",
      Anal.Chem. 47, 674-679 (1975).

102 . McLeod H.A., Coffin D.E., "Pesticide Residue Screening Methods Uti-
      lizing Multidedector Configurations", Water Quality Parameters,
      ASTM STP 573, American Society for Testing and Materials,
      Philadelphia, 1975, pp. 183-195.

103 . McConnel M.L., Novotny M., "Automated High-Resolution Gas Chromatogra
      phic System for Recording and Evaluation of Metabolic Profi-
      les", J.Chromatogr. 112, 559-571 (1975).

104 . Verga G.R., Poy F., "Gas Chromatography of Nitrogen- and Phosphorus-
      -Containing Compounds. A New, High Sensitivity, Variable Se-
      lectivity Detector", J.Chromatogr. 116, 17-27 (1976).

105 . Bertsch W., Hsu F., Zlatkis A., "Hearth Cutting Technique in High
      Resolution Gas Chromatography Applied to Sulfur Compounds in
      Cigarette Smoke", Anal.Chem. 48, 928-931 (1976).

106 . Hrivnac M., Frischknecht W., Cechova L., "Gas Chromatographic Multi-
      detector Coupled to a Glass Capillary Column", Anal.Chem. 48,
      937-940 (1976).

107 . Bruner F., Ciccioli P., Bertoni G., "Analysis of Sulfur Compounds in
      Environmental Samples with Specific Detection and Selective
      Columns", J. Chromatogr. 120, 200-202 (1976).

108 . Mellor N., "Thermionic Detectors in Gas Chromatography. Selective De-
      tection of Phosphorus, Nitrogen and Sulphur Compounds", J.
      Chromatogr. 123, 396-399 (1976).

109 . Sevcik J., Kaiser R.E., Rieder R., "A Semi-specific Flame-Ionization
      Detection System", J.Chromatogr. 126, 263-269 (1976).

110 . Ciccioli P., Bertoni G., Brancaleoni E., Fratarcangeli R., Bruner F.,
      "Evaluation of Organic Pollutants in the Open Air and Atmo-
      spheres in Industrial Sites Using Graphitized Carbon Black
      Traps and Gas Chromatographic-Mass Spectrometric Analysis
      with Specific Detectors", J.Chromatogr. 126, 757-770 (1976).

111 . Giger W., Reinhard M., Schaffner C., Zürcher F., in Keith L. H. edi-
      tor, Identification and Analysis of Organic Pollutants,
      Ann Arbor Science Publishers, Ann Arbor, 1976, pp. 433-452.

112 . Bächmann K., Emig W., Rudolph J., Tsotsos D., "Die gleichzeitige Ver-
          wendung von FID und ECD im Anschluss an eine Kapillarsäule",
          Chromatographia 10, 684-685 (1977).

113 . Adams R.F., Vandemark F.L., Schmidt G.J., "Ultramicro GC Determination
          of Amino Acids Using Glass Open Tubular Columns and a Nitrogen-
          -Selective Detector", J.Chromatogr. Sci. 15, 63-68 (1977).

114 . Baker J.K., "Identification and Chemical Classification of Drugs Based
          on the Relative Response of a Nitrogen Selective Detector and
          a Flame Ionization Detector in Gas Chromatographic Analysis",
          Anal. Chem. 49, 906-908 (1977).

115 . Bertsch W., Anderson E., Holzer G., "Two Dimensional High Resolution
          GLC Environmental Analysis, Preliminary Results", Chromato-
          graphia 10, 449-454 (1977).

116 . Södergren A., "Simultaneous Detection of Halogenated and Other Com-
          pounds by Electron-Capture and Flame-Ionization Detectors
          Combined in Series", J. Chromatogr. 160, 271-276 (1978).

117 . Anderson E.L., Bertsch W., "Practical Aspects of Pt/Ir Effluent
          Splitters for Multidetector GC and Pneumatic Solute Swit-
          ching", HRC&CC 1, 13-17 (1978).

118 . Albert D.K., "Determination of Nitrogen Compound Distribution in Pe-
          troleum by Gas Chromatography with a Thermionic Detector",
          Anal.Chem. 50, 1822-1829 (1978).

119 . Bjorseth A., Eklund G., "Analysis of Polynuclear Aromatic Hydrocar-
          cons by Glass Capillary Gas Chromatography Using Simultane-
          ous Flame Ionization and Electron Capture Detection", HRC&CC
          2, 22-26 (1979).

120 . Frame G.M.II, Flanigan G.A., Carmody D.C., "Application of Gas Chro-
          matography Using Nitrogen Selective Detection to Oil Spill
          Identification", J.Chromatogr. 168, 365-376 (1979).

121 . Bates T.S., Carpenter R., "Determination of Organosulfur Compounds
          Extracted from Marine Sediments", Anal.Chem. 51, 551-554
          (1979).

122 . Poy F., "A New Approach to the Simultaneous Multi-Detection Techni-
          que with Electron Capture and Flame Ionization in Series",
          HRC&CC 2, 243-245 (1979).

123 . Wenzel B., Aiken R.L., "Thiophenic Sulfur Distribution in Petroleum
          Fractions by Gas Chromatography with a Flame Photometric
          Detector", J.Chromatogr.Sci. 17, 503-509 (1979).

124 . Gross D., Gutekunst H., Blaser A., Hamböck H., "Peak Identification
          in Capillary Gas Chromatography by Simultaneous Flame Ioni-
          zation Detection and ^{14}C-Detection", J.Chromatogr. 198,
          389-396 (1980)

125 . Lopez-Avila V., "Analysis of Sludge Extracts by High Resolution GC
          with Selective Detectors", HRC&CC 3, 545-550 (1980).

126 . Becher G., "Glass Capillary Columns in the Gas Chromatographic Sepa-
          ration of Aromatic Amines. II Application to Sample from
          Workplace Atmospheres Using Nitrogen-Selective Detection",
          J. Chromatogr. 211, 103-110 (1981).

127 . McCarthy L.V., Overton E.B., Maberry M.A., Antoine S.A., Laseter J.L.,
"Glass Capillary Gas Chromatography with Simultaneous Flame
Ionization (FID) and Hall Element-Specific (HECD) Detection",
HRC&CC 4, 164-168 (1981).

128 . Bjorseth A., Carlberg G.E., Gjos N., Moller M., Tveten G., in L.H.
Keith editor, "Advances in the Identification and Analysis of
Organic Pollutants in Water", Ann Arbor Science Publishers,
Ann Arbor, 1981.

129 . Marriot P.J., Cardwell T.J., "Chromatographic Parameters Derived from
the Non-linear Response of a Flame Photometric Detector",
Chromatographia 14, 279-284 (1981).

130 . Gagliardi P., Verga G.R., Munari F., "A High Resolution Head Space
Gas Chromatographic System with Multi-Detector Capabilities",
Carlo Erba Application Sheets (1981).

131 . Drozd J., "Chemical Derivatization in Gas Chromatography", Chromatogr.
Revs. 19, 303-356 (1975).

132 . Poole C.F., Zlatkis A., "Derivatization Techniques for the Eletron-
Capture Detector", Anal. Chem. 52, 1002A-1016A (1980).

133 . Knapp D.R., "Handbook of Analytical Derivatization Reaction",
J.Wiley & Sons, New York, 1979.

134 . Blau K., King G. editors, "Handbook of Derivatives for Chromatography", Heyden & Son, London, 1980.

135 . Hartvig P., Ahnfelt N.-O., Hammarlund M., Vessman J., "Analysis of
Nicotine as a Trichloroethyl Carbamate by Gas Chromatography
with Electron-Capture Detection", J.Chromatogr. 173,
127-138 (1979).

136 . Yamaguchi T., Yoshikawa S., Sekine Y., Hashimoto M., "Determination
of (E)-(2,3-dichloro-4-methoxyphenyl)-2-furanylmethanone-O-
-(2-diethylaminoethyl)oxime Methanesulphonate (ANP-4364) in
Plasma Using Gas Chromatography with Electron-Capture Detection", J.Chromatogr. 173, 147-154 (1979).

137 . Doshi P.S., Edwards D.J., "Use of 2,6-dinitro-4-trifluoromethylbenze-
nesulphonic Acid as a Novel Derivatizing Reagent for the Analysis of Catecholamines, Histamines and Related Amines by Gas
Chromatography with Electron-Capture Detection", J.Chromatogr.
176, 359-366 (1979).

138 . Hoshika Y., Muto G., "Sensitive Gas Chromatographic Determination of
Phenols as Bromophenols Using Electron-Capture Detection",
J.Chromatogr. 179, 105-111 (1979).

139 . Tanaka A.,Nose N., Yamada F., Saito S., Watanabe A., "Determination
of Nitrite in Human, Cow and Market Milks by Gas-Liquid Chromatography with Electron-Capture Detection", J.Chromatogr.
206, 531-540 (1981).

140 . Farwell S.O., Gage D.R., Kagel R.A., "Current Status of Prominent Selective Gas Chromatographic Detectors: A Critical Assessment",
J.Chromatogr.Sci. 19, 358-376 (1981).

PROGRESS IN COLUMN TECHNOLOGY

A. Venema
Akzo Research
Corporate Research Department
Postbox 60, ARNHEM
The Netherlands

Summary

In this paper the importance of capillary GC for environmental analyses
is emphasized.
A number of properties of GC columns affecting the general
applicability is discussed.
A survey is given of the progress made during the past few years with
respect to each of these properties.

1. INTRODUCTION

The importance of gas chromatography in the field of environmental
pollution analyses is demonstrated by the huge amount of literature
published during the past years.
Due to the complexity of the samples which have to be separated
capillary gas chromatography has become indispensable to obtain reliable
results. Only a few problems, for instance the determination of relatively
high concentrations of pollutants, can be solved with packed columns. To
achieve the desired "selectivity" these columns are often operated in combination with a specific detection system like FPD, Thermionic detection or
a mass spectrometer.
In view of their general applicability this paper will only deal with
the progress made in the field of capillary columns.
In this scope the steel capillary column will not be discussed as this
column type is only useful in some very restricted areas of capillary gas
chromatography.

2. CAPILLARY GAS CHROMATOGRAPHY .

Since the first papers dealing with glass capillary gas chromatography
were published (1,2) the applicability of this technique has been improved
tremendously.
When literature on capillary column preparation became more explicit
with respect to the experimental details, especially after the first
"Hindelang" Conference in 1975, the application of glass capillary chromatography was not restricted anymore to a few, highly skilled, chromato-
graphers.
Moreover the improved situation during the past few years with respect
to the commercial availability of both GC apparatus dedicated to capillary
GC and high quality capillary columns gave rise to a rapid increase of the
use of capillary GC.
In addition, due to its high separation power, the use of capillary
columns is stimulated by the still growing demands for (environmental)
analyses.

Though in the past capillary GC was thought to be unsuitable for trace analyses, a large amount of literature dealing with this subject has disapproved this opinion.
On the contrary, the use of the right injection technique and the optimal conditions will result in a lower detection limit for capillary GC as compared with an optimized packed column system.
So it will be clear that if an analytical problem dealing with trace analyses of organic compounds has to be solved with gas chromatography, in general a capillary column will be the best choice.

3. REQUIRED CAPILLARY COLUMN PROPERTIES

To obtain good results we need high quality capillary columns; basic requirements are:
3.1 - high separation efficiency;
3.2 - high liquid film stability;
3.3 - high temperature stability;
3.4 - low adsorptivity.
Although some interrelation exists between these requirements they wil be discussed separately.

3.1 High separation efficiency

In the early days of capillary GC a high separation efficiency was the major topic of investigation. As early as 1962 Giddings pointed out that an evenly distributed stationary phase film was a condition to obtain a high resolution (3).
From the technical point of view the preparation of capillary columns with evenly distributed stationary phase films is no problem anymore.
Two different techniques are available:

3.1.1 Static coating procedure

Basically this method has not been altered since its description in literature (4,5).
The most recent experimental details have been published by Grob (6). Although several closing techniques have been published (7,8,9), the original waterglass method is still very useful (10).

3.1.2 Dynamic coating procedure

The procedure has been substantially improved by Schomburg (11,12) since its introduction (13). Some factors affecting the obtained film thickness have been discussed (14,15).

Both methods give very good results with respect to separation efficiency, although slightly better results are obtained with the static method when gum phases are used, while the dynamic method is slightly favourable for stationary phases with a low viscosity.

3.2 High liquid film stability

In order to be of any practical value the even film distribution in the capillary column must last under practical conditions, for instance during temperature programming.
Factors affecting film stability were recently discussed by Grob (16). The following possibilities can be distinguished:

3.2.1 - film stabilization due to intermolecular forces between support and stationary phase;

3.2.2 - film stability due to surface geometry (surface roughness) (3);
3.2.3 - film stabilization by the formation of crosslinked networks.

Re 3.2.1

Film stabilization by intermolecular forces is observed for a-polar polysiloxanes like OV 1 and SE 54. These stationary phases spread spontaneously on glass surfaces.

In general medium-polar or polar stationary phases do not spread on glass. This situation can be improved by changing the character of the glass surface, for instance by chemical modification or pretreatment of the glass with surface active agents.

As in the past this principle was not very successful much more attention has been paid to surface roughening techniques (see 3.2.2). Recently, however, Grob and Blomberg succeeded in coating a number of polar stationary phases on chemically treated glass (17,18).

This promising technique has the advantage of producing real smooth films, in contrast with the surface roughening techniques, thus giving the highest achievable resolution.

Re 3.2.2

Film stabilization by surface roughening is a circumstantially described technique for the preparation of semi-polar and polar capillary columns. A number of techniques and materials have been described, for instance HCl-etching, $BaCO_3$, SiO_2, NaCl-sol, etc.

All these methods, however, affect the resolution attainable, the thermostability and/or the adsorptivity of the capillary column (see 3.3 and 3.4).

If surface roughening has to be used, those methods yielding NaCl on the surface have to be preferred in view of the inertness of the deposited solid (19). Especially the method described by de Nijs et al. (20) is a very elegant one; besides for AR-klar this method is also applicable to Pyrex- and duran glass or fused silica.

3.2.3 Film stabilization by the in situ formation of crosslinked stationary phases

A very new topic is the possibility of network formation in the stationary phase by the introduction of crosslinks between the polymer chains (19). When such a reaction takes place in the stationary phase of a coated capillary the liquid film will be stabilized due to its altered physical properties.

Moreover it seems to be possible to clean capillary columns by solvent rinsing without damaging the coating.

The chemistry of the crosslink reaction and the requisite crosslink density are under investigation in several research institutes.

3.3 High temperature stability

When glass capillary columns were introduced in the early sixties this material was supposed to be completely inert. Indeed the use of glass instead of steel was an improvement. However, during the past few years it has become clear that the chromatographic properties and the thermostability of glass capillary columns strongly depend on the type of glass used and/or the treatment of the glass.

Recently it has been demonstrated (20,22,23,24) that metal ions present in glass are detrimental to the thermal stability of polysiloxane phases. Very probably the same holds for some polyethyleneoxide/polypropyleneoxide phases.

As shown by Dandeneau (24) the use of fused silica, an SiO_2 with a metal ion content << ppm level, indeed tremendously improves the thermal stability of a-polar siloxane and C-20 M coated capillary columns.
The same results have been obtained for glass after removal of the metal ions from the column wall surface by leaching (26, and references cited therein).

With our present knowledge it is now possible to prepare glass- or fused silica capillary columns, coated with a-polar siloxanes, which can be used up to 340-350°C in the isothermal mode, or even up to 380°C in the temperature programmed mode.

It is only 2 or 3 years ago that these temperatures would have destroyed our capillary columns completely.

Unfortunately the progress in the preparation of capillary columns, coated with medium-polar or polar stationary phases, with an increased thermostability is not so obvious.

This point has to be investigated in future research.

3.4 Low adsorptivity

As the main objective in the start of capillary chromatography was the separation of complex hydrocarbon mixtures adsorptivity was not a real problem.
In our today's practice, however, we have to deal with polar compounds too. Moreover these components must be determined on a ng or even a pg level.
In these cases adsorptivity is of tremendous importance. Where reversible adsorption is easily observed by peak tailing, irreversible adsorption only results in a diminished peak height. So, as a result of irreversible adsorption traces of polar compounds in samples of unknown composition might be overlooked.

Rapid progress has been made with respect to the column wall adsorptivity since the introduction of leaching- and silylation techniques for glass capillary columns (27,28,29) and the introduction of fused silica as a column material (25,30).

All these methods are based upon the idea that adsorptivity is caused by the presence of metal ions, silanol groups or polar siloxane bridges. Hence, the methods applied to lower adsorptivity are: removal of metal ions, silylation of SiOH groups, hydrothermal treatment to remove polar siloxane bonds or application of an intermediate polymer layer on the column wall surface.

However, still unexplained results are obtained (31); moreover critical remarks can be made on these origins of adsorption (24).

Even with the best capillary column prepared according to the most advanced methods still some adsorption of very polar compounds (prim. amines) is observed in the sub ng range (31).

Although a lot of research still has to be done on column wall adsorption the present state of art allows the separation of numereous classes of organic compounds without spoiling adsorption effects.
Sometimes derivatization of the sample became superfluous (18,25,32).

4. CONCLUDING REMARKS

Due to a rapidly growing understanding of the fundamentals, which determines the capillary column properties, a very rapid progress has been made in column technology during the past few years, enabling the production of high quality capillary columns.

Though not yet perfect the present state of art of capillary column preparation allows separations to be done which were impossible only a few years ago. With respect to temperature stability, resolution and adsorptivity these types of capillary columns are much better than their packed equivalents.

The commercial availability of these capillary columns will result in a rapid growth of the application of capillary chromatography.

Looking to the research which is being done in numereous places one can expect that further progress will be made in the near future, especially in the field of film stability and adsorptivity.

REFERENCES

1. Desty, D.H. and Goldup, A. in Gas Chromatography 1960, R.P.W. Scott (ed.), Butterworths, London (1960)162.

2. Scott, R.P.W. and Hazeldean, G.S.F., ibid (1960)144.

3. Giddings, J.C., Anal. Chem. 34(4)(1962)458.

4. Bouche, J. and Verzele, M., J. Gas Chromatogr. 6(1968)501.

5. Ilkova, E.L. and Mistryukov, F.A., J. Chrom. Sci. 9(1971)569.

6. Grob, K., HRC & CC 3(1980)525.

7. Cueman, M.K. and Hurley, R.B., HRC & CC 1(1978)92.

8. Sandra, P. and Verzele, M., Chromatographia 11(1978)102.

9. Anders, G., Rodewald, D. and Welsch, Th., HRC & CC 3(1980)298.

10. Rutten, G.A.F.M. and Rijks, J.A., HRC & CC 1(1978)279.

11. Schomburg, G., Husmann, H. and Weeke, F., J. Chromatogr. 99(1974)63.

12. Schomburg, G. and Husmann, H. in Proceedings of the first Int. Symp. on Glass Cap. Chrom., R.E. Kaiser (ed.), Hindelang (1975)61.

13. Dijkstra, G. and de Goey, J. in Gas Chromatography 1958, D.H. Desty (ed.), Butterworths (1958)56.

14. Alexander, G. and Lipsky, S.R. in Proceedings of the 2nd Int. Symp. on Glass Cap. Chrom., R.E. Kaiser (ed.), Hindelang (1977)9.

15. Venema, A. and van der Ven, L.G.J., ibid (1977)385.

16. Grob, K., HRC & CC 2(1979)599.

17. Grob, K. and Grob, G., HRC & CC 3(1980)197.

18. Blomberg, L., Markides, K. and Wännman, T. in Proceedings of the 4th Int. Symp. on Cap. Chrom., R.E. Kaiser (ed.), (1981)73.

19. Grob, K. and Grob, G., J. Chromatogr. 213(1981)211.

20. Venema, A., van der Ven, L.G.J. and van der Steege, H. HRC & CC 2(1979) 405.

21. de Nijs, R. et al., ibid 2(1979)447.

22. Borwitsky, H. and Schomburg, G., J. Chromatogr. 170(1979)99.

23. Lee, M.L. et al., ibid, 199(1980)355.

24. Venema, A. in Proceedings of the 4th Int. Symp. on Gap. Chrom., R.E. Kaiser (ed.) (1981)91.

25. Dandeneau, R., Beute, P., Rooney, T. and Hiskes, R., Amer. Lab. 11(9)(1979)61.

26. Lee, M.L. and Wright, B.W., J. Chromatogr. 184(1980)235.

27. Grob, K., HRC & CC 3(1980)493, and references cited there.

28. Schomburg, G., Husmann, H. and Behlau, H., J. Chromatogr. 213(1981)211.

29. Godefroot, M., Roelenbosch, M., Verstappe, M., Sandra, P. and Verzele, M., HRC & CC 3(1980)337.

30. Dandeneau, R., Stark, T. and Mering, L., paper presented at the Pittsburg conference, Atlantic City, 1980.

31. Verzele, M. et al. in Proceedings of the 4th Int. Symp. on Cap. Chrom., R.E. Kaiser (ed.) (1981)239.

32. Jennings, W., Gas Chromatography with Glass Capillary Columns (2nd edition) (1980) Acad. Press.

RECENT DEVELOPMENTS IN CAPILLARY COLUMN PREPARATION

P. SANDRA, M. VAN ROELENBOSCH, I. TEMMERMAN and G. REDANT

Laboratory of Organic Chemistry, State University of Ghent
Krijgslaan, 271 (S.4), B-9000 GENT (Belgium)

Summary

In the last years capillary column technology has developed
very rapidly. Leached conventional glass and fused silica
columns both, through high temperature silylation, have
given access to very high quality columns, particularly
with regard to inertness and temperature stability; at
least with the polysiloxanes OV-1, SE-30, OV-101, OV-73,
SE-52 and SE-54.
For apolar coatings both glass and fused silica column types are equally effective. Glass columns will continue to
be used for many years to come but due to the ease of handling, fused silica columns will be the most popular column
type.
Coating fused silica columns with more polar polysiloxanes
(OV-17, OV-225, Silars) is in a state of flux and at present glass is the material of choice for these phases. Recently Ohio Valley introduced a new polar silicone gum
which can be coated on the smooth fused silica surface.
The new phase possesses good chromatographic properties and
will extend the use of fused silica columns. For the polyethylene glycols Carbowax 20M and the Superox series the
coatability of leached glass and fused silica is equally
excellent. The temperature stability however of PEG
coatings is better on roughened glass surfaces. By appropriate mixing of OV-1 and Superox 20M any desired polarity
between the two phases can be obtained, particularly increasing the versatility of fused silica columns. Cross-
linking of apolar phases into the capillary columns results in non-extractable films. Columns can easily be re-
generated by rinsing with solvents if the deposition of
high boiling and non volatile substances, which often occurs in environmental analyses, leads to severe losses in
resolution. The introduction of these new technologies
has resulted in scope and application expansion, often
showing that there are still problems which have to be
overcome.

1. COATING WITH APOLAR POLYSILOXANE GUMS

The last twenty years many scientists have been involved in
developing procedures to prepare high quality capillary columns.
This was finally achieved by leaching conventional glass, high
temperature silylation and static coating with an apolar gum
of the polysiloxane type (1,2,3). Such columns are highly
inert, the coating efficiency is high and the temperature stability (>350°) corresponds to the intrinsic thermal stability
of the phase. In 1979 fused silica capillary columns were in-

troduced (4). These columns promised to be highly inert, highly
efficient, thermally stable, mechanically strong and easy to install. Especially the two last characteristics were very favourably accepted by the non expert chemical public. It is now
clear however that several pretreatment steps are required to
obtain, with fused silica columns, the same inertness as attainable with HTS glass columns. The procedures used in the
authors laboratory for making apolar glass and fused silica columns'are summarized in Table 1. The chromatogram (Fig. 1) of
a Round Robin test on a 26 m x 0.32 mm glass column coated with
SE-54 (df : 0.3 μm) after HCl-HF-PSD treatment illustrates the
inertness of a HTS column.

Pretreatment	Material	Conditions	Aim
LEACHING	Gsl	18.0 % HCl-S-150°-24 h	Removes metal ions
		0.1 % HF-S-150° -24 h	Creates SiOH rich
	Gb	18.0 % HCl-S-180°-24 h	surface
		0.1 % HF-S-150° -24 h	Opens active
			Si-O-Si bridges
	FS	3.0 % HCl-D-350°-24 h	Creates SiOH rich
			surface
			Opens active
			Si-O-Si bridges
SILYLATION	Gsl Gb FS	HMDS or PSD (∿ 400°)	SiOH → Si-OTMS
COATING	Gsl Gb FS	Static OV-1, OV-73 SE-52, SE-54 in n-pentane	To obtain a homo- geneous film 0.1 - 1 μm

Table 1 : Procedure for making apolar columns.

Abbreviations : Gsl : glass soda lime, Gb : glass borosilicate,
                FS : fused silica, S : static, D : dynamic,
                HMDS : hexamethyldisilazane, PSD : polysiloxane
                degradation.

2. COATING WITH POLAR POLYSILOXANE PHASES

A great deal of effort is presently directed towards the
coating of fused silica and leached glass with polar polysiloxanes (OV-17, OV-210, OV-225, Silar 5CP, Silar 10C). Modification and surface deactivation by means of methylphenyldisila-
zanes (5), baking with the stationary phase itself (PSD) (2),
deactivation with cyclic siloxanes and silazanes (6,7) and in
situ polymerization of prepolymers containing the functional
groups of the stationary phases (8,9) have been advanced for
glass capillary columns. Some of these treatments have been
tried out on fused silica capillary columns but results are
still in a state of flux in the sense that some of the columns
were quite good and some useless. In fact, to retain liquid
stationary phases, roughening of the silica wall has to be
applied. Several attempts have been made to roughen fused silica columns but all failed. Roughening glass surfaces on the

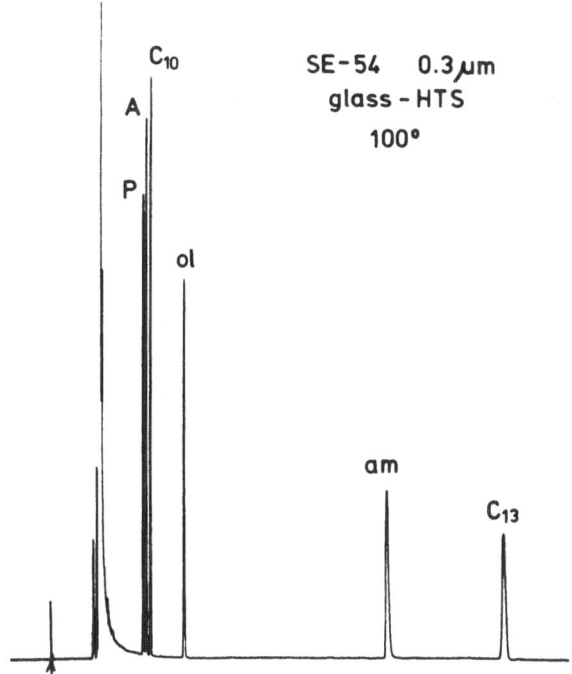

Fig. 1 : Round Robin test organized on the occasion of the IV
Hindelang symposium 1981 on a 26 m x 0.3 mm HTS-SE-54
column. Column temperature 100°, split injection
1/200 4 ng per compound; hydrogen carrier gas 0.2 kg/
cm^2. Peaks : 1) aniline, 2) phenol, 3) n-decane,
4) 1-octanol, 5) n-decylamine, 6) n-tridecane.

contrary causes no problem. For the phases mentioned above,
whisker surfaces are used in our laboratory. The introduction
of new polar silicone gum phases will in the near future extend
the versatility of fused silica columns. Recently Ohio Valley
synthesized a medium polar silicone phase (OV-1701) with interesting chromatographic characteristics. This polar gum phase
can be applied in fused silica columns. The analysis of pesticides on a 20 m x 0.3 mm FS-OV-1701 column is shown in fig. 2.

3. COATING WITH POLAR PHASES OF THE PEG TYPE

Polyethylene glycols (Carbowax 20M and the Superoxes) can be
coated on smooth fused silica surfaces. With film thicknesses
from 0.1 to 0.3 µm columns remain acidic (amines are reversibly
and/or irreversibly adsorbed). The temperature stability varies between 220-240°C depending on the Carbowax 20M batch and
the purity of the carrier gas. At high temperatures the lifetime is limited. Better results are obtained with Superox 20M
coatings with a long-term stability around 240°-250°C. The
adhesion of polyethylene glycols to alkali glass leached with
water and hydrochloric acid is better than to the smooth silica

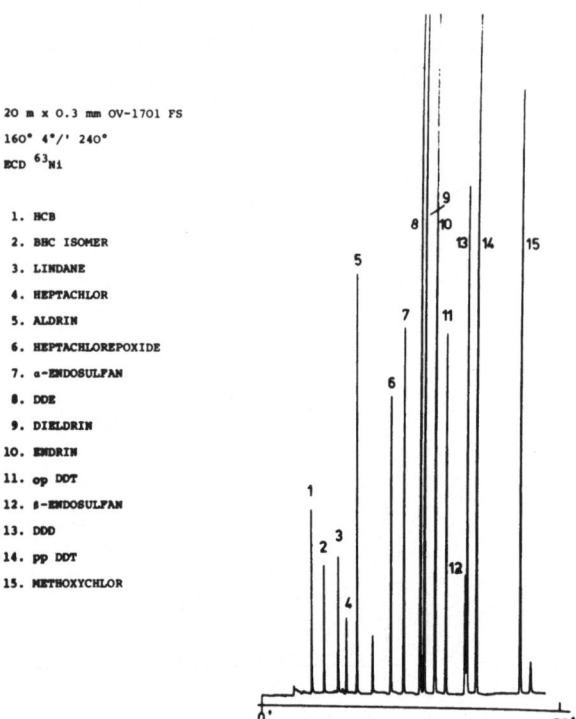

20 m x 0.3 mm OV-1701 FS
160° 4°/' 240°
ECD ^{63}Ni

1. HCB
2. BHC ISOMER
3. LINDANE
4. HEPTACHLOR
5. ALDRIN
6. HEPTACHLOREPOXIDE
7. α-ENDOSULFAN
8. DDE
9. DIELDRIN
10. ENDRIN
11. op DDT
12. β-ENDOSULFAN
13. DDD
14. pp DDT
15. METHOXYCHLOR

Fig. 2 : Analysis of a pesticide mixture on a 20 m x 0.3 mm
FS-OV-1701. Column temperature programmed 160° to
240° at 4°/min. ECD detection, hydrogen carrier gas
1 ml/min. Solid state injection, 10 to 100 pg per
compound.

surfaces. The long life time even at temperatures of 250° is
attributed to the roughened surface created by water leaching.
The inertness of a H_2O-HCl leached alkali glass column is superior to that of a fused silica column (10).

4. COATING WITH MIXED PHASES OV-1-SUPEROX 2OM

The interesting polar phases Ucons and Pluronics cannot be
coated on the smooth silica surfaces. A solution to this problem is the use of binary mixtures of OV-1 and Superox 2OM.
By appropriate mixing, any desired polarity between the two
phases can be realized, approaching the polarity and the selectivity of the Ucons and Pluronics. Excellent columns with regard to efficiency (> 95 %), inertness and thermal stability
(ca 250°C) have been obtained (11).

5. CROSS-LINKING OF APOLAR SILICONE PHASES

The main advantage of cross-linked phases over conventional
stationary phases is their insolubility in solvents. As a con-

sequence phase stripping because of on-column injection at low
temperatures of large sample volumes no longer occurs. Moreover, stationary phase displacements by the injection of highly
polar compounds is eliminated. The deposition of high boiling
and non-volatile sample compounds, often occurring during real
sample analysis (biological, environmental and natural samples)
ultimately leads to a severe loss of resolution. Therefore pre-
purification of the sample often is needed. With cross-linked
phase columns this is not the case as they can easily be rege-
nerated by rinsing the column with organic solvents.
Cross-linking can be obtained by adding peroxides to the stationary phase used for static coating of the capillary column.
After evaporation of the solvent the polymer layer contains the
peroxide which by subsequent heating initiates the cross-linking
into a three dimensional polymer which no longer is soluble.
Most of the experiments carried out in our laboratory are done
with SE-54 (12). Recently Grob reported that cross-linking
also is possible with OV-1 to OV-61 (13). The non-extractabi-
lity of the film is illustrated in figure 3, showing the temperature programmed analysis of an Arochlor 1260 mixture on the
same column after rinsing with 10 ml n-pentane (A) and 100 ml
n-pentane (B) and after extraction overnight with dichloromethane (C) and water (D).

CONCLUSION

In the last years important progress in column preparation has
been made. At present fused silica and leached glass columns
coated with the polysiloxanes OV-1, OV-101, OV-73, SE-54 and
the polyethylene glycol Superox 20M provide the best choice for
everyday capillary GC. The exploitation of the selectivity of
the mixed phases (OV-1 and Superox 20M) coated on fused silica
columns can be of great help in practice.
At present efforts are directed towards cross-linked columns
and towards the evaluation of new polar silicone gums.

ACKNOWLEDGEMENT

M. Van Roelenbosch wishes to thank the Institute of Scientific
Research in Industry and Agriculture (I.W.O.N.L.) for a grant.
We thank Professor M. Verzele for his advice and helpful dis-
cussions.

REFERENCES

1. K. Grob and G. Grob; JHRC & CC, 2 (1979) 31.
2. G. Schomburg, H. Husmann and H. Borwitzky, Chromatographia,
   12 (1979) 651.
3. M. Godefroot, M. Van Roelenbosch, M. Verstappe, P. Sandra
   and M. Verzele, JHRC & CC, 3 (1980) 337.
4. R. Dandeneau and E. Zerenner, JHRC & CC, 2 (1979) 351.
5. K. Grob and G. Grob, JHRC & CC, 3 (1980) 197.
6. L. Blomberg, K. Markides and T. Wännman, JHRC & CC, 3
   (1980) 527.
7. L. Blomberg, K. Markides and T. Wännman, paper presented at the IV Int. Symp. Capillary Chromatography,
   Hindelang, Mai 1981.

8. C. Madani and E. Chambaz, Chromatographia, 11 (1978) 725.
9. L. Blomberg, K. Markides and T. Wännman, J. Chromatogr. 203 (1981) 217.
10. M. Van Roelenbosch, P. Sandra and M. Verzele, in preparation.
11. P. Sandra and M. Van Roelenbosch, Chromatographia, 14, 6 (1981) 345.
12. P. Sandra, G. Redant, E. Schacht and M. Verzele, JHRC & CC, 4 (1981) 411.
13. K. Grob, G. Grob and K. Grob Jr., J. Chromatogr., 211 (1981) 243.

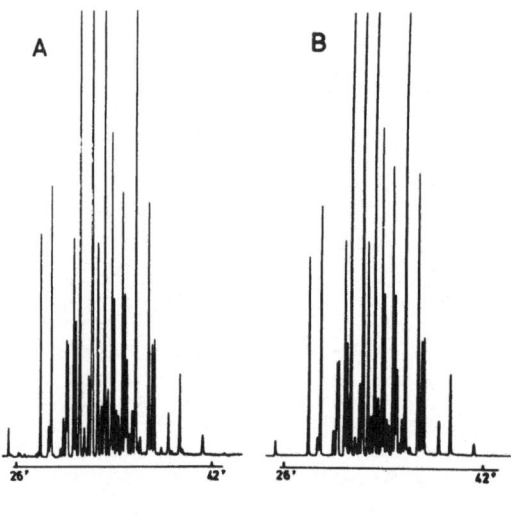

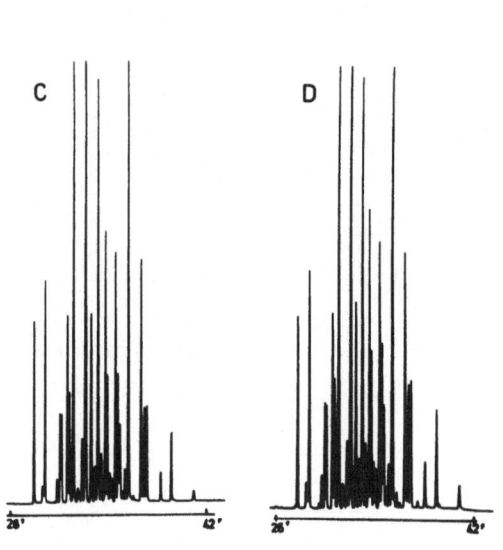

Fig. 3 :

Analysis of Arochlor 1260 on a 20 m x 0.32 mm fused silica column coated with SE-54 and cross-linked as described in ref. 12.
Column temperature 120°C; 10 min isothermal, then programmed to 250°C at 5° min^{-1}. Sensitivity 16 (Carlo Erba 4160). A) After flushing the column with 10 ml n-pentane; B) after flushing the column with 100 ml n-pentane; C) after extraction (column filled) overnight with dichloromethane and flushing with 20 ml dichloromethane; D) after extraction (column filled) overnight with water and flushing with 3 ml methanol and 20 ml dichloromethane.

TECHNIQUES FOR QUANTITATION AND IDENTIFICATION OF ORGANIC MICROPOLLUTANTS
BY HIGH RESOLUTION GAS CHROMATOGRAPHY AND ELEMENT SPECIFIC EMISSION
SPECTROSCOPY

L. Stieglitz, G. Zwick
Institut für Heisse Chemie
Kernforschungszentrum Karlsruhe

Abstract

A combination of a gas chromatograph-microwave plasma detector was modified for the use of high-resolution capillary columns.With the instrument described eleven elements (C, H, D, N, P, O, S, F, Cl, Br,
J) can be monitored simultaneously in the effluent from a gas
chromatograph. To date a detailed study of the response of the
elements C, H, D, F, Cl, Br, J and S was undertaken. The relative
molare response of the elements in compounds varies only by ca. $\pm 10\%$.
The detection limits were determined and found to be in the range of
nanograms per sec. From the elemental ratios measured an identification of the various compounds in possible.

1. EINLEITUNG

Unter den Mikroverunreinigungen des Wassers befindet sich eine Vielzahl von Verbindungen, die Heteroelemente wie Halogene, Phosphor, Schwefel
und Stickstoff enthalten. Aus diesem Grund haben elementspezifische Detektoren innerhalb der letzten Jahre gesteigertes Interesse gefunden. Eine Anwendung selektiver Detektoren bietet vor allem in der Schadstoffanalytik
bedeutende Vorteile, da mit ihnen die Fülle unspezifischer Daten auf eine
problem-orientierte Information reduziert wird. Der Analytiker sieht nur
das, was er unmittelbar zur Lösung der gestellten Fragen benötigt. Ein Detektor mit bedeutendem Anwendungspotential ist in dieser Hinsicht der Mikrowellen Plasmadetektor (MPD). Hier werden die gaschromatographisch getrennten Substanzen in einem durch Mikrowellen induziertem Plasma in die
Elemente zerlegt und zur Emission charakteristischer Strahlung angeregt,
die durch ein optisches Gitter zerlegt und schließlich über Photokathoden
gemessen werden kann.

Der Detektor wurde in Kombination mit Gaschromatographie erstmals von
McCormack zur Analyse von Pesticiden eingesetzt /1/ und hat seitdem erfolg-
reiche Anwendung gefunden, so zur Bestimmung von organischen Quecksilber-
verbindungen /2/, von As, Se und Sb in Umweltproben /3,4,5/ sowie von
flüchtigen Bleialkylen /6/. Eine allgemeine Untersuchung von organischen
Schadstoffen aus Rheinwasser wurde von J. Bonnekessel et al. /7/ mithilfe
des MPD durchgeführt. Trihalomethane wurden von B.D. Quiumby /8/ im Trinkwasser quantitativ bestimmt. In diesem Betrag wird eine Kombination mit
einer Hochleistungskapillarsäule beschrieben, die eine simultane Bestimmung der Elemente Kohlenstoff, Wasserstoff, Deuterium, Fluor, Chlor, Brom
und Schwefel und eine Ermittlung der Elementverhältnisse ermöglicht.

2. Experimentelles

Instrumentierung: Ein Perkin-Elmer Gaschromatograph, Mod. F22, wurde
an einen komerziellen Mikrowellen Plasmadetektor, Mod. MPD-850, (Applied
Chromatography Systems, Ltd) angeschlossen. Gegenüber dem bereits früher
beschriebenen System /9/ wurden folgende Änderungen vorgenommen: Der Gaschromatograph wurde für den Einsatz von Glaskapillarsäulen umgebaut

(Abb.1). Der Einspritzblock (1) wurde durch einen splitlosen On-column-Injektor (Fa. DANI) ersetzt. Als Interface zwischen Trennsäule (5) und Plasmakapillare (17) wurde eine Quarzkapillare verwendet, die bis 300°C geheizt werden konnte. Im beheizten Plasmakopf (16) wurde Scavengergas (O_2/N_2), (26) sowie Spülgas (He), (23) mit je 30 ml/min zudosiert. Über das Ventil (25) und (24) und die Umgehungsleitung (22) konnte der Solventpeak rückgespült und ausgeblendet werden. Das zur Aufrechterhaltung des Plasmas notwendige Vakuum (1-3 Torr) wurde durch eine Drehschiebepumpe (Fa. Edwards) - (20) erreicht. Das Plasma wird in der Kapillare (17) durch einen Mikrowellengenerator mit 2450 MHz mit einer maximalen Leistung von 200 W erzeugt. Das Licht von dem Plasma wird in einem Gitterspektrometer zerlegt (in Abb.1 nicht gezeigt) und bei fest eingestellten Wellenlängen die charakteristische Emissionsstrahlung von elf Elementen (C, H, D, N, O, P, S, F, Cl, Br, J) mit Photomultipliern gemessen. Die analoge Registrierung erfolgt über Vielkanalschreiber und liefert sog. Element-Chromatogramme. Jeder Elementkanal kann zur gleichzeitigen Datenerfassung über A/D-Wandler an ein Hewlett-Packard Labordatensystem Mod.3353, angeschlossen werden, das als Multikanalintegrator Retentionszeiten und Flächenwerte der einzelnen Elemente ausdruckt.

Ghost-Korrektur: Durch die Emission des Kohlenstoff-Kontinuums besteht eine gewisse Querempfindlichkeit zu anderen Elementen. Die Größe dieses "Ghost"-Signals ist proportional der vorhandenen Kohlenstoffmenge und kann elektronisch weitgehend eliminiert werden

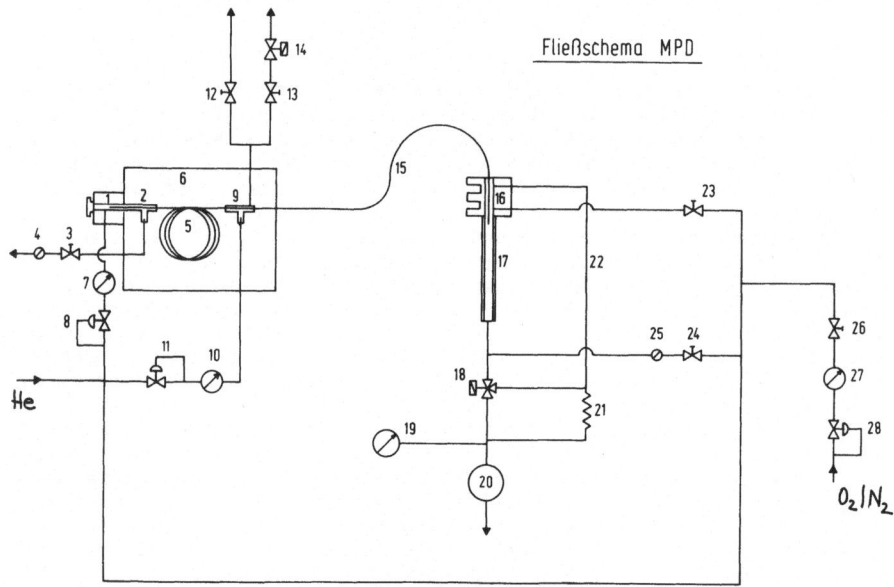

Abb.1: Fließschema des Mikrowellenplasmadetektors
1 On-Column-Einspritzblock, 2 Splitter, 3 Nadelventil, 4 Auf-Zu-Splitausgang, 5 Kapillarsäule, 6 Ofen, 7 Manometer, 8 Druckregler, 9 Kopplungs-T-Stück, 10 Manometer Vordruck T-Stück, 11 Druckregler Vordruck T-Stück, 12 Nadelventil für ständige T-Spülung, 13 Nadelventil für teilweise T-Stückspülung, 14 Magnetventil, 15 Kapillare, beheizt (250°C), 16 Plasmakopf beeizt (250°C), 17 Quarzkapillare (Plasma), 18 Dreiwege-Magnetventil, 19 Vakuum Manometer, 20 Vakuumpumpe, 21 Strömungswiderstand, 22 Umgehungsleitung, 23 Nadelventil für Kopfspülgas, 24 Nadelventil für Rückspülung, 25 Auf-Zu-Ventil, 26 Nadelventil für Scavengegas (O_2/N_2), 27 Manometer für Scavengegas, 28 Druckregler für Scavengegas

Gaschromatographische Bestimmungen: Säule OV-101, 50 m lang, 0,3 mm i.d.; Trägergas Helium, 6.0, Vordruck 1.35 bar. Temperaturprogramm 20°C (6 min isotherm) programmiert mit 8°/min auf 225°C.

Chemikalien: Folgende Verbindungen handelsüblicher Reinheit wurden als Lösungen in Pentan eingesetzt: Dibromchlormethan, Dichlorbenzol, Thioanisol, Benzoesäuremethylester Trichlorbenzol, Hexachlorbutadien, Dibromoctan, Methylthiophen, Methylmerkaptobenzthiazol, Hexadecan, Hexachlorbenzol, Fluorbenzol, m-Fluorbenzotrifluorid, p-Fluortoluol, Tetrafluor 1,2, dibrombenzol, Tetrafluor 1,4 dibrombenzol, Fluorbenzoesäureäthylester. Weiter wurden folgende deuterierte Verbindungen gemessen: D_5-Chlorbenzol, d_{10}-Xylol, d_5-Brombenzol. Die Konzentrtionen der Eichlösungen lagen im Bereich von 200-400 ng/µl. Als interne Standardsubstanzen zur quantitativen Bestimmung wurden eingesetzt: für C, H, Br, Cl Bromdichlortoluol, für S 2-Bromthiophen, Für F Chlortrifluorxylol und für D d_8-Naphtalin.

Aufbereitung der Wasserproben: Als Anwendungsbeispiele wurden Wasserproben des Unterrheins, Trinkwasserproben des Kernforschungszentrums Karlsruhe sowie Proben eines Kläranlagenablaufs untersucht. Die Anreicherung der flüchtigen Verbindungen erfolgte nach K. Grob /10/. Hierzu wurden 0.1 bis 2.0 l der Wasserprobe bei 23°C im geschlossenen Kreislauf ausgegast und die flüchtigen Komponenten auf einem Mikro-Aktivkohlefilter adsorbiert. Nach Elution mit insgesamt 14 µl CS_2 und Zusatz des internen Standards wurden die Eluate direkt "on-column" eingespritzt (1-2 µl).

3. Ergebnisse und Diskussion

In Abb. 2 sind die Kapillarchromatogramme für Kohlenstoff, Schwefel, Chlor und Brom gezeigt. Nach der Injektion (on-column) wurde der Lösungsmittelpeak vollautomatisch ausgeblendet und über die Umgehungsleitung /22/ rückgespült. Im Vergleich zu Messungen mit einem Flammen-Ionisationsdetektor wurde keine Verschlechterung der Trennleistung und der Peakform festgestellt. In Abb.3 sind weitere Elementchromatogramme dargestellt, wobei zusätzlich die Elemente Fluor, Wasserstoff und Deuterium gezeigt sind. Die spezifische Bestimmung von Wasserstoff (10 mV Schreiberempfindlichkeit) erfolgt ohne Probleme. Der Anstieg der Null-Linie bei höheren Temperaturen wird durch Desorption von Spuren Wasser erklärt. Der Deuteriumkanal, ebenfalls 10 mV Schreiberempfindlichkeit, zeichnet sich durch besonders geringes Rauschen aus. Bei den volldeuterierten Verbindungen 5,6,7 und 12 treten kleine Wasserstoffpeaks auf. Eine mögliche Erklärung dafür ist a) Verunreinigung der Präparate durch partiell oder nicht deuterierte Verbindungen b) Beeinflussung des H-Kanals infolge einer Querempfindlichkeit durch den D-Kanal. Die Ursache wird gegenwärtig noch untersucht.

Quantitative, elementspezifische Bestimmungen

Eichlösungen der oben angegebenen Verbindungen werden gaschromatographisch mit dem MPD analysiert und die relative Anzeigeempfindlichkeit (RMR) pro Mol Kohlenstoff, Wasserstoff, Deuterium, Fluor, Chlor, Brom und Schwefel errechnet. Die Werte wurden dabei auf geeignete interne Standardsubstanzen bezogen. Die Ergebnisse sind mit ihren Standardabweichungen in Tab.I gezeigt. Wie ersichtlich, sind die RMR-Werte für ein bestimmtes Element innerhalb geringer Schwankungsbreiten unabhängig von der Verbindung, in der es vorliegt. Damit werden bei Verwendung eines entsprechenden internen Standards quantitative Bestimmungen der jeweiligen Elemente ohne weitere Eichungen ermöglicht. Eine Variation des Gasstroms in der Plasmakapillare (Scavenger + Make-up) von 30 bis 60 ml/min brachte keine Änderung der Werte.

Die Menge des Elementes A errechnet sich als

$$m_A = \frac{m_{STD}}{F_{STD}} \cdot F_A \cdot f$$

wobei m_{STD} die eingesetzte Menge des internen Standards, F_A und F_{STD} die Flächenwerte des Substanz- und Standardpeaks und f der Standardkorrekturfaktor aus Tab.I ist.

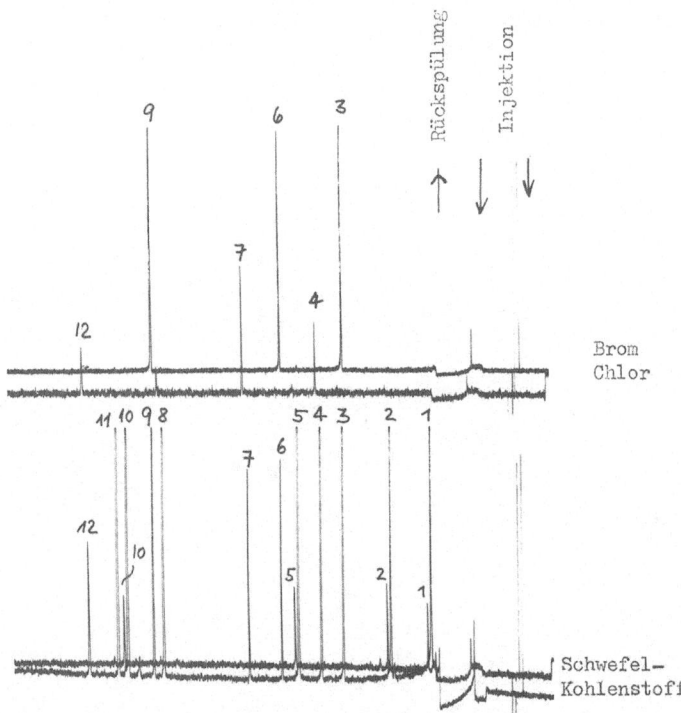

Abb.2: Kapillar-Elementchromatogramme für Kohlenstoff, Schwefel, Chlor und Brom; experimentelle Details siehe Text; Peakzuordnung: 1. Thiophen, 2. 3-Methylthiophen, 3. Brombenzol, 4. 1,3 Dichlorbenzol, 5. Thioanisol, 6. Tetrafluor-1, 4-Dibrombenzol, 7. Hexachlorbutadien, 8. 1-Chlordodekan, 9. 1,8-Dibromoctan, 10. 2-Methyl-1merkaptobenzthiazol, 11. Hexadecan, 12 Hexachlorbenzol.

Die Möglichkeit der spezifischen Messung von Wasserstoff und Deuterium ergibt neue Perspektiven bei der Auswahl von Substanzen als interne Standards. Es wird vorgeschlagen, den Wasserproben vor der Durchführung von Anreicherungs- und Trennverfahren repräsentativ für die zu untersuchenden Substanzen oder Substanzklassen eine volldeuterierte Verbindung als Leitsubstanz zuzusetzen. Über die Messung des D-Signals kann im Konzentrat eine substanzspezifische Bestimmung der Wiederfindungsraten unabhängig von der quantitativen Analyse erfolgen.

Aus der Empfindlichkeit (Impulse /g Element) und dem Rauschen wurde die Nachweisempfindlichkeit errechnet. Die Werte sind in Tab.II gezeigt. Für die untersuchten Elemente liegen die Nachweisgrenzen im Bereich von 0.2 bis 8 Nanogramm /sec. Die Selektivität gegenüber Kohlenstoff ist ebenfalls in Tab.II enthalten. Die Zahlenwerte bedeuten das Vielfache, um das ein Element besser angezeigt wird, als die gleiche Menge Kohlenstoff.

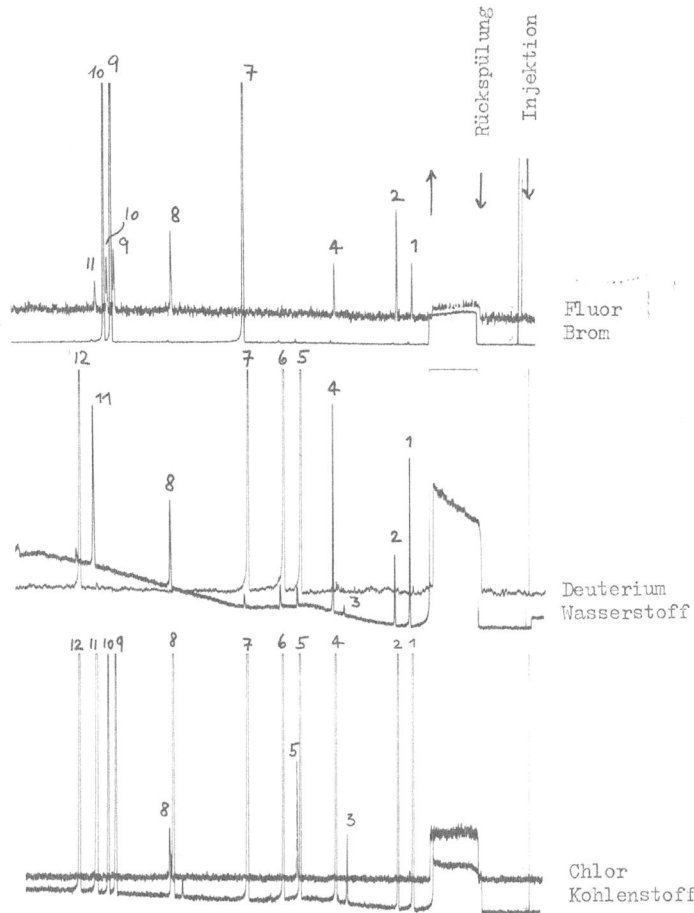

Abb.3: Kapillar-Elementchromatogramme für Kohlenstoff, Fluor, Chlor, Brom, Wasserstoff und Deuterium. Experimentelle Details siehe Text; Peakzuordnung: 1. Fluorbenzol, 2. m-Fluorbenzotrifluorid, 3. KW, Verunreinigung des Lösungsmittels, 4. p-Fluortoluol, 5. Deuterochlorbenzol (d_5), 6. p-Deuteroxylol (d_{10}),7. Deuterobrombenzol (d_5), 8. α'-Chlor-$\alpha, \alpha, \alpha,$- -Trifluor-m-xylol, 9. Tetrafluor- 1,2-dibrombenzol, 10. Tetrafluor 1,4 dibrombenzol, 11. o-Fluorbenzoesäureäthylester, 12. Deuteronaphthalin (d_8)

Elementverhältnisse. Mit der beschriebenen Versuchsanordnung können die Elementchromatogramme simultan aufgenommen werden. Bezieht man nun die relative molare Anzeigeempfindlichkeit eines Elementes A auf die von Kohlenstoff C, so ergeben sich Verhältnisfaktoren f_A, mit denen aus den Flächenwerten der jeweiligen Elementchromatogramme die Elementverhältnisse A/C der unbekannten Substanzen ermittelt werden können. Die anhand der Eichlösungen bestimmten Werte von F_A für einen Röhrenstrom von 125 mA sind: für Wasserstoff 0.26 $\pm$ 0.02; für Chlor 1.25 $\pm$ 0.04; für Brom 9.62 $\pm$ 0.22 und Schwefel 0.70 $\pm$ 0.03. Die Elementverhältnisse A/C ergeben sich nach der Beziehung

$$\frac{A}{C} = \frac{F_A}{F_C} \cdot \frac{1}{f_A}$$

Tab. I   Relative molare Anzeigeempfindlichkeit

Element	Wellenlänge Å	interner Standard	relative molare Anzeige RMR
Kohlenstoff	2478	Bromdichlortoluol	$1,05 \pm 0,13$
Wasserstoff	4861	"	$0,998 \pm 0,18$
Chlor	4794	"	$1,03 \pm 0,11$
Brom	4705	"	$1,16 \pm 0,14$
Schwefel	5454	Bromthiophen	$0,96 \pm 0,10$
Fluor	6856	Chlor-trifluor-xylol	$0,98 \pm 0,13$
Deuterium	6562	d_8-Naphthalin	$1,043 \pm 0,03$

eingespritzte Elementmenge: 50-1oo ng

Tab. II   Nachweisempfindlichkeit und Selektivität des MPD

Element	Nachweisempfindlichkeit g/s	Selektivität
Kohlenstoff	$0,7 \cdot 10^{-9}$	1
Wasserstoff	$0,24 \cdot 10^{-9}$	nicht bestimmt
Chlor	$8,1 \cdot 10^{-9}$	3oo
Brom	$0,42 \cdot 10^{-9}$	6oo
Schwefel	$5,3 \cdot 10^{-9}$	3oo
Fluor	$1 \cdot 10^{-9}$	5oo

Tab. III   Anwendung des MPD auf Analyse von Wasserproben

Substanzklasse : Organohalogenverbindungen

Retentions-zeit	Konzentration ng/l				Elementverhältnisse gemessen			Elementverhältnisse berechnet			Verbindung
	C	H	Br	Cl	H/C	Br/C	Cl/C	H/C	Br/C	Cl/C	
Rhein (Duisburg)											
11.4	57	18	-	180	3.00	-	1.07	-	-	-	unbekannt
12.36	100	-	-	290	1,43	-	0,94	1,00	-	1,00	CH$_2$Cl$_2$
13.70	75	-	-	660	-	-	2,88	1,00	-	3	CHCl$_3$
15.09	65	-	-	170	2,24	-	0,86	2	-	1	C$_2$H$_4$Cl$_2$
15.65	110	-	-	460	0,65	-	1,35	0,5	-	1,5	C$_2$HCl$_3$
22.31	130	-	-	730	-	-	1,86	-	-	2	C$_2$Cl$_4$
32.52	90	-	-	70	1,15	-	0,32	0,66	-	0,33	C$_6$H$_4$Cl$_2$
32.80	180	-	-	160	0,9	-	0,31	0,66	-	0,33	C$_6$H$_4$Cl$_2$
33.86	490	-	-	460	0,88	-	0,31	0,66	-	0,33	C$_6$H$_4$Cl$_2$
Trinkwasser (KFK)											
12.9	35	-	260	180	-	1,07	1,81	1	1	2	CHBrCl$_2$
13.01	30	-	-	120	-	-	1,27	0,5	-	1,5	C$_2$HCl$_3$
18.03	70	6,5	990	210	1,32	2,25	1,10	1	2	1	CHBr$_2$Cl
21.51	40	-	970	-	1,26	3,32	-	1	3	0	CHBr$_3$
23.70	Standard				0,67	0,25	-	0,75	0,25	-	Bromthiophen
46.05	Standard				0,76	0,14	0,28	0,71	0,14	0,28	Bromdichlortoluol

wobei F_A und F_C die Flächenwerte der jeweiligen GC-Peaks des Elements A und von Kohlenstoff sind. Aus den ermittelten Elementverhältnissen kann unter Miteinbeziehung der Retentionszeit eine Identifikation erfolgen.

Anwendung zur Wasseranalyse. Die Methode wurde auf verschiedene Proben von Oberflächengewässern, Trinkwasser sowie auf Proben einer Kläranlage (Kernforschungszentrum) angewandt. Die bisherigen Unteruschungen beschränkten sich dabei auf leicht flüchtige Schadstoffe, die mit den Ausgasverfahren erfaßbar sind. Die Ergebnisse sind in Tab.III zusammengefaßt. Neben den Elementen Kohlenstoff, Wasserstoff, Chlor und Brom wurde ebenfalls Fluor und Schwefel detektiert. In den untersuchten Proben wurden dabei keine Signale für F registriert. Schwefel trat nur in den Klärwerksabläufen (als Dimethyldisulfid) auf. Flüchtige Verbindungen mit F und S können in den Proben von Fluß- und Trinkwässern damit nur unterhalb der Nachweisgrenze von ca. 0.5-1 ng/l (bei 2 l Wasserprobe) vorhanden sein. Mithilfe der zugesetzten internen Standards wurden die vorliegenden Konzentrationen von C, Br, Cl über die relativen molaren Responsefaktoren (Tab.I) berechnet. In einem weiteren Versuch wurde die Methode auch zur Analyse von Proben aus dem Ablauf der Kläranlage des Kernforschungszentrums eingesetzt. Als einzige Chlorverbindung wurde dabei p-Dichlorbenzol in Konzentrationen von 8 ug/l gefunden, das anhand seiner Retentionszeit und des Cl/C-Verhältnisses von 0.33 identifiziert wurde. In Tab.III sind ebenfalls die gemessenen Elementverhältnisse aufgeführt. Anhand dieser Werte sowie der Retentionszeiten konnten die wesentlichsten Organohalogenverbindungen identifiziert werden. Vor allem die Verhältnisse Cl/C und Br/C zeigten dabei mit den theoretisch errechneten Werten gute Übereinstimmung. Sie variierte um 10-15% um die Soll-Werte. Teilweise größere Variationen zeigten die H/C Werte, die somit nur als zusätzliche Identifizierungshilfe herangezogen werden können.

Zusammenfassung

Der Mikrowellenplasmadetektor stellt unter den spezifischen Detektoren eine besonders interessante Alternative dar und weist folgende Vorzüge auf: a) Die relative molare Anzeigewahrscheinlichkeiten sind weitgehend ($\pm$ 10%) unabhängig von der Art der Verbindung. Damit können mit einer ausreichenden Genauigkeit die Elementkonzentrationen auch zunächst unbekannter Verbindungen ermittelt werden. b) Die einzelnen, in der Schadstoffanalytik interessierenden Elemente (C, H, O, N, P, F, Cl, Br, J, S) können unabhängig voneinander detektiert werden. Dies ist besonders interessant für Verbindungen mit F, Br und Jod, für die kein geeigneter, genügend spezifischer Detektor existiert. c) Aus den simultan gemessenen Elementchromatogrammen lassen sich die Elementverhältnisse ermitteln, die zusammen mit den Retentionszeiten zur Identifizierung herangezogen werden können.

5. Literatur

/ 1/ A.J. McCormack, S.C. Tong, W.D. Cooke; Anal. Chem. 37, /1965/, 1470-1476

/ 2/ C.A. Bache, D.J. Lisk; Anal. Chem. 37, /1965/, 1477-1480

/ 3/ Y. Talmi; Anal. Chim. Acta, 74. /1975/ 107-117

/ 4/ Y. Talmi, A.W. Andren; Anal. Chem. 46, /1974/ 2122-2126

/ 5/ Y. Talmi, V.E. Norvell; Anal. Chem. 47 /1975/ 1510-1516

/ 6/ D.C. Reamer, W.H. Zoller, T.C. O'Haver, Anal. Chem. 50 /1978/, 1449-1453

/ 7/ J. Bonnekessel, L. Braunstein, K. Hochmüller; Vom Wasser 50 /1978/ 191-229

/ 8/ B.D. Quimby, M.F. Delaney, P.C. Uden, R.M. Barnas; Anal. Chem. 51 875-880

/ 9/ L. Stieglitz, W. Roth; in Analytical Techniques in Environmental Chemistry, ed J. Albaiges, Pergamon Press, Oxford, New York /1980/ p. 345-352

/10/ K. Grob, F. Zürcher, J. Chromatogr. 117 /1976/ 285-294

APPLICATIONS IN GAS CHROMATOGRAPHY

C. O'DONNELL

Water Resources Division, An Foras Forbartha,

St. Martin's House, Waterloo Road,

Dublin, 4, Ireland.

Summary

These notes describe solutions to some practical problems in gas
chromatography. The first note concerns the use of a second detector
to monitor internal standards, while the second describes a flow-
controlled effluent splitter for use with relatively large
concentrations of trihalomethanes from a purge and trap system.

1. USE OF INTERNAL STANDARDS WITH DUAL DETECTORS

Extracts of environmental water samples contain numerous compounds and
give complex pictures on GC analysis. For this reason it is difficult
using a single detector to select internal standards which will not distort
the chromatograms and possibly affect the results in a particular case.
Nevertheless, internal standards are required for these applications as the
run times tend to be fairly long, 20 to 40 minutes (compared to many routine
assays of 1 to 2 minutes), and repetition of injections is very time-
consuming. The internal standard peak or peaks can give an assurance that
the injection has gone as intended and that problems such as septum leaks
are absent. When used with inlet splitting systems for capillary work the
internal standard gives a check that sufficient sample has gone onto the
capillary column. For these reasons it is desirable to find a way of
using internal standards with environmental extracts without obscuring the
chromatograms or risking interference with the internal standards
themselves from components of the extract which may have similar retention
times.
    As a general approach to the analysis of surface water extracts, an
effluent splitter is used dividing the column effluent between ECD and FID
detectors. At typical concentrations the FID response is very low and
therefore it has been decided to use this detector for internal standards.
Samples are spiked with 100 ppm of C_{21} and C_{23} n-hydrocarbons, which give a
good response on a clean FID baseline without interfering with the ECD
baseline. The ECD sensitivity is illustrated by the chromatogram obtained
from a standard mixture of 0.1 ppm each of lindane, parathion and dieldrin
in methanol containing 100 ppm each of C_{21} and C_{23} n-hydrocarbons (fig. 1).
The first peak on the FID trace is nonadecane, which was added to the
pesticide mixture as an additional label. Any accidental contamination of
samples or syringes by the standard is easily detected, and co-injection of
standard and sample can be monitored by the height of the C_{19} peak relative
to the other hydrocarbons.
    In general, the use of the FID detector with hydrocarbon internal
standards at a higher concentration than the background level in the
samples allows the analysis of complex extracts to be monitored without
interference with the sensitive ECD response. The response of the FID
detector is stable and linear over a wide range, and it is possible to use
this response to study the drift of the ECD detector, as often occurs after
the injection of dirty samples.

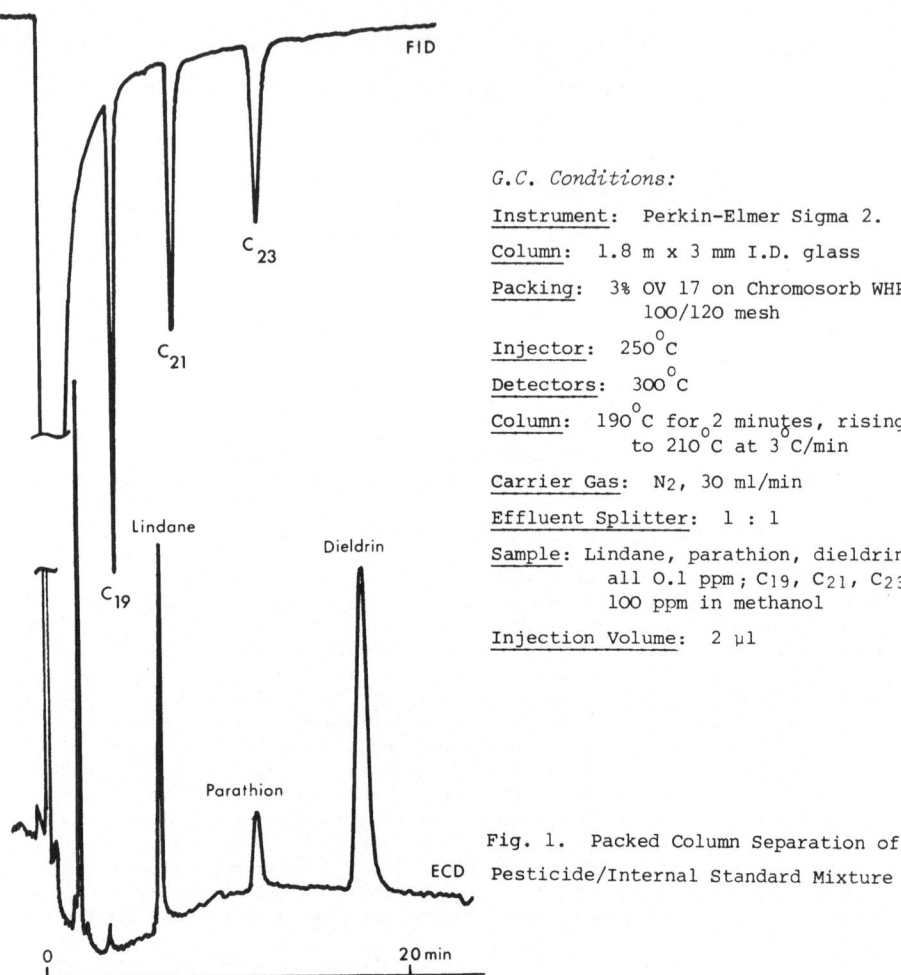

G.C. *Conditions:*

Instrument: Perkin-Elmer Sigma 2.

Column: 1.8 m x 3 mm I.D. glass

Packing: 3% OV 17 on Chromosorb WHP
100/120 mesh

Injector: 250 °C

Detectors: 300 °C

Column: 190 °C for 2 minutes, rising
to 210 °C at 3 °C/min

Carrier Gas: N2, 30 ml/min

Effluent Splitter: 1 : 1

Sample: Lindane, parathion, dieldrin
all 0.1 ppm; C19, C21, C23
100 ppm in methanol

Injection Volume: 2 µl

Fig. 1. Packed Column Separation of
Pesticide/Internal Standard Mixture

This standard system was used initially with packed columns and using a carrier gas flow rate of 30 ml/min. For this purpose a standard Perkin-Elmer splitter kit was employed with a split ratio of 1 : 1 (fig. 1). For use with SCOT columns a make-up gas (N_2, 25 ml/min) is introduced to increase the effluent volume prior to splitting. Peaks in this case are sharper, but there is still no interference between detector signals (fig. 2). An inlet splitter is used with a low split ratio and the FID standards are used as a measure of the amount of sample reaching the column.

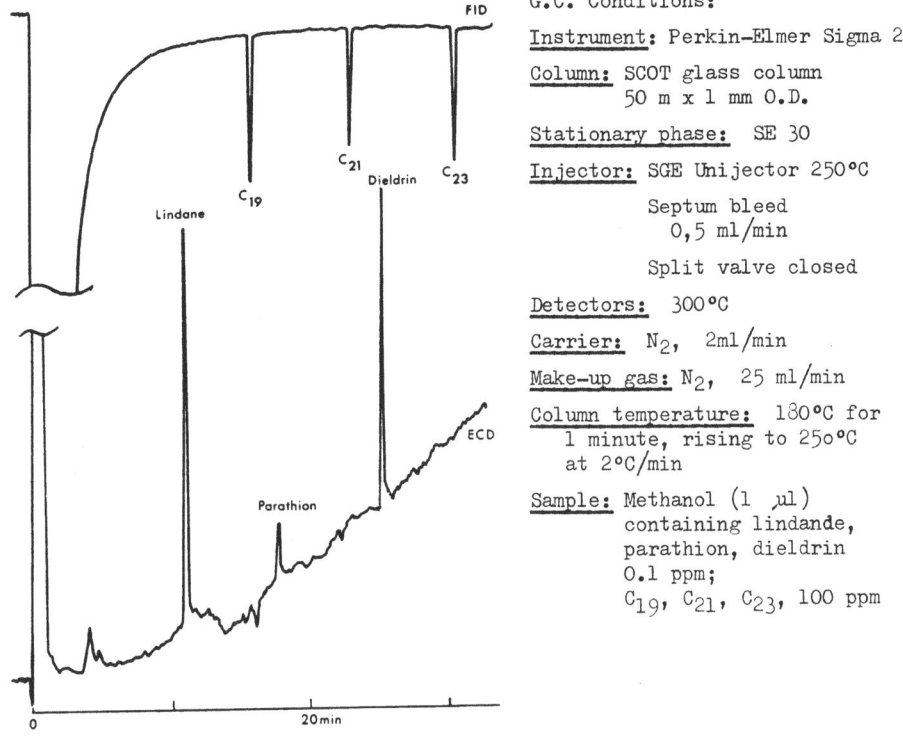

G.C. Conditions:

Instrument: Perkin-Elmer Sigma 2

Column: SCOT glass column
    50 m x 1 mm O.D.

Stationary phase: SE 30

Injector: SGE Unijector 250°C

    Septum bleed
    0,5 ml/min

    Split valve closed

Detectors: 300°C

Carrier: N_2, 2ml/min

Make-up gas: N_2, 25 ml/min

Column temperature: 180°C for
    1 minute, rising to 250°C
    at 2°C/min

Sample: Methanol (1 ,ul)
    containing lindande,
    parathion, dieldrin
    0.1 ppm;
    C_{19}, C_{21}, C_{23}, 100 ppm

Fig. 2   SCOT Column Separation of Pesticide/Internal Standard
Mixture

With WCOT columns of fused silica it is possible to insert the column
end up to the detector avoiding the use of make up gas and also minimising
band broadening in the transfer from column to detector. To take full
advantage of this, a 25 metre column was halved to form two similar 12.5 m
columns which were both installed in the inlet splitter. This system also
appears to work well with one column leading to each detector (fig. 3).
When used with a high split ratio or Grob splitless injection the FID
trace gives a reliable indication that the sample has been transferred onto
the column and is therefore very useful in setting up the conditions for an
analysis. In this case the internal standards are on a different column
and care should be taken to ensure that both columns are performing
comparably.

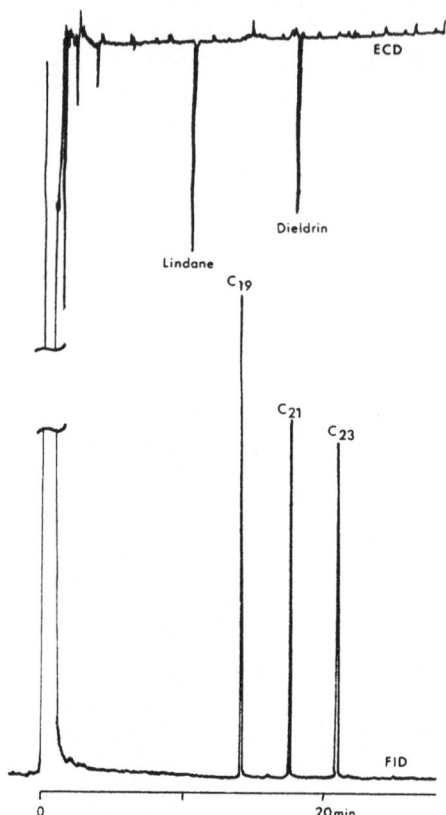

G.C. Conditions:

Instrument: Perkin-Elmer Sigma 2

Column: WCOT fused silica column
12.5 m x 0.22 mm I.D.

Liquid phase: CP-Sil 5

Injector: SGE Unijector 250°C

Septum bleed: 0.5 ml/min

Split flow: 25 ml/min

Split delay: 30 sec

Detectors: 300°C

Carrier: N_2 0.6 bar

Column temperature: 60°C for
1 minute, rising to 120°C at
25°C/min then to 290°C at
8°C/min

Sample: Methanol (1 µl) containing
lindane, dieldrin 0.1 ppm,
C_{19}, C_{21}, C_{23}, 100 ppm

Fig. 3   WCOT Column Separation of
Pesticide/Internal
Standard Mixture

2.   AN EFFLUENT SPLITTER FOR G.C. WITH EXTERNAL CONTROL

The use of dynamic headspace sampling for the removal of volatile organic compounds from water samples gives great improvements in sensitivity over normal practice as most of the volatiles from a large water sample can be introduced into a gas chromatograph. For example, using a purge and trap system with a 5 ml water sample, the amount chromatographed is up to a thousand times greater than would be introduced by direct injection of a 5 µl aliquot. However, where the purge is used as cleanup treatment rather than for preconcentration, this added sensitivity can cause certain problems. In the analysis of trihalomethanes in drinking water using a purge and trap system the volume purged was 5 ml. This gave very large signals on the gas chromatograph which could be handled by attenuating the amplifier of the flame ionisation detector (FID) but which overloaded the electron capture detector (ECD). The U.S. standard of 100 µg/l for trihalomethanes is equivalent to 0.5 µg in a 5 ml sample, so that it was necessary to reduce the amount reaching the ECD.

The column effluent was split between FID and ECD and trials with standard solutions showed that a large split ratio would be needed, with very little going to the ECD. A standard trihalomethane solution (50 µl containing chloroform 500 ng, bromodichloromethane 100 ng, dibromochloromethane 100 ng and bromoform 100 ng) was introduced into the purge apparatus. Bromochloromethane solution (30 µl containing 300 ng) was added as an internal standard and the mixture was purged for 15 minutes at 30°C with nitrogen (30 ml/min). The volatiles were trapped on Porapak Q (80-100 mesh) in a stainless steel tube 5 cm x 4.5 mm I.D. The trap was then transferred to a thermal desorption apparatus (Bendix Flasher) connected to the injection port of a gas chromatrograph (Perkin-Elmer Sigma 2). The desorption temperature was 185°C into a carrier flow of 30 ml/min nitrogen. Separation was on a 1.8 m x 3 mm glass column packed with 80-100 mesh Carbopack c / 0.2% Carbowax 1500, with a temperature gradient from 60°C to 160°C at 8°C per minute after a delay of 3 min at 60°C. The column effluent was split 1 : 1 between ECD and FID detectors using a standard Perkin-Elmer splitter kit. Under these conditions the trihalomethane peaks were so large as to saturate the ECD, which had a flow of 25 ml/min of argon (95%)/methane (5%) as a purge gas.

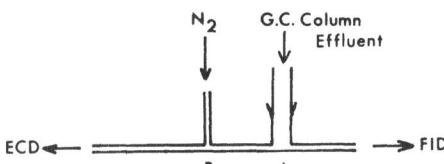

Fig.4 .   T-piece arrangement - diagrammatic

A second T-piece was attached to the ECD inlet downstream from the splitter, as shown diagrammatically in fig.4 . By introducing a flow of nitrogen into the system via this T-piece, sufficient back pressure could be set up in the line A-B to affect the split ratio at A. The FID signal was increased, and the additional gas flow through the ECD also decreased the ECD signal. The correct make-up flow rate could only be found by experiment. The make-up T-piece was connected to the second gas flow controller of the gas chromatograph which gave a steady flow of gas which could be controlled from outside the oven while a run was in progress. As each peak emerged, adjustments to the make-up flow could be made to bring them on scale. At typical make-up flow of 5 to 6 ml/min the trihalomethane peaks were down to 25% of the saturation level (fig. 5). The split ratio- once set, was stable, allowing reproducible operation of the full purge and trap system.

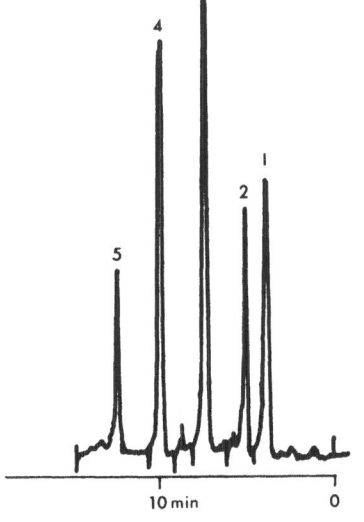

1.   Bromochloromethane 300 ng
                    (Internal Standard)

2.   Chloroform 500 ng

3.   Bromodichloromethane 100 ng

4.   Dibromochloromethane 100 ng

5.   Bromoform 100 ng.

Fig. 5 .   Analysis of trihalomethane standard by purge and trap system

THE USE OF FUSED-SILICA CAPILLARY COLUMNS IN GAS CHROMATOGRAPHY

AND GAS CHROMATOGRAPHY/MASS SPECTROMETRY

D. MEEK and W.J. REID
Water Research Centre, Stevenage Laboratory, Elder Way,
Stevenage, Herts. SG1 1TH

Summary

Fused-silica capillary columns have recently come into common use in
the analysis of environmental samples by gas chromatography (GC) and
gas chromatography-mass spectrometry (GC-MS). A method of installing
a fused-silica column into a GC-MS system, whereby 'dead space' is
minimised has been developed and has been in successful use for a
considerable period. A comparison of the performance of several
makes of fused-silica column has been made with that of a glass
capillary. The response of standard mixtures of pure compounds with
varying polarities, and that of a sample, of a factory effluent were
compared using the two types of column and measurements of column
efficiency made. For most substances tested the performance of the
glass and fused-silica columns was found to by similar but the latter
was much more convenient to use. The useful life of the columns was
variable but in general the glass columns lasted for a longer period
than fused-silica. Fused-silica columns deactivated with Carbowax
20M had a shorter lifetime, when used at temperatures of up to 250°C,
than those deactivated by silicone treatment.

1. INTRODUCTION

In the analysis of trace quantities of the organic components of
environmental samples by gas chromatography (GC) and gas chromatography -
mass spectrometry (GC-MS) the quality of the GC column is a significant
factor. The column should permit the separation of as many compounds as
possible in a reasonable analysis time. To this end it should be efficient, (ie a capillary column) and compounds injected into it should not be
altered or decomposed by the coating or the material from which the column
is made. Similarly the injector should be as inert as possible.
Dandenau[1,2] has described the preparation of use of fused-silica
columns and these are now available from several commercial suppliers.
Lipsky et al[3] have tested these columns and concluded that fused silica-
glass tubing containing less than 1 ppm of metal oxides makes excellent
capillary columns for GC use. This type of column tubing when drawn with·
a very thin wall and externally coated with polyimide is flexible, which
makes handling simpler and also enables a direct link between the GC and
MS to be easily made. It has been the practice with fused silica columns
to deactivate the tubing with Carbowax 20M before coating with the
stationary phase. Whilst this has advantages in reducing 'tailing' of
polar compounds and in producing a more even coating of the stationary
phase, the column eventually produced lacks thermal stability, sometimes
produces distorted retention times and has a relatively short life if used
consistently at the maximum operating temperature. Columns are now
produced which have been undercoated using a silicone deactivation technique and which seem to be more satisfactory than those treated with
Carbowax.

2. EXPERIMENTAL

The chromatograph used for the GC tests was a Carlo-Erba 'Fractovap' 2900 fitted with sub-ambient temperature control, operated under the following conditions:-

Type of injection	- splitless
Injection volume	- 1 μl
Injection temperature	- 300°C
Detector temperature (FID)	- 280°C
Column oven programme	- 4° per min, 30°C-240°C
Carrier gas	- hydrogen
Carrier gas flow rate	- 1-2 ml per min
Make up gas flow rate	- 30 ml per min
Vent delay	- 30 sec.

The chromatograph which was linked to the mass spectrometer was a Hewlett-Packard Model 5710A fitted with a Grob-type injector, Model 18740A. The operating conditions were the same as those described for the Carlo Erba GC except that the flame ionisation detector was replaced by the mass spectrometer. The MS used was a VG 70-70 magnetic sector instrument operating under the following conditions:-

Electron energy	- 70 eV
Emission current	- 200 μA
Scan time	- 1 sec per decade
Accelerating voltage	- 4 kV
Mode	- EI
Resolution	- ~1000.

2.1. GC COLUMNS

The GC columns used in these tests were:-
1. Jaeggi glass 50 m WCOT column coated with OV1
2. Hewlett-Packard 50 m fused-silica WCOT column coated with OV1
3. Quadrex 'Black Knight', 50 m fused-silica WCOT column coated with OV1
4. Chrompak 25 m fused-silica WCOT column coated with CP-sil-5
5. GC^2 50 m fused-silica WCOT column coated with OV1.

Columns 2 and 3 were undercoated with Carbowax 20 m and columns 4 and 5 by silicone deactivation. The Jaeggi column was prepared using Grob's[4,5] technique.

None of the columns used was new; all had been in routine use for a period and had received 20-100 injections prior to the tests of column performance. The Hewlett-Packard fused-silica capillary had received more use than the others. The types of sample injected on to the columns prior to these tests were broadly similar but there is no doubt that the injection of certain types of samples can affect column performance for a considerable time.

2.2. TEST MIXTURES

Two test solutions were used for comparison of the columns together with an ether extract of a sample derived from an effluent containing an industrial waste. The first test solution consisted of a series of n-alkanes from C_6 to C_{22} at a level of 10 ng per μl in hexane and ethyl-benzene solvents. This mixture also contained pyrene, which normally elutes between C_{20} and C_{21} n-alkanes. The other test solution was a Grob[6] standard mixture prepared as shown in Table 1.

Table 1

Compound	Concentration (mg/l)
1. 2,3-butanediol	62.0
2. n-decane	28.3
3. 1-octanol	35.5
4. Nonanal	40.0
5. 2,6-dimethylphenol	32.0
6. n-undecane	28.7
7. 2,6-dimethylaniline	32.0
8. 2-ethylhexanoic acid	38.0
9. Methyl decanoate	42.5
10. Dicyclohexylamine	31.3
11. Methyl undecanoate	41.5
12. Methyl dodecanoate	40.0

2.3. CONNECTIONS

The connection between the MS and the fused silica GC columns was made by passing the end of the column through a piece of stainless steel tubing inside the heated interface into the source of the spectrometer. Escape of gas from the source ion block was prevented by using a mica-impregnated PTFE collar into which the steel tubing was fitted and which was pushed into the source retractor cup. This collar also acted as an insulator preventing electrical leakage between the ion block and the source housing. A $\frac{1}{4}$ inch diameter sleeve was silver-soldered on to the stainless steel tube carrying the column and the system made vacuum-tight by sealing this sleeve on to the re-entrant with a graphite olive. The arrangement is shown in Fig. 1 and had the advantage that the column was in a continuous length from the bottom of the injector to the MS source, without the necessity for a glass jet.

The bottom of the injector was also modified to avoid the use of a glass hanger and shrink-fit connections, which are difficult to make when fused silica columns are used. The arrangement is shown in Fig. 2.

When the glass capillary column was used the end of the column was connected to a length of fused silica tubing which passed into the MS source in the manner described for fused silica columns. This eliminated any dead-space effects which might have biased the comparison had the connections been made in the normal manner using glass-lined stainless steel tubing and a glass jet. Shrink-fit PTFE tubing was used to make the connection between the glass column and the fused-silica tubing.

Splitless injections were used throughout and were made into a glass or fused silica-lined Grob-type injector, with back flushing after 30 seconds. The end of the column in the chromatograph was pushed into the detector until it was level with the top of the jet of the FID.

3. RESULTS AND DISCUSSION

3.1. EFFICIENCY

A measure of the efficiency of each column was obtained by measuring its separation number (n_{sep}) which is the possible number of separated peaks between two n-paraffin peaks resulting from components of consecutive carbon number.

The measurements were made using C_{10}, C_{11}, and C_{12} n-alkanes and $n_{(sep)}$ calculated from:-

$$n_{(sep)} = \left[\frac{d}{Y_1 + Y_2} \right] - 1$$

where d = the distance between successive peaks

Y_1 = the peak width (half height) of first selected peak
Y_2 = " " " " " " second " "

The separation numbers of the five columns tested were

Column	$n_{(sep)}$
1	32
2	12
3	31
4	27
5	32

The shapes of the alkane peaks were satisfactory.
Chromatograms of a complete alkane mixture (C_6-C_{22} using hexane and ethyl benzene as solvents) are shown in Fig. 4a-e. Pyrene was also present in the mixture. The peak eluting between C_9 and C_{10} was acetophenone which was an impurity in the mixture. The Chrompak column[4] gave a separation number only a little less than that of Columns 1, 3 and 5 even though it was only half as long. Experience has indicated that most separations can be carried out satisfactorily using a 25-m rather than 50-m column but when the column has been in use for a prolonged period its injector end may become contaminated with non-volatile material and there is less opportunity for the removal of coils to correct the problem. Column 2 had a lower separation number than the other columns but this column had been used for a considerable period on routine samples before these tests were carried out, and may well have given a better value at an earlier stage. The column deactivation technique using Carbowax 20M is also suspect, as the columns treated in this way deteriorated more rapidly than the silicone pre-treated Carbopak or GC[2] column when used routinely at temperatures up to 250°C.

Chromatograms of the Grob test mixture run on the five columns are shown in Fig. 5a-f. A chromatogram of the mixture run on a conditioned but unused 25-m OV1-coated fused silica column is shown in Fig. 5e. In all the traces, the peaks for the n-alkanes and the methyl esters of the C_{10}-C_{12} acids are well shaped but 2-ethylhexanoic acid in no instance gave a satisfactory peak height or shape although the peak for dicyclohexylamine was well shaped. This suggested that the columns tended to be basic and adsorbed acidic compounds. On the glass column the peak height for 2,6-dimethylphenol was less than that of 2,6-dimethylaniline which also suggested as basic column. Column 2 gave a very small peak for the 2,6-dimethylphenol and Column 3 gave no separation between nonanal and the xylenol so that no comparison could be made with the dimethylaniline peak.

The shape and size of the octanol peak, which gives a measure of adsorption of polar compounds was satisfactory only in the chromatograms obtained from Columns 4 and 5. On the glass and other fused silica columns the peak was broad and tailing and the height of the peak only a small percentage of that expected for complete and undistorted elution. The peak for 2,3-butanediol was satisfactory on Column 2 but with the other columns the peak was small and tended to tail, although this effect is expected when non-polar columns are used[6]. There was a tendency for the diol peak to split into two peaks although it is not known whether this was a chromatographic or injector defect or whether the original material used to prepare the Grob standard contained two isomers of the diol. The mass spectra from the two peaks were identical.

An ether extract of a sample of a trade effluent was injected on to the columns fitted into the GC-MS. The TIC chromatograms produced were compared to detect any differences caused by the column used.

The peaks in the chromatograms were largely alkanes and alkenes and whilst differences were apparent the overall separation of these peaks was very similar. The Chrompak column failed to resolve the trialkyl phosphate from the C15 n-alkane peak although the Quadrex column gave a baseline separation between them and the glass column showed a reasonable separation. The glass column gave a bigger peak for palmitic acid than the fused-silica columns but the latter end of the chromatogram was generally less well resolved on glass compared to fused silica.

3.2. COLUMN INSTALLATION

The installation of all the fused-silica columns was simpler than that of the glass column. With the glass column it was necessary to pass air through the column to burn off the stationary phase at the ends while they were being straightened to fit into the chromatograph; this procedure was unnecessary when fused silica columns were used. The silica columns were also mich less fragile, although scratches on the polyimide coating were found to cause weakness. Cutting these columns was accomplished by heating with a small flame and breaking the column at the point where the coating started to shrivel. This prevented any of the coating finding its way down the column bore. The ferrules used for attaching the columns to the injector were made of graphite or Supeltex M-3 (Supelco Inc). These were fairly soft and could withstand a temperature of 300°C. The harder Vespel olives proved inconvenient as the column had to fit snugly through the hole in the olive, otherwise leaks resulted, and a considerable selection of olives was required in order to find one which would fit any particular column. An important consideration when a fused-silica column is being selected for GC-MS purposes is the way in which the coils are held together. In these tests Columns 3 and 4 were found together without a former by partially fusing the outer coating of the individual coils together. Whilst this resulted in a light, compact column, the coils were difficult to separate. If the end of the column is to pass into the MS source then a considerable length of free column is necessary and this might not be available on columns where the coils are fused. Furthermore, if it becomes necessary to remove the front portion of the column because it has become contaminated with high-boiling material, the length of free column available might be inadequate. Other columns such as those manufactured by Hewlett-Packard or GC[2] are wound on a former and the coils detach easily, making the installation much simpler and facilitating the removal of sections of contaminated column.

3.3. COLUMN LIFE

The life of the columns is variable and the decision to discard is rather subjective, but in general the glass columns last about six months when several routine injections of extracts of water, effluent and sewage sludge are made each day. On the same basis the fused-silica columns deactivated with Carbowax 20M survived for about three months and the silicone pretreated columns for about four or five months. The fused-silica columns were normally programmed up to 250°C whereas the glass columns were not normally operated above 220°C. This difference may account for the differences in column life.

4. CONCLUSIONS

A comparison of the efficiency of glass columns with fused-silica columns did not indicate that either was preferable to the other. Similarly the Grob test mixture and the sample of industrial effluent indicated that whilst glass or silica might be more suitable for some applications than for others, for routine use on aqueous extracts either type of column could be used. The costs of glass and fused-silica columns are very variable and the column performance is by no means related to the price. When convenience of use is considered, however, there is no doubt that fused-silica shows considerable advantages over glass, especially when a mass spectrometer is used as the detector. The system described to connect the end of the column directly into the source of the mass spectrometer has been in everyday use for 2 years and has proved perfectly satisfactory. The elimination of dead space at both the injector and detector ends of the column has resulted in improvements in separation efficiency.

REFERENCES

1. DANDENEAU, R. and ZERENNER, E.H. An investigation of Glass for Capillary Chromatography. J. High Resoln Chromatogr., Chromatogr. Commun. 2, 351 (1979).
2. DANDENEAU, R., BENTE, P., ROONEY, T. and HISKES, R. Flexible Fused Silica Columns, an advance in high resolution chromatography. Am. Lab., September 61 (1979).
3. LIPSKY, S.R., McMURRAY, W.J. and HERNANDEZ, M. Fused Silica Glass Columns for Gas Chromatographic Analyses. J. Chromatogr. Sci. 18, 1 (1980).
4. GROB, K. and GROB, G. A new, generally applicable procedure for the preparation of glass capillary columns. J. Chromatog. 125, 471-485 (1976).
5. GROB, K., GROB, G. and GROB, K. Jr. The barium carbonate procedure for the preparation of glass capillary columns: Further information and developments. Chromatographia, 10, 4, 181-187 (1977).
6. GROB, K. Jr. Comprehensive standardised quality test for glass capillary columns. J. Chromatog. 156, 1-20 (1978).

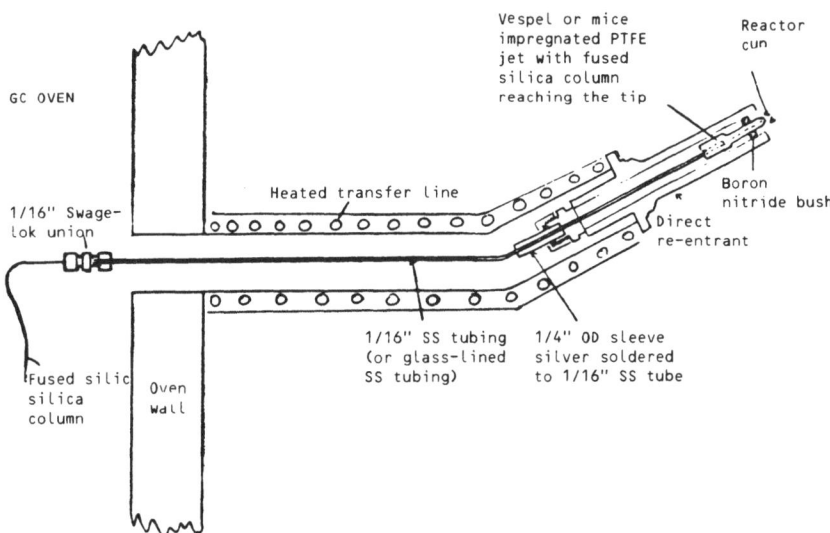

Fig. 1. Connection of a fused silica solumn from GC to the MS source

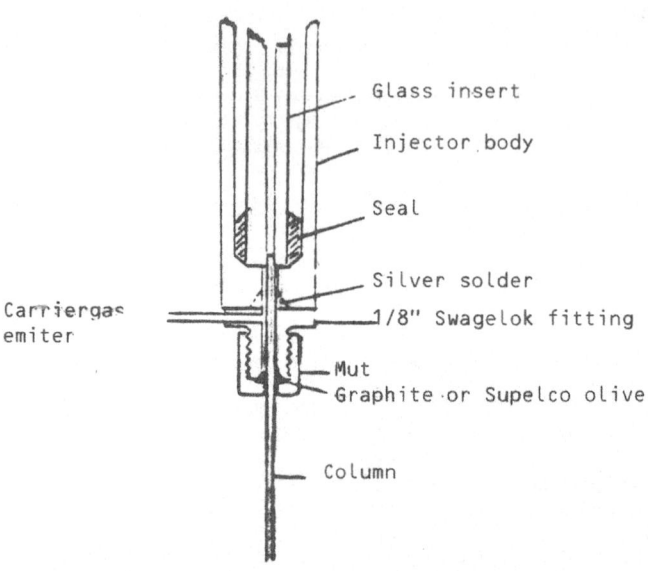

Glass insert

Injector body

Seal

Silver solder

Carriergas emiter

1/8" Swagelok fitting

Nut

Graphite or Supelco olive

Column

Fig. 2. Modified HP injector

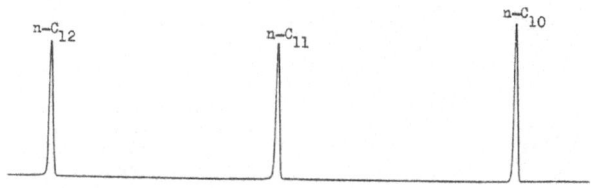

Fig. 3a Glass (Column 1)

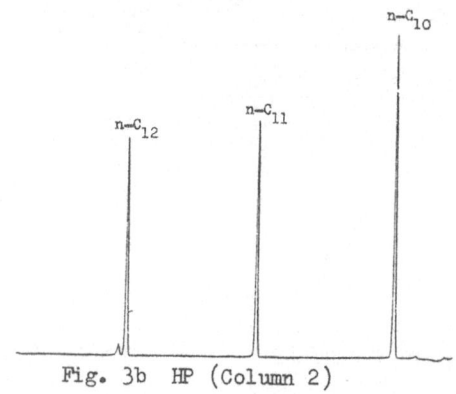

Fig. 3b HP (Column 2)

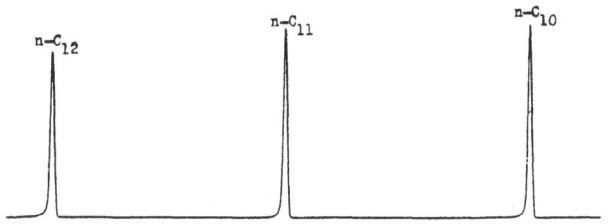

Fig. 3c Quadrex (Column 3)

Fig. 3d Chrompak (Column 4)

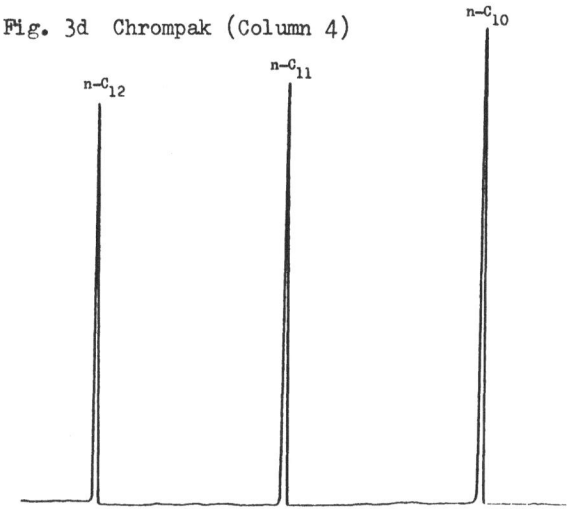

Fig. 3e GC^2 (Column 5)

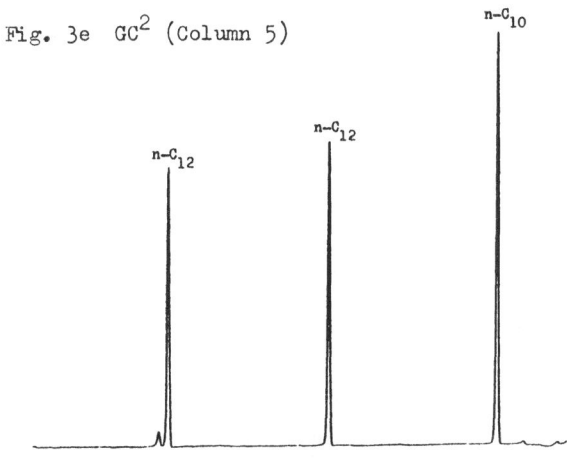

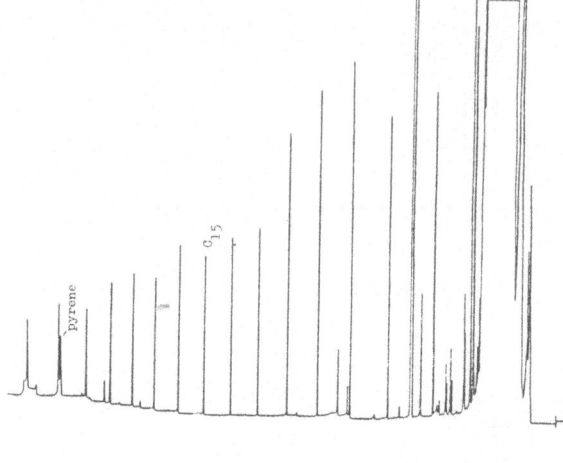

Fig. 4a
C_6-C_{22} s-alkane standard,
glass (Column 1)

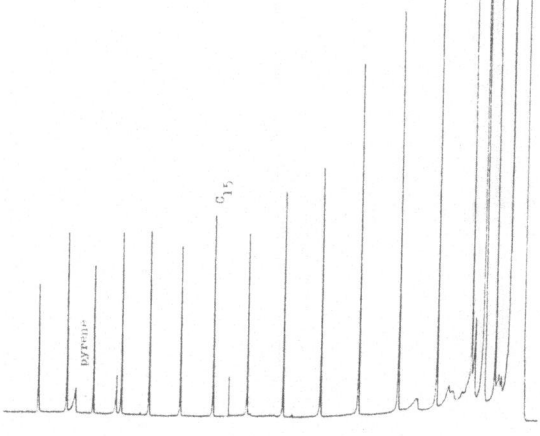

Fig. 4b
C_6-C_{22} n-alkane standard,
NP (Column 2

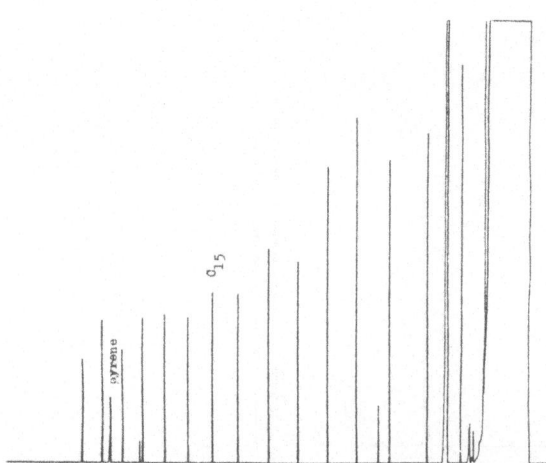

Fig. 4c
C_6-C_{22} n-alkane standard,
Quadrex (Column 3)

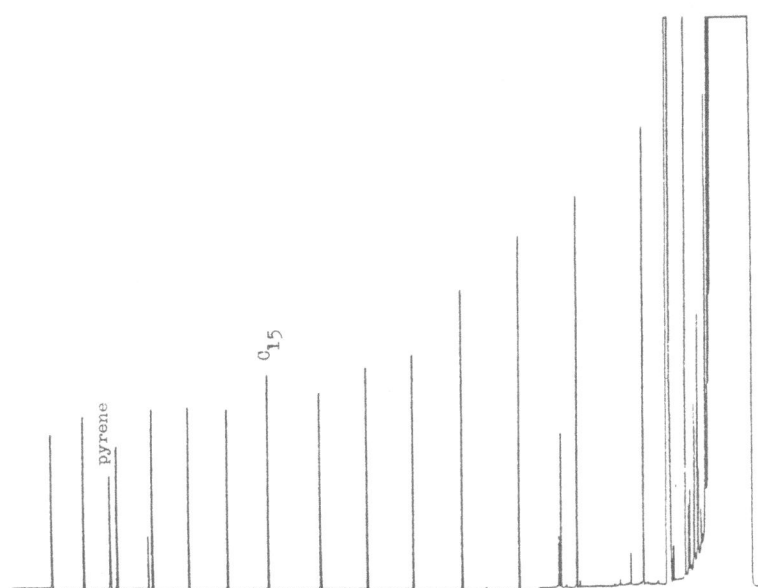

Fig. 4d C_6-C_{22} n-alkane standard, Chrompark (Column 4)

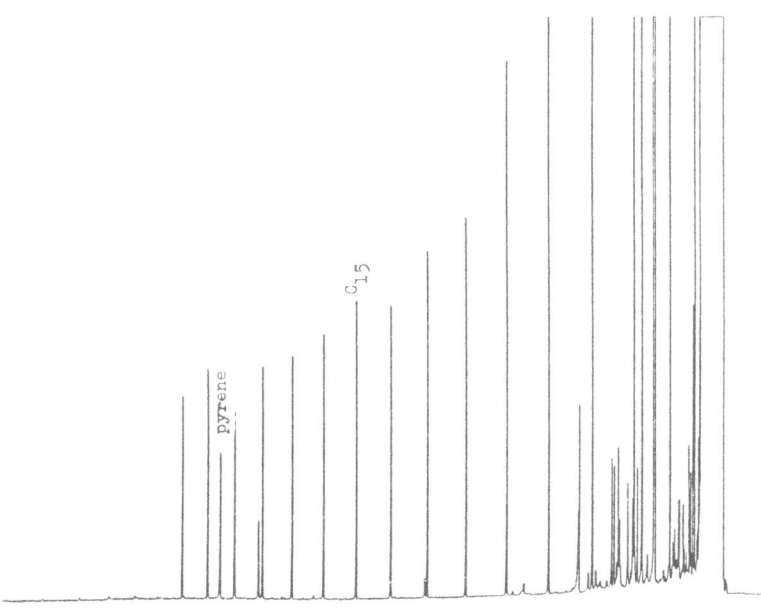

Fig. 4e C_6-C_{22} n-alkane standard, GC^2 (Column 5)

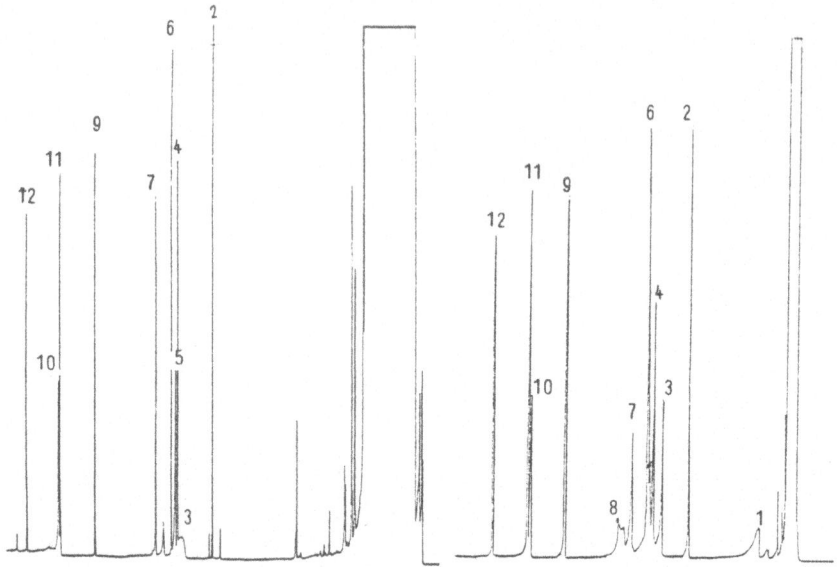

Fig. 5a Grob standard glass
(Column 1)

Fig. 5b Grob standard, NP
(Column 2)

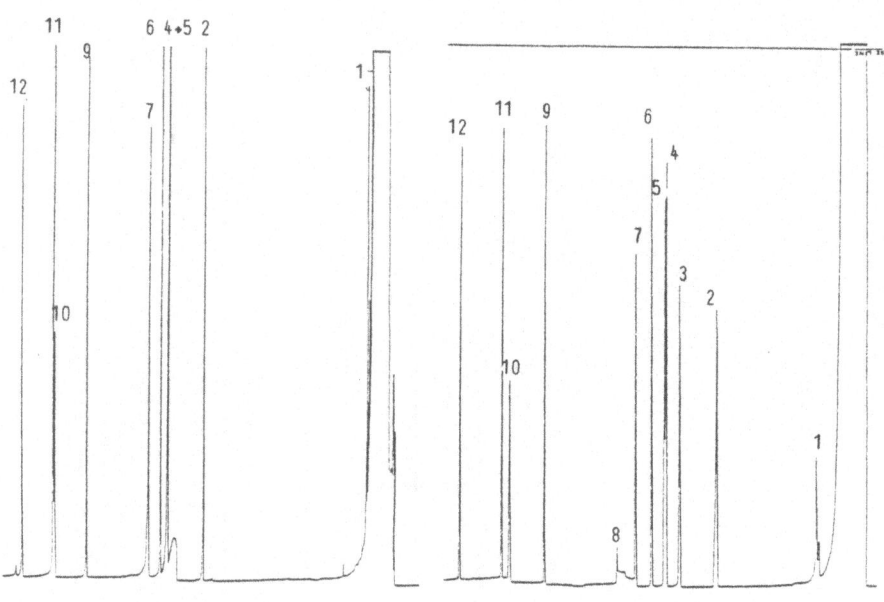

Fig. 5c Grob standard, Quadrex
(Column 3)

Fig. 5d Grob standard, Chrompak
(Column 4)

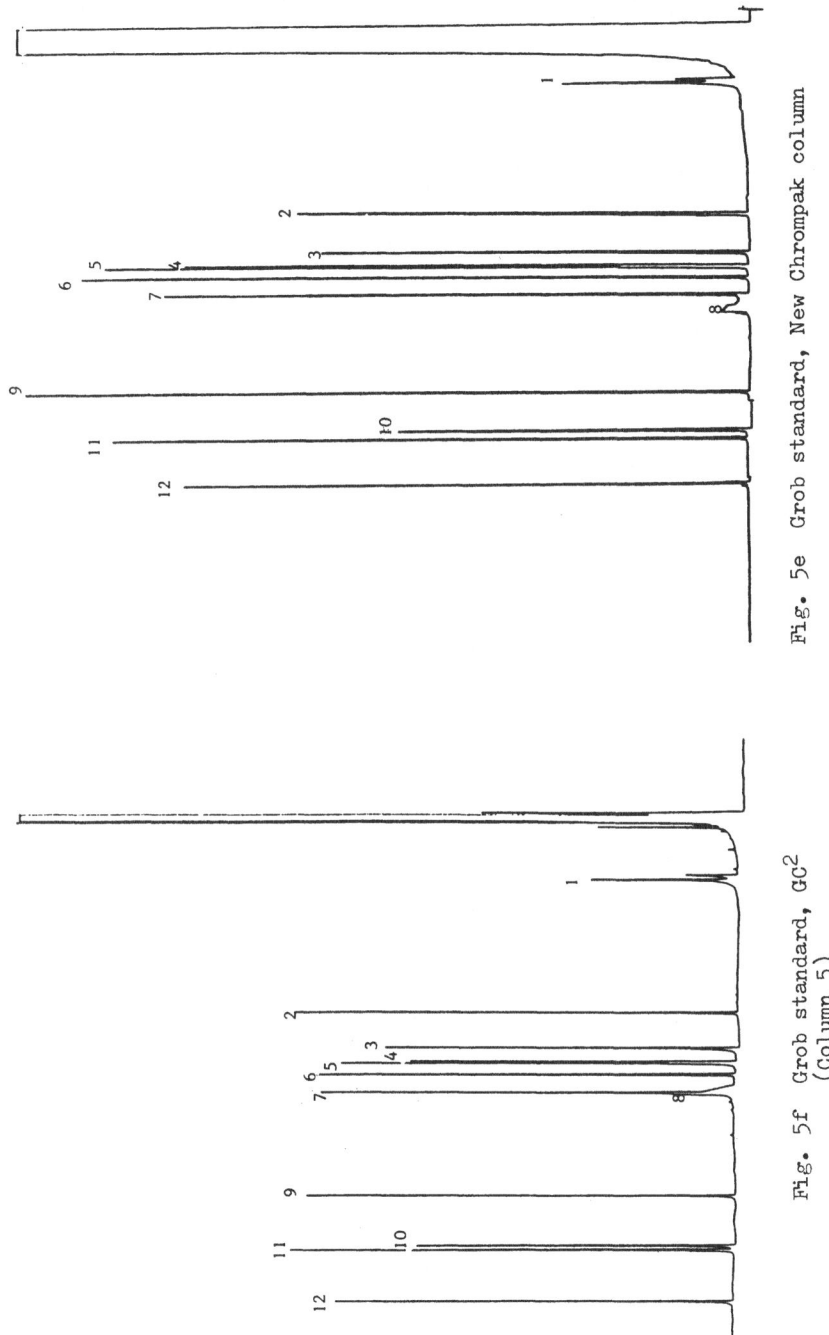

Fig. 5e Grob standard, New Chrompak column

Fig. 5f Grob standard, GC^2
(Column 5)

Fig. 6a. TIC chromatogram of an effluent extract, Glass (Column 1)

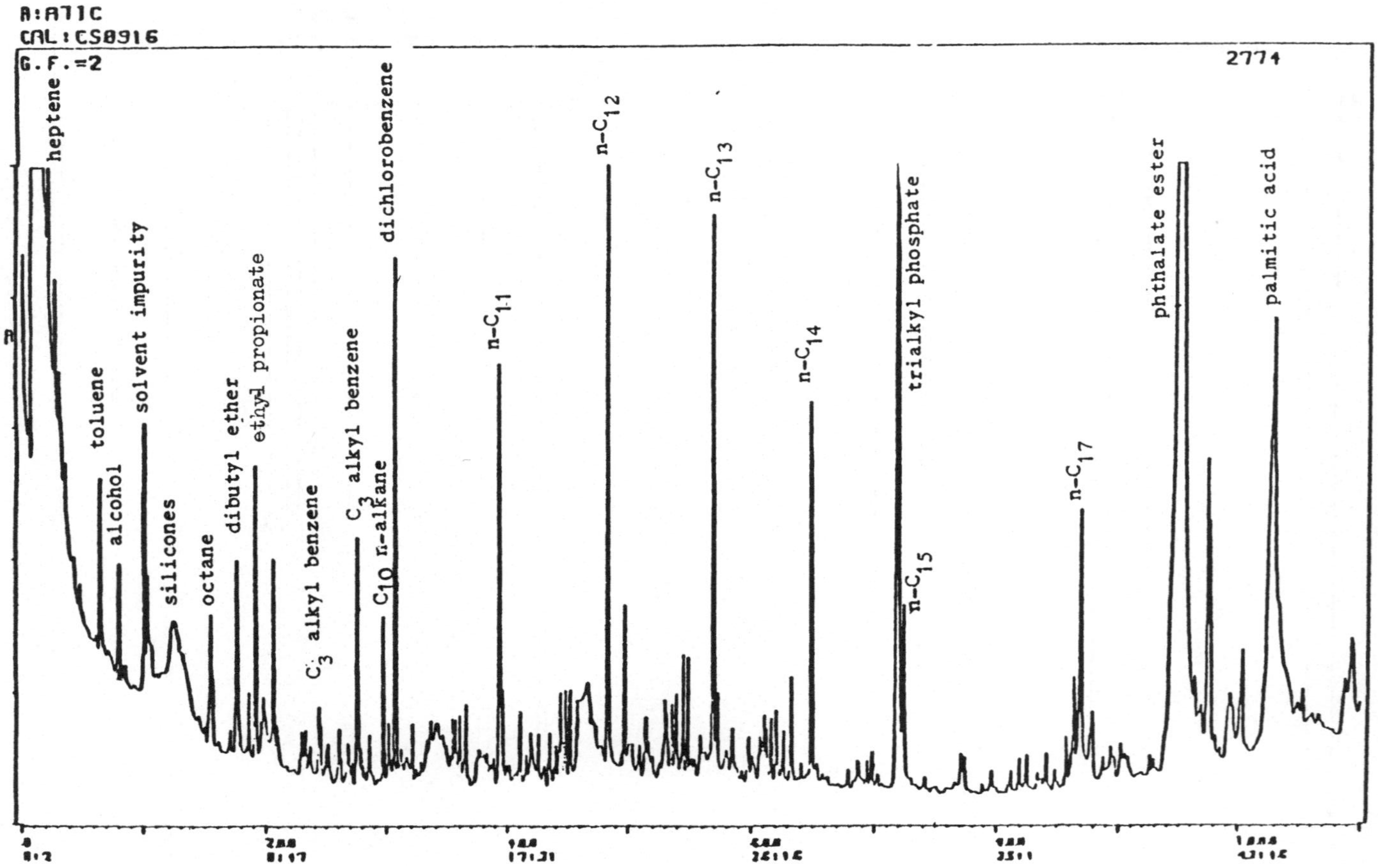

Fig. 6b TIC chromatogram of an effluent extract Quadrex (Column 3)

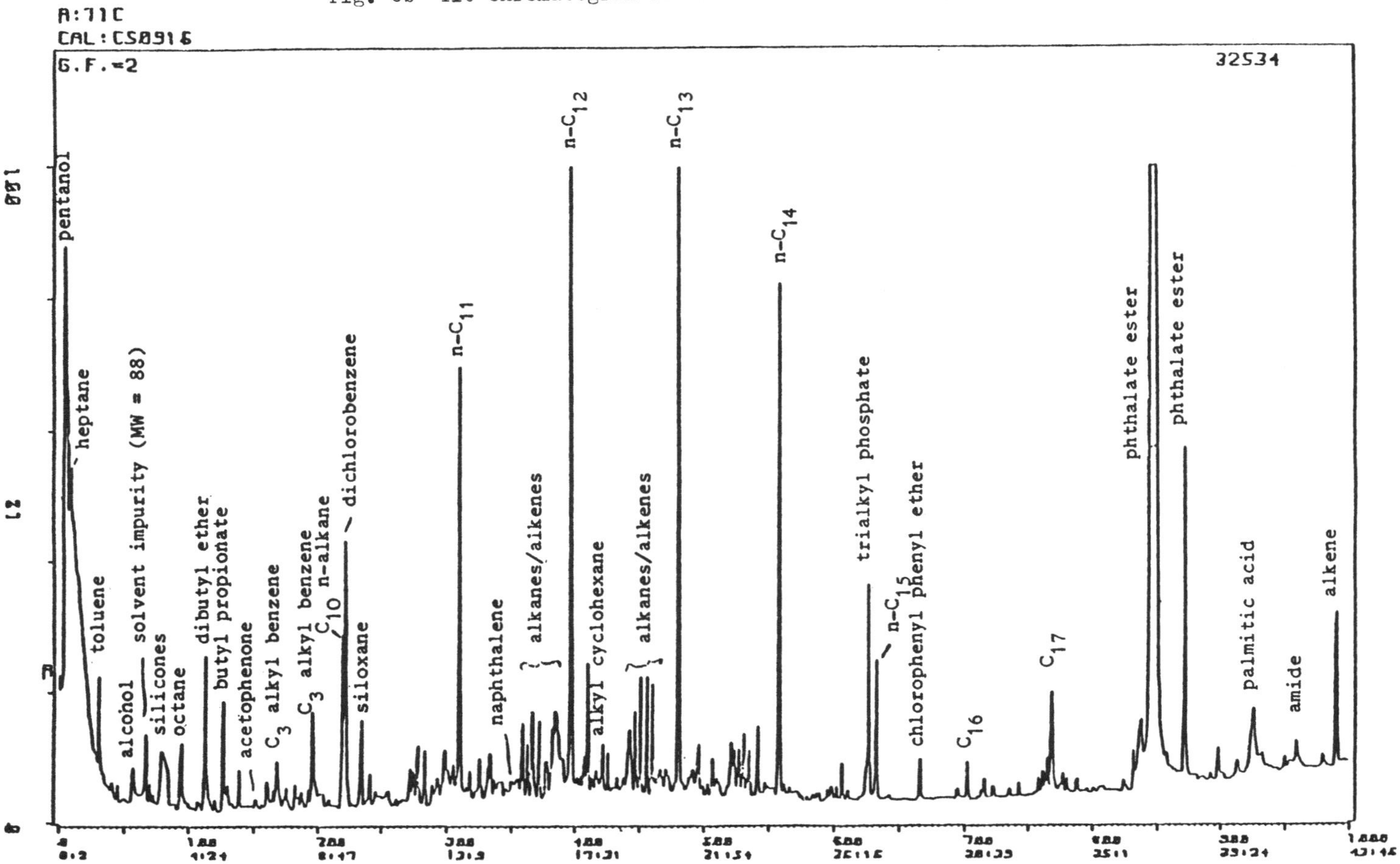

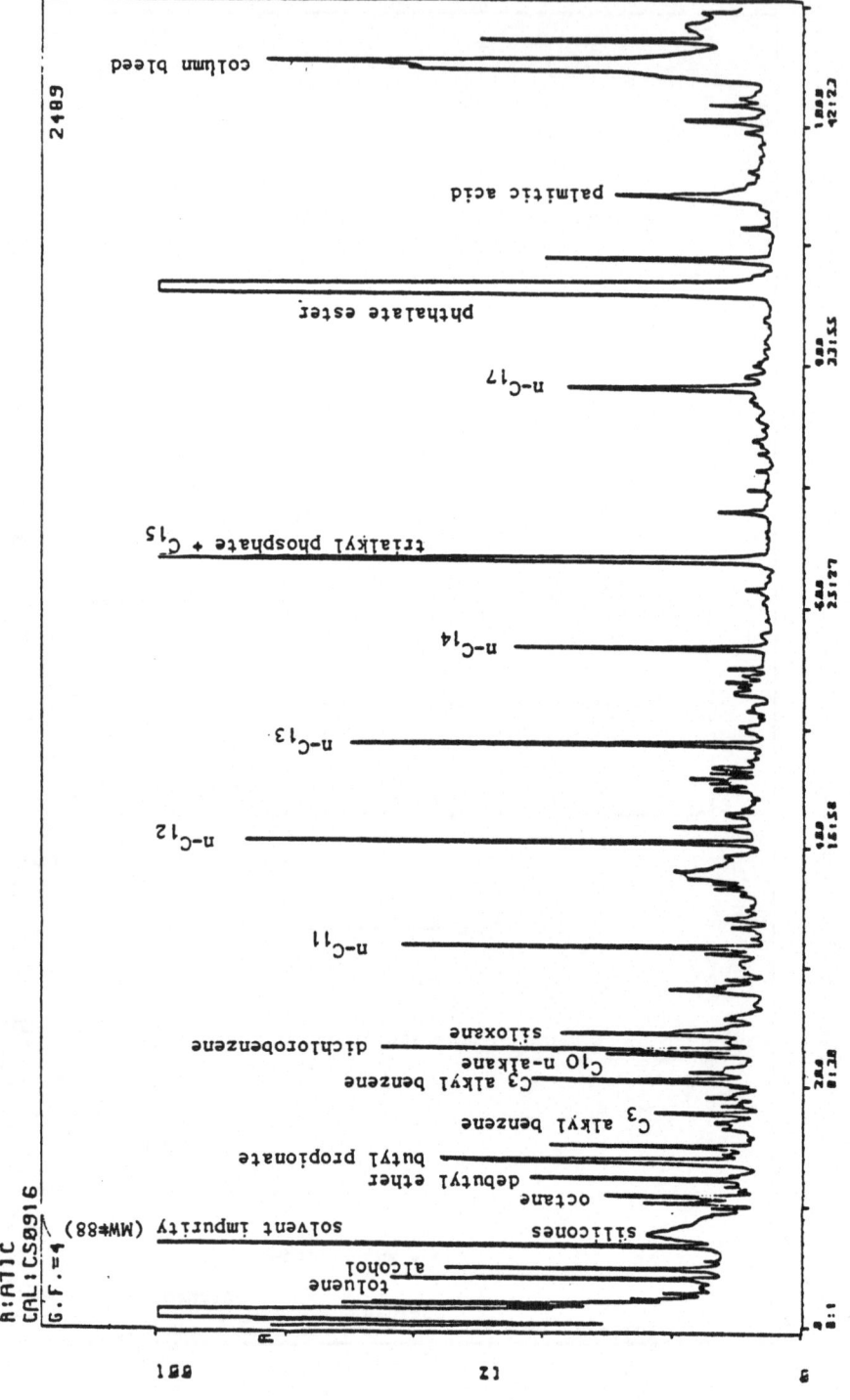

Fig. 6c TIC chromatogram of an effluent extract, Chrompak (Column 4)

THE DETERMINATION OF LINEAR PTGC RETENTION INDICES
FOR USE IN ENVIRONMENTAL ORGANICS ANALYSIS

H. KNÖPPEL, M. DE BORTOLI, A. PEIL, H. SCHAUENBURG AND H. VISSERS

Joint Research Center Ispra
Commission of the European Communities

Summary

The availability of mass spectra is not always sufficient for the identification of environmental organic pollutants. Often GC retention data are
useful to remove ambiguities.

A method is described which allows to obtain relative retention data from
linear temperature programmed GC (PTGC) runs. The method

● is based on polynomial interpolation,

● does not require the presence of a complete series of homolog
  straight chain alkyl compounds in the sample   and

● may use also non-alkane sample compounds as references for the
  calculation of relative retention data.

The reproducibility of indices obtained under constant experimental
conditions using apolar columns has been assessed as well as the influence
of a few experimental parameters.

1. INTRODUCTION

      Linear PROGRAMMED TEMPERATURE GAS CHROMATOGRAPHY (PTGC) using glass
or fused silica capillary columns combined with MASS SPECTROMETRY (MS)
has become a standard method in ENVIRONMENTAL ORGANIC ANALYSIS (EOA).

      GC retention data in addition to mass spectra are required or provide
additional evidence for identification in case of

● isomers with identical or very similar mass spectra

● minor constituents in complex mixtures yielding poor quality
  mass spectra.

      Relative retention data (INDICES) have advantages over retention
times due to

● a better reproducibility at nominally constant GC conditions,

● smaller changes when GC parameters are changed,

● better suitability for reference data storage.

A method for the calculation of retention indices is described which

● is based on polynomial instead of linear interpolation,

● does not require the presence of a complete series of non-alkanes
  (or other homolog series) in a sample,

● offers the possibility to use compounds of different classes
  simultaneously as reference compounds.

2. DESCRIPTION OF THE METHOD

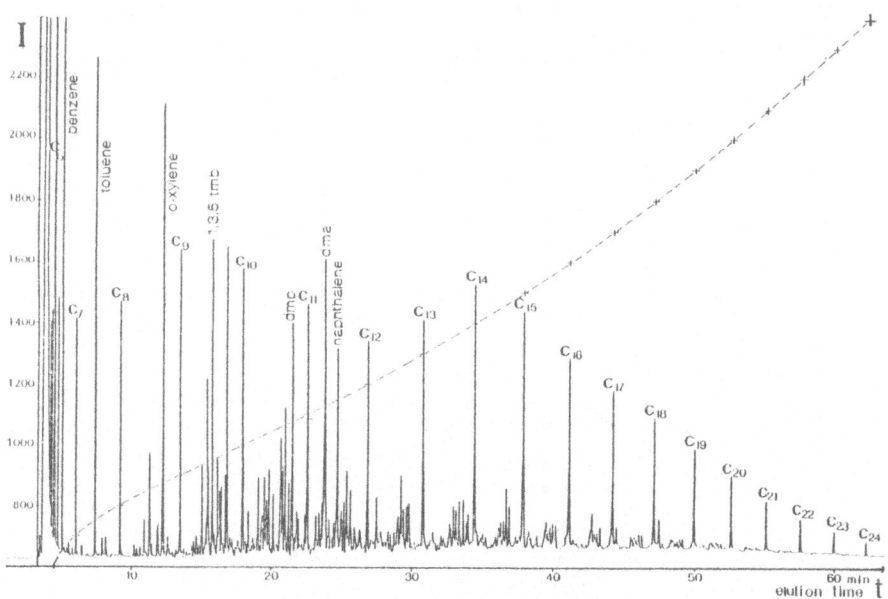

Fig. 1 - GLASS CAPILLARY CHROMATOGRAM OF A REFERENCE MIXTURE CONTAING
N-ALKANES $(C_6 - C_{24})$ AND FURTHER COMPOUNDS LIKE INDICATED
(tmb = trimethylbenzene, dmp = 2,6 dimethylphenol,
dma = 2,6 dimethylaniline)

The calculation of the retention indices for the constituents of an environmental (or other) sample is performed in two steps :

1. ● Separate a reference mixture (see e.g. Fig. 1) containing the **nor**-
alkane series under the same chromatographic conditions as subsequent
samples.

   ● Define a 'master' polynom

$$(1) \quad I = \sum_{i=0}^{m} a_i \cdot t^i$$

giving retention index I as a function of retention time t
(see dashed curve in Fig. 1).

   ● Calculate coefficients a_i using the least squares method for
approximation of polynom (1) to the points $(t_n / 100 \cdot n)$;

t_n = retention time of nor-alkane with n carbon atoms

$I_n = 100 \cdot n$ = retention index of nor-alkane with n carbon atoms
(by definition)

m = degree of master polynom $\leqslant$ number of reference alkanes - 2 = 13

2. • Separate sample.

• Determine retention times and indices of reference compounds contained in the sample which

• may be added or not

• need not to be nor-alkanes.

• Use least squares method in order to adapt the first 1 - 1 coefficients of the 'master' polynom in such a way that the polynom

$$(2) \quad I = \sum_{i=0}^{\ell-2} a'_i \cdot t^i + \sum_{i=\ell-1}^{m} a_i \cdot t^i$$

fits best the points (t_k / I_k);

l = number of reference compounds

t_k = retention times of reference compounds in the sample $(k = 1,2, \ldots , l)$

I_k = retention indices of the reference compounds

a'_i = adapted coefficients.

• Calculate retention indices of sample compounds using eq. (2).

A more detailed description of the index calculation method has been given elsewhere (1).

(1) H. KNÖPPEL, M. DE BORTOLI, A. PEIL, H. SCHAUENBURG and H. VISSERS Proceedings of the Second European Symposium "Physico-chemical Behaviour of Atmospheric Pollutants", Varese (Italy), 29 September - 1 October 1981, H. OTT and B. VERSINO (eds.), D. Reidel Publishing Company, Dordrecht, Holland, 1982, p. 99.

3. RESULTS

COMPARISON OF PTGC RETENTION INDICES CALCULATED BY

LINEAR AND BY POLYNOMIAL INTERPOLATION

Alkane Carbon Number	Retention Time (Min)	RETENTION INDICES		
		By Definition	Calculated by Linear Interpolation	Calculated by Polynomial interpolation
9	14.905	900	900.0 900.0	899.99
10	19.558	1 000	1 003.7 1 003.7 .	999.97
11	23.879	1 100	1 100.0 1 100.0	1 100.00
12	27.904	1 200	1 203.5 1 200.0	1 199.99
13	31.656	1 300	1 300.0 1 309.6	1 299.96
14	35.164	1 400	1 403.0 1 412.1	1 400.02
15	38.469	1 500	1 500.0 1 508.7	1 500.06
16	41.593	1 600	1 602.7 1 600.0	1 600.00
17	44.555	1 700	1 700.0 1 686.6	1 700.00

EXPLANATION

Underlined values have been used as references and are those given by definition in the case of linear interpolation. In the case of polynomial interpolation the corresponding retention time values have been used for the adaptation of the master polynomial.

RESULTS

● Linear interpolation needs a <u>complete series of n-alkanes</u> as references.

● Polynomial interpolation gives good results ($\pm$ 0,06 Index Units) with <u>four references in the range</u> $C_9 - C_{17}$.

COMPARISON OF THE REPRODUCIBILITY OF RETENTION TIMES AND INDICES UNDER CONSTANT EXPERIMENTAL
CONDITIONS

COMPOUND NAME	MEAN RETENT. TIME (MIN)	DMAX (%)	MEAN RETENTION INDICES I AND MAXIMUM DEVIATIONS DMAX OBTAINED BY ADAPTATION TO THE UNDERLINED VALUES					
			I	DMAX (%)	I	DMAX (%)	. I	DMAX (%)
n-hexane	2,564	0,18	600,00	0,02	600,15	0,81	600,37	0,12
BENZENE	3,611	0,25	646,16	0,93	646,06	0,57	646,15	0,00
n-heptane	5,367	0,32	700,00	0,01	700,01	0,43	699,99	0,04
TOLUENE	7,641	0,45	751,83	0,37	751,89	0,35	751,82	0,00
n-octane	9,994	0,52	799,99	0,00	799,99	0,23	799,94	0,02
o-XYLENE	13,691	0,57	874,94	0,04	874,93	0,03	874,93	0,02
n-nonane	14,905	0,56	900,00	0,00	899,99	0,00	900,00	0,02
1,3,5-TRIMETHYLBENZENE	17,425	0,58	953,23	0,07	953,23	0,03	953,24	0,05
n-decane	19,559	0,51	999,99	0,00	1000,00	0,04	1000,00	0,05
N,N-DIMETHYLANILIN	22,301	0,60	1062,62	0,05	1062,63	0,05	1062,62	0,06
2,6-DIMETHYLPHENOL	22,958	0,58	1078,07	0,04	1078,09	0,05	1078,07	0,06
n-undecane	23,879	0,58	1100,00	0,00	1100,01	0,03	1099,99	0,00
NAPHTHALENE	26,211	0,67	1157,10	0,06	1157,11	0,06	1157,10	0,07
n-dodecane	27,902	0,62	1199,98	0,00	1199,97	0,02	1199,99	0,00
n-tridecane	31,654	0,74	1300,01	0,00	1300,00	0,06	1300,13	0,10
n-tetradecane	35,159	0,82	1399,97	0,01	1399,98	0,08	1400,24	0,22
n-pentadecane	38,460	0,78	1500,02	0,00	1500,05	0,04	1500,32	0,25
n-hexadecane	41,586	0,95	1599,98	0,00	1600,00	0,00	1599,99	0,00
n-heptadecane	44,548	1,04	1700,00	0,00	1700,02	0,06	1699,24	0,71
ANTHRACENE	46,208	1,21	1758,43	0,14	1758,44	0,10	1756,88	1,45
n-octadecane	47,360	1,09	1799,99	0,00	1799,99	0,16	1797,69	2,13

EXPLANATIONS

- THE REPORTED RETENTION TIME AND INDEX VALUES ARE AVARAGES OF TEN GC RUNS,
- TEMPERATURE WAS 35°C FOR 5 MIN AND SUBSEQUENTLY PROGRAMMED UP TO 250°C AT 4°C/MIN
- DMAX IS GIVEN IN % OF THE ELUTION TIME RESP. RETENTION INDEX DIFFERENCE OF ADJACENT N-ALKANES
  (FOR N-ALKANES HALF THIS DIFFERENCE IS TAKEN), IN THE CASE OF INDICES THIS DIFFERENCE IS 100
  BY DEFINITION , I.E. DMAX IS ALSO THE ABSOLUTE MAXIMUM DEVIATION.
- INDICES IN THE FIRST COLUMN HAVE BEEN OBTAINED BY CALCULATING 'MASTER' POLYNOMIALS FOR ALL
  TEN RUNS. INDICES IN THE SECOND AND THIRD COLUMN HAVE BEEN CALCULATED BY ADAPTING ONE OF THE
  MASTER POLYNOMIALS TO THE OTHER NINE RUNS.

RESULTS

- RELATIVE MAXIMUM DEVIATIONS OF RETENTION INDICES ARE IN GENERAL 8 - 10 TIMES SMALLER THAN
  THOSE OF THE RETENTION TIMES.
- ADAPTING THE MASTER POLYNOMIAL WITH APPROPRIATELY CHOSEN REFERENCE COMPOUNDS, A REPRODUCIBILITY
  OF THE INDICES BETTER THAN 0.1 INDEX UNIT CAN BE ACHIEVED.

COMPARISON OF PTGC RETENTION INDICES OBTAINED WITH DIFFERENT OV-1 COATED CAPILLARY COLUMNS

COMPOUND NAME	COLUMN							
	1	2	3	MEAN OF COL. 1-3	4	5	6	MEAN OF COL. 4-6
o-XYLENE	874.94	874.90	875.33	875.06 ± 0.27	874.60	874.82	874.80	874.74 ± 0.14
1,3,5-TRIMETHYL BENZENE	953.23	953.18	953.43	953.28 ± 0.15	953.15	953.08	953.11	953.11 ± 0.04
N,N-DIMETHYL ANILINE	1062.62	1062.55	1062.62	1062.60 ± 0.05	1062.09	1062.24	1062.23	1062.19 ± 0.10
2,6-DIMETHYL PHENOL	1078.07	1078.22	1077.92	1078.07 ± 0.15	1077.42	1077.45	1077.46	1077.44 ± 0.02
NAPHTHALENE	1157.10	1156.91	1157.19	1157.07 ± 0.16	1156.83	1156.92	1156.94	1156.90 ± 0.07
ANTHRACENE	1758.43	1757.79	1758.18	1758.13 ± 0.34	1757.79	1757.05	1758.07	1757.97 ± 0.18

EXPERIMENTAL PARAMETERS OF THE DIFFERENT COLUMNS

COLUMN NUMBER	COLUMN MATERIAL	COLUMN LENGTH (m)	INT. DIA (mm)	FILM THICKNESS (μm)	TEMPERATURE PROGR. RATE r (°C/min)	CARRIER FLOW RATE F (ml/min)	INLET PRESSURE (psi)	$\frac{r \cdot W}{F \cdot \varrho}$ *)
1	SOFT GLASS	24	0.3	1.13	4.0	3.7	16	0.055
2	"	25	0.3	1.13	4.0	3.8	11.5	0.056
3	"	30	0.3	1.13	3.0	3.1	16.2	0.062
4	DURAN	25	0.31	1.16	4.0	3.9	13	0.058
5	"	25	0.31	1.16	4.0	3.9	13	0.058
6	"	25	0.31	1.16	4.0	3.8	13	0.059

*) W = weight, ϱ = densitiy of the stationary phase

EXPLANATIONS

- ALL COLUMNS HAVE BEEN RUN ISOTHERMALLY FOR 5 MIN AND WERE SUBSEQUENTLY PROGRAMMED AT A RATE INDICATED IN THE ABOVE TABLE TO 250°C.
- ALL COLUMNS HAVE BEEN PERSILYLATED ACCORDING TO THE GROB PROCEDURE AND HAVE BEEN STATICALLY COATED.
- COLUMNS 1 - 3 HAVE BEEN PRODUCED USING AN OLD GLASS DRAWING MASHINE. THEREFORE INTERNAL DIAMETERS SHOW AT THE SAME NOMINAL VALUE CONSIDERABLE VARIATIONS AS CAN BE VERIFIED COMPARING INLET PRESSURE, LENGTH AND FLOW RATE VALUES.
- COLUMNS 4 - 6 HAVE BEEN KINDLY PROVIDED BY M. GALLI, MEGA CAPILLARY COLUMN LABORATORY.

RESULTS

- APOLAR CAPILLARY COLUMNS COATED WITH THE SAME STATIONARY PHASE CAN BE PRODUCED IN A WAY TO REPRODUCE PTGC RETENTION INDICES WITHIN BETTER THAN 0.5 INDEX UNITS.
- WITH COLUMNS PRODUCED IN THE SAME LABORATORY ACCORDING TO THE BEST STATE OF THE ART, A REPRODUCIBILITY OF PTGC RETENTION INDICES OF 0.1 - 0.2 INDEX UNITS AND BETTER CAN BE OBTAINED.
- USING APOLAR CAPILLARY COLUMNS OF THE SAME TYPE AND MODERN GC EQUIPPMENT INTERLABORATORY REPRODUCIBILITY OF PTGC RETENTION INDICES WITHIN 0.5 INDEX UNITS OR BETTER SHOULD BE OBTAINABLE.

SESSION III - SEPARATION AND ANALYSIS OF NON-VOLATILE COMPOUNDS

Chairman: M. FIELDING, Water Research Centre,
Medmenham Laboratory, United Kingdom

Review papers :

- Developments in selective detectors for HPLC

- Kopplung eines Hochleistungschromatographen mit einem Massenspektrometer

- Identification of non-volatile organic compounds in water

Short papers :

- Routine HPLC of polynuclear aromatic hydrocarbons

- Evaluation of a mass detector for HPLC determination of organic compounds in water

Poster papers :

- Determination of polycyclic aromatic hydrocarbons (PAH's) at the low ng/1 level in the Biesbosch water storage reservoirs (Neth.) for the study of the degradation of chemicals in surface waters

- Assessment of a moving belt type HPLC-MS interface with respect to its use in organic water pollution analysis

DEVELOPMENTS IN SELECTIVE DETECTORS FOR HPLC

H.POPPE

Laboratory for Analytical Chemistry,University of Amsterdam,
Nieuwe Achtergracht 166,1018 WV,Amsterdam,The Netherlands.

Summary

The position of the selectivity of the detection in the
approach of the selectivity of the complete analytical
procedure is discussed.Various instrumental solutions for
the attainment of selectivity of detection are reviewed.
Special emphasis is laid on the combination of post column
reaction with selective measuring techniques such as flu-
orimetry and electrochemical methods.

1 Introduction

The analytical application of high pressure ·liquid chromatography depends to a high degree on the availability of
suitable detectors.These should fulfil a number of requirements,of which the main ones are (1,2,3):sufficiently low contribution to peak width,adequately low detection limit and
adequate selectivity in the detection of the analytes under
study.

The latter requirement is generally much more important
than in gas chromatography,and it is often the first to consider in organic trace analysis applications of HPLC.In biological,medical and environmental analysis the gas chromato-
grapher at least can be sure that high molecular weight compounds will not generate peaks in the chromatogram of the
analytes of lower molecular weight.However,in liquid chromatography the separation criterion is nearly always polarity
rather than volatility,and compounds of any molecular weight
may emerge and interfere in any position in the chromatogram.
As the indicated samples invariably contain a large number
and often large amounts of high molecular weight material,
interference is an omnipresent problem in trace analysis with
HPLC.

Selectivity of detection is one of the potential solutions
for the interference problem.However,it will be clear that all
the susequent steps in the analytical process,as given in Fig.1
allow for suitable measures for the improvement of analytical
selectivity.The use of selective detection therefore has to be
considered in the perspective of the alternatives.Table I gives
a general,crude and often simplifying overview of the drawbacks
and advantages met with when one tries to enhance the selectivity in a particular step of the process.

From this it is clear that the use of selective detection
principles allows to avoid other measures which could lead to
significant losses in analytical performance with respect to
speed,precision,accuracy and development costs.The only penalty
for the introduction of selective detection is the increase in
equipment costs.

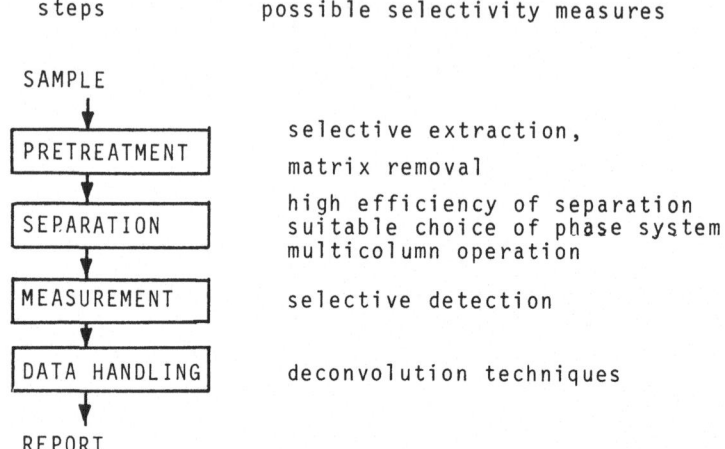

Figure 1

The analytical process and means for the
enhancement of analytical selectivity

steps   possible selectivity measures

SAMPLE

PRETREATMENT   selective extraction,
                       matrix removal

SEPARATION   high efficiency of separation
                       suitable choice of phase system
                       multicolumn operation

MEASUREMENT   selective detection

DATA HANDLING   deconvolution techniques

REPORT

Table I

Costs of the improvement of analytical selectivity in various steps of the analytical process			
steps			
PRETREATMENT	SEPARATION	MEASUREMENT	DATA HANDLING
diminished accuracy and precision	time (N)	equipment	precision
increased development	equipment and development (multicolumn)		equipment
			development (software)

There are various physical and chemical principles on
which a selective detector for HPLC can be founded. However, before discussing these another general problem should be indicated. These principles, as fluorescence or electrochemical activity, etc., are chosen in the hope that the compounds of interest
are especially responsive. However, the fortunate case that the
responsive group coincides with the group of analytical interest is only seldomly encountered. In most cases one has a difficult choice (if any) between a too selective detector, which
does not respond to some important compounds of the group of
interest and a device of too low selectivity with the inherent
interference problems.
    Highly selective detectors are only useful when their de-
tection limits are also low, in the ppb range or lower. This

follows from the limited loadability of the HPLC phase systems.
Above a certain concentration compounds will elute from columns
as excessively broadened,distorted peaks,by which the chromatographic selectivity (resolution) will be destroyed.For the very
popular phase systems consisting of alkyl-modified silica as
the stationary phase and mixtures of water and an organic solvent as the stationary phase,this concentration limit is about
300 ppm (4).It follows that the eluting concentration of a
trace,occurring in a q-fold lower concentration in the sample
than some major constituent,cannot be higher than approximately 300/q if we want to preserve the resolution.It should be
noted that this holds true also when the major constituent is
fully invisible with the detection system used;the overloading
compound will affect the chromatographic behaviour of the traces as well.The factor q may greatly exceed 1000 in contemporary analytical problems.Therefore selective detection principles
are only useful if their detection limits are in the ppb range
or preferably lower.

2 Overview of principles for selective HPLC detection

Table II summarizes the most important principles that are
in use for selective HPLC detection.

Table II

Principles for selective HPLC detection

e = excellent ; m = moderate ;i = insufficient ; - = does
not apply

principle	performance		
	detection limit	selectivity	scope
wavelength choice UV	m	m/i	e
simultaneous multi-wavelength	m	m	e
fluorescence	e	e	i
anodic electrochem.	e	e	i
cathodic electrochem. (mercury elctrodes)	m	e	i
atomic spectrometry	e	e	m/e
electron capture	e	e	i
homogeneous post column reaction	e	e	e
post column reaction with extraction	e	e	e

A few remarks can be added to table II.
The free choice of the measuring wavelength,available after the introduction of UV absorbance detectors with a mono-
chromator,is in many cases a suitable approach for the enhancement of selectivity.However,because of the very broad bands

in the UV spectra of organic molecules,the selectivity ratios never are very high.

Much more powerful in this respect is in principle the use of simultaneous multiwavelength detection by means of the so called image sensing devices such as the photodiode arrays.In combination with the UV multicomponent analytical scheme,known and applied already for some decades, a system of tunable selectivity is in principle available.The main bottleneck in the introduction of this concept is not the UV system itself but the price of the latter and the development of data handling procedures which can be applied "on the fly" and which are transparent to the user.It can be expected that the near future will bring substantial progress in this field.

Fluorescence is used nowadays in many laboratories for routine analysis in biochemical,clinical and environmental applications and we can refer to standard textbooks for practical and theoretical guidelines for the use of fluorescence (2,5). A development for (probably the near) future is the introduction of laser excitation (6) in commercial equipment.This will lower the detection limit and will make the system more compatible with small diameter columns.An important application field for the fluorimetric detection is in the combination with post column reaction detection systems (see below).

The electrochemical method using oxidation at solid electrodes (anodic -), especially amperometry (7) and coulometry (8), now plays an important role in the chomatographic analysis of oxidizable compounds occurring in trace amounts in samples of biomedical origin, especially the catecholamines.The combination of low detection limit and suitable selectivity is again decisive here.

Cathodic (reductive) electrochemical processes have been applied for detection purposes only in analytical research laboratories until the last years.The main reason for this is probably that the detection limits obtainable with the proposed devices were not low enough.This in turn is connected with the fact that the only reasonable choice for the electrode material is mercury and this is commonly used as the dropping mercury electrode (DME) which is difficult to apply in a flow through detection device.However,during the last years interesting results have been presented which show that various measures, such as the use of short dropping times (9) , pulse methods and alternating current methods (10) can bring the detection limit down to well below 1 ng and can enhance the selectivity. Also better instrumentation for flow through application of polarography is now becoming commercially available.These developments could also make the tensammetric mode,selective for surfactants (11) an attractive approach.

A competitor for the mercury drop electrode is the electrode consisting of a thin film of mercury on a solid electrode which of course has the advantage that it is easily incorporated in a flow through design.It is not yet clear what the practical value of this technique will be.From static electroanalysis it is known that the stability of such electrodes can present difficulties.

In any case,the interference by oxygen and other reducible impurities in the mobile phase has to be coped with.As nitrogen sparging is often found to be not effective enough,an electrochemical scrubber (12) was proposed.In view of the results of all these efforts it may be expected that cathodic electrochemistry will develop into a valuable technique for selective HPLC detection.The widespread occurrence of reducible

compounds as micropollutants,e.g. pesticides with nitro groups
,enhances the significance of this development in the context
of this symposium.

Atomic spectrometry, especially atomic absorption,has been
interfaced to liquid chromatography by many workers,of which
a few are cited (13-16).The interfacing is not an easy matter,
especially in the case of the so called flameless AA.Also atomic fluorescence (17) and emission spectrometry with the in-
ductively coupled argon plasma (18) have been proposed.In
these cases the interfacing presents no problem.However,in most
cases the excessive investment costs of the equipment will be
prohibitive for many laboratories.Nevertheless,for special
studies, especially those on metal speciation in the environment,the use of such element specific devices in combination
with HPLC has a unique potential.

Electron capture detection has been proposed for HPLC detection a number of years ago.However,the system that relies on
the total vaporization of the mobile phase is not compatible
with recent trends in the choice of phase systems for HPLC,in
which aqueous solvents,often with the addition of buffers and
ion pair formers,are used.

Summarizing the above it can be noticed that there is a
natural but annoying correlation between the degree of selectivity and the scope of a measurement principle.In most of the
selective detectors we can obtain a signal only from few compounds;selectivity is high.However,as we are mostly not able
to change the nature of the selctivity,the scope of the devices
is very narrow.

The application of post column reaction detection circum-
vents this problem to a large extent.Therefore we devote a separate section to this technique.

3 Post column reaction detection

By carrying out chemical reactions on the eluting compounds,we can convert non-detectable or ill detectable material into products which are highly suitable for the detection
instrumentation at hand.The products may consist of derivatives
of the original compounds,as is e.g. the case in the conversion
of amino acids with ninhydrin or o-phtaldehyde into coloured
or fluorescent products,but they may also be just a reaction
product,as is the case in the method for the detection of phenols by means of their oxidation with cerium(IV).The product
cerium(III) can be well measured in a fluorimeter (19).Concen-
trations in the ppb-range can be measured in this way.

In post column reaction detection the selectivity is mainly controlled by the chemistry involved.This fact adds a very
important degree of freedom in the design of the analytical
system.The selectivity can be tuned by choosing the suitable
chemical reactions and by influencing these by the choice of
the conditions such as temperature,reagent concentration and
pH. It is this flexibility and the fact that it is so easily
exploited because of the vast knowledge on selective organic
reactions,gathered during the times that reactions were the
main asset of organic analysis,which makes the post column reaction detection technique so important in organic trace analysis.

The general principles and problems of the method have
been thoroughly discussed (20-24).After addition of the reagent
mixture to the column effluent in a low dead volume mixing
T-piece,conditions should be such that:

a) adequate mixing of reagent and column effluent occurs (the
    mixing device) and
b) the mixture is delayed before detection long enough in or-
    der to allow for a sufficient extent of reaction (the delay
    device).
The requirement a) is discussed in ref.(22) .It generally presents no big difficulties,because a suitable mixing device can
be easily constructed while avoiding extra peakbroadening to
any appreciable extent.

Requirement b) on the other hand leads to more problems.
When the reaction rate is so low that a long delay time is necessary,the system has to be carefully designed as the imple-
mentation of the delay or residence time t_r invariably leads to
a finite residence time distribution.The standard deviation of
the latter is denoted by σ_{tr} .The design of the reactor is
governed by the compromise between t_r and σ_{tr}.The latter is
better known in chromatography as the contribution to peak
width.The compromise,necessary to avoid loss of chromatographic resolution,is also controlled by external constaints such
as the available pressure drop,construction techniques etc.,
and leads to the following three types of design.

The first is a simple open tube,which should be preferably helically coiled because this diminishes peak broadening
(24).It can be used (22,23) for delay times t_r up to about 1-2
minutes.Above that limit it is impossible to keep σ_{tr} down
without recourse to inpractical high pressures over the reaction system.

Somewhat better in this respect is the packed bed reactor
(22,23),essentially a chromatographic column filled with non-
sorbing particles.It can be used with t_r values up to about
5-10 minutes,when a pressure drop of 30-50 bar is found acceptable.

It should be noted that it is often possible to avoid larger reaction times by a suitable choice of conditions.Higher
temperatures are an obvious approach.In this context it should
be noted that it is entirely unnecessary to carry the reaction
to virtual completion in these systems.The required reproducibility is obtained by the fact that the reaction time is perfectly constant in a flow through system.Also the stability
of the formed product is not critical.The benefit of the increased temperature was shown for the ninhydrin case,where it
was found that 140 ^{o}C as a reaction temperature allowed for
a 2 minutes reaction time with optimal signal to noise ratio
(25).

If reaction times larger than 5-10 minutes are unavoidable
the third reactor type should be used.This type relies on segmentation of the liquid stream for the elimination of peak
broadening,as it is known from the Autoanalyzer[R] system.In
this way very long delay times (in excess of 1 hour) can be
applied and the resulting peak broadening even in the extreme
cases remains relatively low and mainly determined by the
effects in the so called debubbler,the phase separator necessary for delivering a homogeneous mixture to the measuring instru
ment(20,21).

The classical gas segmentation can be replaced by solvent
segmentation with an inmiscible liquid (26).This technique
leads to a significant advantage over the homogeneous reaction
systems:extractions of products or reagents in flow through
mode can be carried out.In homogeneous reactors the reagents
which give an appraciable response in the measuring equipment
cannot be used.With the two phase systems,and provided the

product and the reagent move to different phases,we can use this type of detection reactions as well.This brings a large number of classical labelling techniques and ion pairing within reach of post column reaction detection (27).For surfactants the potential of this method using ion pair extraction has been demonstrated (28).

4 Conclusion

Selective detection in HPLC is presently the subject of concentrated research in a number of laboratories.The use of many kinds of physico-chemical phenomena for this purpose yields methods of good detection limit and often excellent selectivity.However,the scope of these methods is often rather limited.

Combination of these classical selective detectors with post column reaction techniques offers the combination of tunable selectivity and low detection limit and will probably be the method of choice for many of the complicated problems encountered in environmental analysis.It is therefore a fortunate circumstance that equipment for post column reaction is no becoming commercially available.The inventiveness and endeavour of the analytical chemists should do the remainder of the work.

References

1   J.F.K.Huber,J.Chromatog.Sci.,7 (1969) 172
2   L.R.Snyder and J.J,Kirkland,Introduction to Modern Liquid Chromatography,2nd Ed.,Wiley,New York 1979
3   H.Poppe,Anal.Chim.Acta,114 (1980) 59
4   A.W.J.de Jong,J.C.Kraak,H.Poppe and F.Nooigedacht,J.Chromatog.,193 (1980) 193
5   R.P.W.Scott,Liquid Chromatography Detectors,Elsevier,Amsterdam,1977
6   G.J.Diebold and R.N.Zare,Science 196 (1977) 1439
7   P.T.Kissinger,Anal.Chem., 49 (1977) 447A
8   J.Lankelma and H.Poppe,J.Chromatog., 125 (1976) 375
9   H.B.Hanekamp,P.Bos and R.W.Frei,J.Chromatog., 186 (1979) 489
10 H.B.Hanekamp,W.H.Voogt and P.Bos,Anal.Chim.Acta, 118 (1980) 73
11 J.Lankelma and H.Poppe,J.Chromatog.Sci.,14 (1976) 310
12 H.B.Hanekamp,W.H.Voogt,P.Bos and R.W.Frei,Anal.Chim.Acta 118 (1980) 81
13 T.M.Vickrey,M.S.Buren,H.E.Howell,Anal.Lett., A11 (1978) 1075
14 N.Yoza,K.Kouchiyama and S.Ohashi,At.Absorpt.Newl.,18 (1979) 39
15 W.Slavin and G.J.Schmidt,J.Chromatog.Sci.,17 (1979) 610
16 H.Koizumi,R.D.McLaughlin and T.Hadeishi,Anal Chem., 51 (1979) 387
17 D.D.Siemer,P.Koteel,D.T.Haworth,W.J.Taraszewski and S.R.Lawson,Anal.Chem., 51 (1979) 575
18 C.H.Gast,J.C.Kraak,H.Poppe and F.J.M.J.Maessen,J.Chromatog., 185 (1979) 549
19 A.W.Wolkoff and R.H.Larose,J.Chromatog.,111 (1975) 472
20 L.R.Snyder,J.Chromatog., 125 (1976) 287
21 L.R.Snyder,Anal.Chem.,48 (1976) 1017
22 J.F.K.Huber,K.M.Jonker and H.Poppe,Anal.Chem.,52 (1980) 2
23 R.S.Deelder,M.G.F.Kroll,A.J.B.Beeren and J.H.M. van den Berg, J.Chromatog.,149 (1978) 669
24 R.Tijssen,Anal.Chim.Acta, 114 (1980) 71

25 K.M.Jonker,H.Poppe and J.F.K.Huber,Chromatographia,11 (1978)
   123
26 A.H.M.T.Scholten,U.A.Th.Brinkman and R.W.Frei,J.Chromatog.,
   205 (1981) 229
27 J.F.Lawrence,U.A.Th.Brinkman and R.W.Frei,J.Chromatog., 185
   (1979) 473
28 C.P.Terwey-Groen,J.C.Kraak,W.M.A.Niessen,J.F.Lawrence,C.E.
   Werkhoven-Goewie,U.A.Th.Brinkman and R.W.Frei,Int.J.Environ.
   Anal.Chem.,9 (1981) 45

KOPPLUNG EINES HOCHLEISTUNGSCHROMATOGRAPHEN MIT EINEM
MASSENSPEKTROMETER

K. LEVSEN
Institute of Physical Chemistry, University of Bonn

Summary

The current state of on-line coupling of a liquid chromatograph with a mass spectrometer (LC/MS) is reviewed. The
two commercially available systems are discussed in detail:
(a) With the movingbelt system the LC effluent is transported on a kapton belt through two vacuum locks into the
ion source of the mass spectrometer. The solvent is
removed from the belt by evaporation within the vacuum
locks. The belt is heated up inside the ion source leading
to a flash evaporation of the sample into the ionization
chamber. (b) With the direct liquid introduction (DLI)
system the LC effluent is split in the interface and ~ 1%
of the effluent is directly introduced into the ion source
operating in the chemical ionization (CI) mode. The LC
solvent acts as CI reagent gas. If a micro HPLC is used
the total effluent can be introduced into the ion source.
Moreover, a recently developped jet-interface is described.
The potential of newer mass spectrometric ionization
methods such as plasma desorption, laser induced mass
spectrometry, secondary ion mass spectrometry and fast
atom bombardment for future HPLC-MS interfaces will be
discussed.

1. EINLEITUNG

      Die Kopplung Gaschromatograph-Massenspektrometer (GC-MS)
stellt heute die wichtigste Methode zur Analyse organischer
Verunreinigungen im Wasser dar. Diese Methode verbindet die
hohe Trennleistung der Kapillargaschromatographie mit der Sub-
stanzspezifität und hohen Nachweisempfindlichkeit der Massenspektrometrie. Bedauerlicherweise werden mit dieser Methode jedoch nur die flüchtigen Verbindungen erfaßt, die vermutlich
lediglich 20-30% der in Oberflächenwassern vorhandenen organischen Verunreinigungen ausmachen. Deshalb ist es verständlich,
daß intensive Bemühungen unternommen werden, um mit Hilfe der
Hochleistungsflüssigchromatographie (HPLC) eine Auftrennung sowie einen Nachweis auch der weniger flüchtigen Wasserinhalts-
stoffe zu erreichen. Für diese Untersuchungen ist eine on-line
HPLC-MS Kopplung aus zwei Gründen wünschenswert: (a) Die heute
verfügbaren HPLC-Detektoren sind weder substanzspezifisch noch
annähernd ähnlich empfindlich wie GC-Detektoren. Ein Massen-

spektrometer wäre ähnlich wie bei der GC-MS ein idealer wenn
auch teurer Detektor. (b) Eine direkte HPLC-MS Kopplung würde
eine rasche und im Idealfall eindeutige Identifizierung schwe-
rer flüchtiger Verunreinigungen im Wasser erlauben.

Auf der anderen Seite ist eine direkte HPLC-MS Kopplung
weniger zwingend als die direkte GC-MS-Kopplung, da sich die im
HPLC aufgetrennten Komponenten relativ einfach aufsammeln und
off-line massenspektrometrisch untersuchen lassen. Ein solches
Verfahren ist jedoch zeitaufwendig. Außerdem verzichtet man auf
den Einsatz des Massenspektrometers als direkten Detektor.

Unglücklicherweise ist eine Kopplung eines HPLC mit einem
Massenspektrometer ungleich schwieriger als eine GC-MS-Kopplung,
so daß bei der Entwicklung von HPLC-MS-Kopplungen in der Regel
ein Kompromiß beim Betrieb des HPLC oder des Massenspektrometers oder bei beiden gemacht werden muß. Trotz einer Vielzahl
von Schwierigkeiten sind in den zurückliegenden zehn Jahren
eine Reihe von HPLC-MS Interface entwickelt worden, deren Vor-
und Nachteile in diesem Artikel dargestellt werden sollen (1,2).
Ein Schwerpunkt wird auf der Beschreibung jener Systeme liegen,
die heute kommerziell erhältlich sind.

2. GRUNDLEGENDE BETRACHTUNGEN

Es ist wünschenwert, daß ein Interface weder die Arbeits-
weise des HPLCs noch die des Massenspektrometers beeinträchtigt.
Wodurch sollte demnach eine optimale HPLC-MS Kopplung ausge-
zeichnet sein? Die wünschenswerten Eigenschaften der drei Komponenten einer HPLC-Kopplung, des Flüssigchromatographen, des
Massenspektrometers und des Interfaces sind in der folgenden
Tabelle zusammengefaßt (1):

Tabelle I. Wünschenswerte Eigenschaften einer HPLC-MS Kopplung

HPLC:
a. Keine Beschränkung bei Lösungsmittelwahl
b. Verwendung von Pufferlösungen
c. Gradienteneluierung
d. Fließraten bis zu 2 cm^3/min ≈ 2000 cm^3/min (Gas)

MS:
a. Zur Aufrechterhaltung des Vakuums sollten die Fließraten
   < 20 cm^3/min (Gas) sein
b. Keine Beeinträchtigung der Empfindlichkeit durch den HPLC
c. Registrierung eines Totalionenchromatogramms
d. Neben konventionellen Ionisierungsmethoden (EI/CI) sollten
   Methoden zur Ionisierung thermisch labiler, stark polarer
   Verbindungen verfügbar sein.

Interface:
a. Hohe Anreicherung der Probe relativ zum Lösungsmittel
b. Maximaler Transfer der Probe vom HPLC zum Massenspektrometer
c. Keine Peakverbreiterung durch das Interface
d. Verdampfung auch von Proben geringer Flüchtigkeit

Die Hauptschwierigkeiten bei der Konstruktion eines HPLC Interfaces liegen auf der einen Seite darin, daß zur Aufrechterhaltung des Vakuums in der Ionenquelle des Massenspektrometers (0.5 mbar im CI Betrieb, < 0.1 µbar im EI Betrieb) maximal eine Gasmenge von 20 cm^3/min (im CI Betrieb) in das Spektrometer eingelassen werden kann, während die vom HPLC anfallende Lösungsmittelmenge einer Gasmenge von bis zu 2000 cm^3/min entspricht. Die zweite Schwierigkeit liegt darin, nicht flüchtige Verbindungen zu verdampfen. Im folgenden wird dargestellt, inwieweit eine Lösung dieser Probleme bei den einzelnen Interfacen gelungen ist.

3. "ATMOSPHERIC PRESSURE IONIZATION" QUELLE (API)

Von Horning (3) ist eine Ionenquelle entwickelt worden, die unter normalem Atmosphärendruck arbeitet (API Quelle) und deshalb besonders zur Kopplung an einen HPLC geeignet sein sollte. Hierzu wird ein Teil des Eluats verdampft und der Dampf aus Lösungsmittel und gelöstem Stoff direkt in die API Quelle eingelassen. Die Ionisierung erfolgt entweder mit Hilfe einer radioaktiven Quelle, die Elektronen emittiert, oder aber mit Hilfe einer Coronaentladung. Um das für das Massenspektrometer erforderliche Vacuum aufrecht zu erhalten, sind Ionenquelle und Massenanalysator lediglich durch ein sehr dünnes Loch (wenige µm) verbunden. Der Einsatz einer solchen Kopplung hat trotz der berichteten guten Empfindlichkeit (im Nanogrammbereich) wenig Verbreitung gefunden, was nicht nur daran liegt, daß API Quellen kommerziell nicht erhältlich sind. Negativ auf den Einsatz der Kopplung hat sich auch die Tatsache ausgewirkt, daß Lösungsmittelverunreinigungen stark stören.

4. SEMIPERMEABLE MEMBRANE

Jones und Yang (4) haben versucht, das Lösungsmittel von der gelösten Probe mit Hilfe semipermeabler Siliconmembrane abzutrennen. Es wurde jedoch keine hohe Probenanreicherung erzielt. Außerdem mußte eine erhebliche Peakverbreiterung in Kauf genommen werden, so daß auch dieses System keine weitere Verbreitung gefunden hat.

5. BEWEGLICHES BAND (MOVING BELT)-INTERFACE

Von Scott et al. (5) stammt die Idee, das HPLC Eluat auf einen beweglichen Draht tropfen zu lassen, der Lösungsmittel und Probe zur Ionenquelle transportiert, wobei das Lösungsmittel durch ein differentielles Pumpsystem verdampft wird. Dieses System wurde von McFadden und Schwartz (6) weiterentwickelt und wird heute von zwei Firmen (Finnigan, VG Micromass) angeboten. Abb.1 zeigt eine schematische Darstellung des Interfaces. Anstelle eines Drahtes wird ein Kaptonband verwandt, auf das das HPLC Eluat auftropft. Auf dem Band bildet sich ein dünner Film des Eluates, der kontinuierlich zur Ionenquelle transportiert wird. Auf dem Weg dorthin wird das Eluat zunächst aufge-

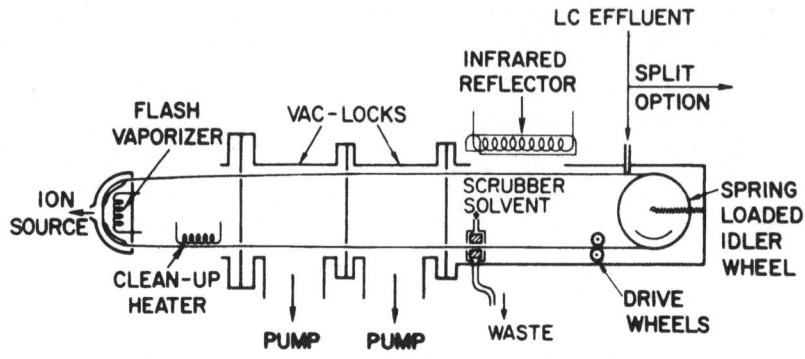

Abb. 1. "Moving Belt" - Interface (Finnigan)

heizt und passiert dann zwei Vakuumschleusen, in denen das Lösungsmittel weitgehend verdampft wird, bevor das Band in die
Ionenquelle eintritt. Unmittelbar vor der Ionenquelle wird das
Band noch einmal kurz erhitzt, wobei die Probe in die Ionisierungszone hinein verdampft wird und entweder durch Elektronen-
stoß (EI) oder Chemische Ionisation (CI) ionisiert wird. Beim
Rücklauf passiert das Band noch eine weitere Heizzone zum Rei-
nigen sowie bei einigen neueren Modellen eine Waschzelle, in
der sich mit Lösungsmittel getränkte Gaze befindet. Beide Vor-
richtungen sollen Memory-Effekte verhindern.
    Werden nicht-polare, leicht flüchtige Lösungsmittel (z.B.
Hexan) verwandt, so können Lösungsmittelmengen bis zu 1 cm^3/min
verkraftet werden. Bei einem hohen Wassergehalt des Lösungsmittels sinkt die Kapazität des Bandsystems auf $\sim$ 0.1 cm^3/min, so
daß ein Split vor dem Interface angebracht werden muß.
    Ein Nachteil des Interfaces liegt darin, daß die Probe vor
der Ionisierung verdampft werden muß, was die Untersuchung extrem polarer, thermisch labiler Verbindungen ausschließt, obwohl die Hochleistungsflüssigchromatographie gerade zur Auftrennung solcher Verbindungen geeignet ist. Für diese Problem-
stellung ist die Entwicklung der Desorptions-Chemischen Ionisation ("desorption chemical ionization", DCI) von Bedeutung, die
auf eine Idee von McLafferty und Baldwin (7) zurückgeht. Bei
dieser Methode wird die Probe auf einer möglichst inerten
Oberfläche aufgebracht und direkt in das Plasma einer CI Quelle
eingeführt. Mit dieser Technik können auch relativ polare,
thermisch labile Moleküle ionisiert und unzersetzt in die Gas-
phase transferiert werden. Es liegt nahe, diese Methode auch
auf das "moving belt" System anzuwenden. Hierzu ist es lediglich
nötig, das Kaptonband bis in die Ionisierungszone hineinzufüh-
ren. Eine solche Modifikation des "moving belt" Systems wurde
z.B. von der Firma Varian MAT (jetzt Finnigan MAT) durchgeführt.
Mit diesem Interface gelang die Analyse von Nucleotiden und
Polysacchariden, die als thermisch extrem labile Verbindungen

bekannt sind (8). Obwohl der Ionenentstehungsmechanismus noch
nicht völlig verstanden wird, spielt die inerte Oberfläche des
Kaptonbands und die rasche Aufheizung des Bandes in der Quelle
eine wichtige Rolle. Komplette Massenspektren lassen sich mit
10^{-8} g Probe erhalten. Das Interface verursacht keine nennens-
werten Verbreiterungen der chromatographischen Peaks.

6. DIREKTEINLASS ("direct liquid introduction", DLI)

Der einfachste Weg einer Kopplung HPLC-MS besteht darin,
das Eluat ohne vorherige Abtrennung des Lösungsmittels direkt
in das Massenspektrometer einzulassen. Da das Vakuumsystem die
gesamte Lösungsmittelmenge selbst unter CI Bedingungen nicht
verkraftet, ist ein Split des Eluats vor dem Einlaß in das
Spektrometer erforderlich. Dieser Weg wurde zuerst von
McLafferty beschritten (9,10), der den Flüssigchromatographen
und das Massenspektrometer durch eine Glaskapillare verband,
die an der Spitze auf ca. 10 µm ausgezogen war. Durch eine
Splitanordnung wurde dafür gesorgt, daß nur ca. 1% des Lösungsmittels in das Spektrometer eintritt. Um die Pumpleistung des
Spektrometers zu erhöhen, wurde die Ionenquelle von einem kryo-
genen Pumpsystem umgeben, d.h. von Kühlfingern, die mit flüssi-
gem Stickstoff gekühlt werden. Die entscheidende Idee war, das
Lösungsmittel selbst direkt als Reaktandgas für die Chemische
Ionisation zu verwenden. Ein Problem bei dem Betrieb dieses im
Prinzip sehr einfachen Interfaces ergab sich dadurch, daß sich
die Kapillare in kurzen Abständen zusetzte und sich häufige
Druckschwankungen ergaben. Arpino et al. (11) konnten in einer
jüngeren systematischen Untersuchung zeigen, daß die Verstopfung der Kapillare dadurch zustande kommt, daß das Lösungsmittel nicht erst in der Ionenquelle, sondern bereits in der Kap-
pillare verdampft. Die Autoren zeigten durch theoretische und
experimentelle Untersuchungen auf, daß sich das Problem lösen
läßt, wenn man den Flüssigchromatographen und das Massenspektrometer nicht durch eine Kapillare, sondern durch eine dünne
Membrane, die ein 1-5 µm Loch enthält, miteinander verbindet.
Unter diesen Bedingungen verdampft das Lösungsmittel nicht unmittelbar beim Eintritt in das Vakuum. Vielmehr bildet sich ein
Flüssigkeitsjet, der aus ca. 5 µ großen Flüssigkeitströpfchen
besteht. Die Flüssigkeitströpfchen verlieren auf dem Weg zum
Ionisierungsort bevorzugt Lösungsmittelmoleküle, so daß eine
Probenanreicherung erfolgt. Die Bildung eines Flüssigkeitsjets
ist zugleich eine Voraussetzung dafür, daß auch thermisch labile Verbindungen intakt ionisiert werden können. Von Vorteil ist
es weiterhin, wenn nicht nur das Lösungsmittel als Reaktandgas
für die Chemische Ionisation dient, sondern ein weiteres Reaktandgas, z.B. Ammoniak, zugegeben wird. Auf diese Weise konnten
die Autoren intakte Disaccharide untersuchen. Schon vor den
theoretischen Untersuchungen von Arpino et al. hatte Melera
(12) von der Firma Hewlett-Packard ein Interface entwickelt,
dem die gleichen Überlegungen zugrunde liegen. Der Aufbau wird
in Abb.2 gezeigt. Das HPLC Eluat strömt an einer Stahlmembrane
vorbei, die ein 5 µm Loch hat. Ungefähr 1% des Eluats tritt in
die Ionenquelle ein. Die Ausbildung eines Flüssigkeitsjets in
der Ionenquelle wird durch eine Kühlung des Interfaces begün-

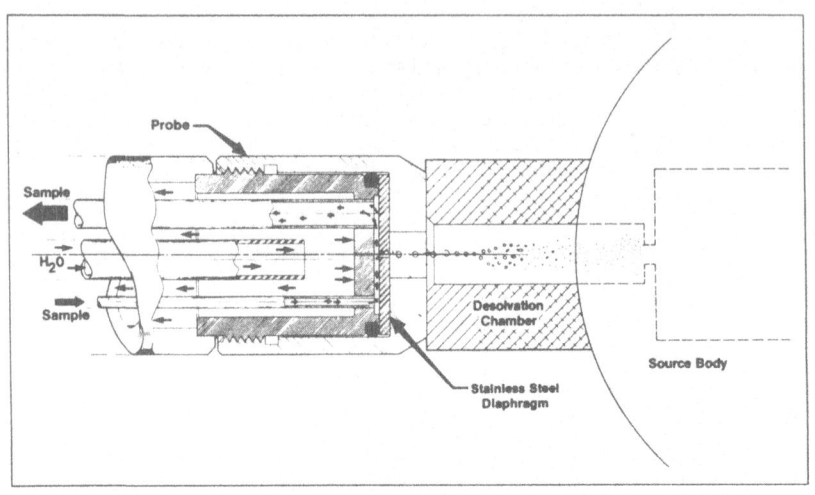

Abb. 2, "Direct Liquid Introduction" - Interface (Hewlett-Packard)

stigt. Dieses seit zwei Jahren kommerziell vertriebene Interface
hat sich im Dauerbetrieb bewährt. Es kommt zu keiner Peakverbreiterung. Wird das Massenspektrometer in der "single ion
monitoring" (SIM) Mode betrieben, so lassen sich injizierte Probenmengen von 50-500 μg nachweisen.
    Das Interface hat drei Nachteile:
a. Es können nur CI Spektren aufgenommen werden.
b. Durch Ionisierung des Lösungsmittels und nachfolgende
    Ionen-Molekül-Reaktionen enthalten die Massenspektren einen
    störenden Untergrund im unteren Massenbereich. In der Pra-
    xis läßt sich lediglich der Massenbereich oberhalb m/z 110
    zur Identifizierung von Proben ausnutzen.
c. Durch die Splitanordnung gelangt nur ∿ 1% der Probe in das
    Massenspektrometer.

7. EINSATZ VON MIKROFLÜSSIGCHROMATOGRAPHEN

    Seit einigen Jahren werden Mikroflüssigchromatographen
(Firma JASCO) kommerziell angeboten, die Trennsäulen mit einem
Innendurchmesser von nur 0.5 mm verwenden. Die hierbei anfallende Lösungsmittelmenge ist 100 mal kleiner (∿ 10 μl/min) als
bei üblichen HPLC-Geräten, wodurch die Kopplung mit einem Massenspektrometer vereinfacht wird. Solche Mikro-HPLC-Geräte sind
für das "Direct Liquid Introduction"-Interface geeignet, da die
gesamte anfallende Lösungsmittelmenge in das unter CI-Bedingungen operierende Massenspektrometer eingelassen werden kann, was
die Empfindlichkeit des Systems beträchtlich erhöht. Eine Kopplung eines Mikro-HPLC mit einem Massenspektrometer wurde von
Henion et al. (13) sowie von uns (14) erprobt. Die Kopplung
kann sehr einfach realisiert werden: HPLC und Massenspektrometer werden über eine Stahlkapillare mit relativ großem Innen-

durchmesser (0.1 mm) als Strömungswiderstand verbunden. Durch
den relativ großen Innendurchmesser tritt keine Verstopfung der
Kapillare auf. Probenmengen von $\leq 10^{-10}$ g lassen sich im SIM-
Betrieb nachweisen. Allerdings kommt es bei der Verwendung
einer relativ weiten Kapillare zu keiner Jet-Bildung, so daß
sich thermisch labile Verbindungen bisher nicht erfassen lassen. Von Nachteil ist, daß die Trennleistung der Mikro-HPLC-Geräte merklich schlechter ist als die der konventionellen Geräte.
Eine Verbesserung der Trennleistung in naher Zukunft zeichnet
sich noch nicht ab.

8. JET-INTERFACE

     Bei der GC-MS-Kopplung hat sich unter der Vielzahl der ent-
wickelten Separatoren der Jet-Separator besonders bewährt. Es
liegt nahe, ein Jet-Interface zur Probenanreicherung auch für
die HPLC-MS Kopplung zu entwickeln. Tatsächlich sind eine Reihe
solcher Systeme erprobt worden (15,16). So benutzt das im Ab-
schnitt 6 erwähnte DLI-Interface eine Jet-Anordnung, auch wenn
es ursprünglich nicht als Jet-Interface konzipiert war. Von den
in den vergangenen Jahren entwickelten Jet-Interfacen soll lediglich das von Vestal et al. (17,18) entwickelte ausführlicher
diskutiert werden. Mehrjährige Versuche von Vestal et al. (19),
einen Flüssigkeits-Jet mit Hilfe eines Lasers aufzuheizen, um
so eine rasche Verdampfung des Lösungsmittels zu erreichen,
zeigten nur einen bescheidenen Erfolg. Die Arbeitsgruppe fand
schließlich eine sehr viel einfachere Anordnung, die in Abb.3
schematisch dargestellt ist.

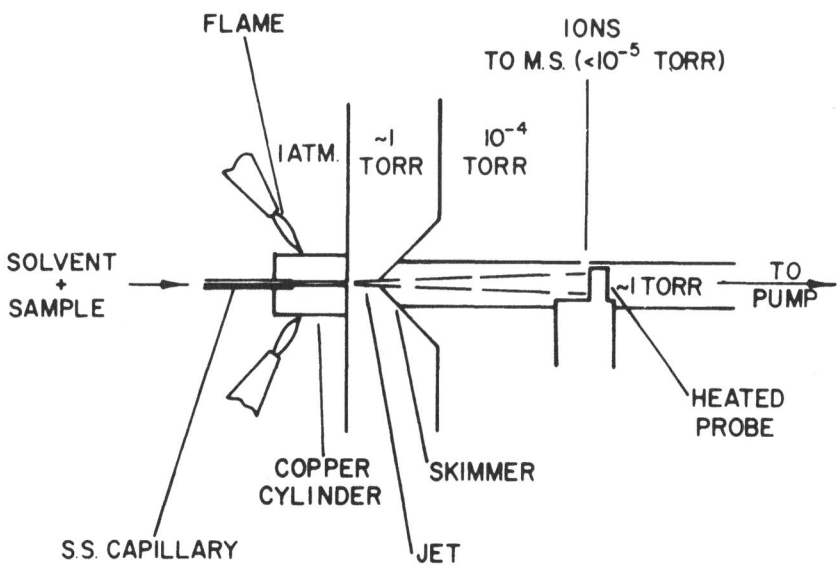

Abb. 3. Jet - Interface

Das HPLC-Eluat durchströmt eine dünne Kapillare (0.15 cm i.D.), die in einen Kupferblock eingebettet ist. Der Kupferblock wird mit vier kleinen Wasserstoff-Sauerstoffflammen auf ca. 1000 °C erhitzt. Das HPLC-Eluat wird dadurch sehr rasch (in wenigen Millisekunden) erhitzt und teilweise verdampft, wobei sich ein Jet aus Dampf und Flüssigkeitströpfchen nahe dem Ende der Kapillare bildet. Bei der anschließenden adiabatischen Expansion in eine Vakuumkammer wird nach Angaben der Autoren 95% der Flüssigkeit verdampft, während der Rest in Form kleiner Tröpfchen auf Schallgeschwindigkeit beschleunigt und dabei stark kollimiert wird. Der Jet-Strahl tritt durch einen Skimmer in die Ionenquelle, wo er auf eine auf 250 °C geheizte Kupferplatte auftritt. Durch den Aufprall auf die heiße Platte werden die Flüssigkeitströpfchen vollständig verdampft und unter CI-Bedingungen ionisiert. Es zeigte sich, daß unflüchtige Probenmoleküle bevorzugt in den Flüssigkeitströpfchen bleiben, so daß eine starke Anreicherung der Probe relativ zum Lösungsmittel erfolgt. Während nur 3-5% der Lösungsmittelmoleküle die Ionenquelle erreichen, werden bei der Untersuchung unflüchtiger Substanzen bis zu 50% der Probenmoleküle in den Ionisierungsraum transferiert.

Überraschenderweise stellten die Autoren fest, daß es auch bei Abschaltung des ionisierenden Elektronenstrahls zur Ionenbildung kommt (18). Der genaue Mechanismus der Ionenbildung ist noch unbekannt. Für die analytische Nutzung dieser neuen Ionisierungsmethode ist die Tatsache von besonderem Interesse, daß die intakte Ionisierung thermisch sehr labiler Verbindungen, wie z.B. Dinucleotide, gelingt.

Obwohl sich die Kopplung noch im Entwicklungsstadium befindet, kommt ihr möglicherweise in der Zukunft eine größere Bedeutung zu.

9. SCHONENDE IONISIERUNGSVERFAHREN

Die vorangehenden Abschnitte haben gezeigt, daß es durch geeignete Modifikationen der bereits bestehenden HPLC-MS-Kopplungen gelingt, auch thermisch labile Verbindungen bis zu einem gewissen Grade zu identifizieren. Speziell zur massenspektrometrischen Identifizierung von stark polaren, thermisch sehr labilen Verbindungen sind in den zurückliegenden Jahren eine Reihe schonender Ionisierungsmethoden wie Felddesorption (FD) (20), Californium-Plasmadesorptions-Massenspektrometrie (PDMS) (21), Laser-induced-Massenspektrometrie (LIMS) (22), Sekundärionenmassenspektrometrie (SIMS) (23) und "Fast Atom Bombardment" (FAB) (24) entwickelt worden, von denen die Felddesorption die älteste und etablierteste, die "Fast Atom Bombardment" die jüngste und vielversprechendste Methode ist. Während aus technischen Gründen eine on-line Kopplung, HPLC-FD nicht realisierbar zu sein scheint, sollte der Einsatz der anderen Ionisierungsmethoden in HPLC-MS Kopplungen im Prinzip zu verwirklichen sein. So sind im zurückliegenden Jahr erste Vorschläge für eine Kopplung HPLC-PDMS (25), HPLC-LIMS (26) sowie HPLC-SIMS (27) gemacht worden. Hierbei wird überwiegend ein Metallband zum Transport des Eluats zum Ionisierungsort benutzt (26,27). Obwohl mit diesen Systemen sehr polare Verbindungen nachgewiesen werden können, bedarf es noch erheblicher Anstrengungen, um die vorgeschlagenen Kopplungen so weiter zu entwickeln, daß sie im Routinebetrieb einsetzbar sind.

10. SCHLUSS

In den vorangehenden Abschnitten wurde gezeigt, daß in den letzten Jahren HPLC-MS Kopplungen entwickelt wurden, die sich im Routinebetrieb bewährt haben. Mit diesen Kopplungen gelingt auch die Identifizierung mäßig polarer Verbindungen. Kopplungen zum Nachweis extrem polarer, thermisch sehr labiler Verbindungen sind in der Entwicklung, doch werden möglicherweise noch mehrere Jahre vergehen, bis die Systeme technisch ausgereift sind.

HPLC-MS Kopplungen wurden bisher fast ausschließlich zur Auftrennung und Identifizierung von Proben aus dem biochemischen bzw. medizinischen Bereich eingesetzt. Eine Verwendung solcher Kopplungen zur Auftrennung und Identifizierung schwerflüchtiger, organischer Inhaltsstoffe im Wasser bietet sich an. Nach unserer Erfahrung bedarf dies jedoch in der Regel einer Vorreinigung bzw. Vortrennung der Probe.

LITERATUR

1. P.J. Arpino und G. Guiochon, Anal. Chem. 51, 682A (1979).

2. W.H. McFadden, J. Chromatogr. Sci. 18, 97 (1980).

3. D.I. Carroll, I. Dzidic, R.N. Stillwell, K.D. Haegele und E.C. Horning, Anal. Chem. 41, 2369 (1975).

4. P.R. Jones und S.K. Yang, Anal. Chem. 47, 1000)1975).

5. R.P.W. Scott, C.G. Scott, M. Munroe und J. Hess Jr., J. Chromatogr. 99, 395 (1974).

6. W.H. McFadden und H.L. Schwartz, J. Chromatogr. 122, 389 (1976).

7. M.A. Baldwin und F.W. McLafferty, Org. Mass Spectrom. 7, 1353 (1973).

8. C. Brunnée, L. Delgmann, G. Dielmann, W. Meyer und P. Thorenz, 29. Meeting of the American Society for Mass Spectrometry, Minneapolis, 1981, Poster TPA 4.

9. M.A. Baldwin und F.W. McLafferty, Org. Mass Spectrom. 7, 1111 (1973).

10. P.J. Arpino, B.G. Dawkins und F.W. McLafferty, J. Chromatogr. Sci. 12, 574 (1974).

11. P.J. Arpino, P. Krien, S. Vatja und G. Devant, J. Chromatogr. 203, 117 (1981).

12. J.W. Serum und A. Melera, 29th Pittsburgh Conference on Analytical Chemistry and Applied Spectroscopy, Cleveland, Ohio (1978). Paper No. 589.

13. J.D. Henion und G.M. Maylin, Biom. Mass Spectrom. 7, 115 (1980).

14. K.H. Schäfer und K. Levsen, J. Chromatogr. 206, 245 (1981).

15. T. Takeuchi, Y. Hirata und Y. Okumura, Anal. Chem. 50, 659 (1978).

16. S. Tsuge, Y. Hirata und T. Takeuchi, Anal. Chem. 51, 166 (1979).

17. C.R. Blakley, J.J. Carmody und M.L. Vestal, Anal. Chem. 52, 1636 (1980).

18. C.R. Blakley, J.J. Carmody und M.L. Vestal, J. Am. Chem. Soc. 102, 5933 (1980).

19. C.R. Blakley, M.J. McAdams und M.L. Vestal, J. Chromatogr. 158, (1978).

20. H.D. Beckey, Int. J. Mass Spectrom. Ion Phys. 2, 500 (1969).

21. R.D. McFarlane und D.F. Torgerson, Science 191, 920 (1976).

22. R.O. Mumma und F.J. Vastola, Org. Mass Spectrom. 6, 1373 (1972).

23. A. Benninghoven und W.K. Sichterman, Anal. Chem. 50, 1180 (1978).

24. M. Barber, R.S. Bordoli, R.D. Sedgwick und A.N. Tyler, J. Chem. Soc.,Chem. Comm. 325 (1981).

25. H. Jungclas, H. Danigel und L. Schmidt, Org. Mass Spectrom., eingereicht.

26. E.D. Hardin und M.L. Vestal, Anal. Chem. 53, 1492 (1981).

27. A. Benninghoven, A. Eiche, M. Junack, W. Sichtermann, J. Krizek und H. Peters, Org. Mass Spectrom. 15, 459 (1980).

IDENTIFICATION OF NON-VOLATILE ORGANIC COMPOUNDS IN WATER

B. CRATHORNE and C.D. WATTS

Water Research Centre, Henley Road, Marlow,
Buckinghamshire, England.

Summary

A review of research into the isolation, separation and identification of non-volatile organic compounds in water is presented.

The review has been restricted to the most commonly used methods for the isolation of non-volatile organics; *viz* freeze-drying, reverse-osmosis, activated carbon adsorption and XAD resin adsorption. It is shown that each of these techniques is capable of isolating a range of non-volatile organics from water but it is apparent that no one method is capable of extracting all organics present in a water sample.

Separation methods for non-volatiles invariably use high-performance liquid chromatography and several chromatographic methods have been employed. Undoubtedly reversed-phase chromatography has received the most widespread usage although efficient separations of certain types of non-volatile compound have been obtained on normal-phase chromatography, size-exclusion chromatography and ion-exchange chromatography.

Identification of separated non-volatile compounds is difficult but feasible by the application of state-of-the-art mass spectrometric techniques. In particular the use of field desorption mass spectrometry with accurate mass measurement and collisonal activation with linked scanning is shown to be highly promising.

1. INTRODUCTION

This paper presents a short review of research into the isolation, separation and identification of non-volatile organic compounds in water. Results of recent work at the Water Research Centre (WRC) will be used to illustrate certain areas of research.

For the purposes of this review a 'non-volatile' organic compound is defined as a compound which is not generally amenable to analysis by gas chromatography (GC). This may be by virtue of high molecular weight, high polarity or thermal instability. Naturally this concept is to some extent flexible since advances in the use of derivatisation or custom-made columns are extending the range of compounds able to be analysed by GC. However the techniques reviewed in this paper are capable of analysing inherently non-volatile compounds in addition to some compounds which can be analysed by GC.

One particular need for methods of identification for non-volatile organics stems from the rapid growth of interest over the last few years in organics in drinking water[1,2]. This interest arises from concern over the possible health effects resulting from long-term exposure to organics in drinking water, especially halogenated compounds[1]. Gas chromatography-mass spectrometry (GCMS) is almost universally used to obtain information on the range and type of organic compounds present in treated water. It is now generally accepted that approximately 80% of the organic matter present in both raw and drinking water is not amenable to analysis by GCMS and there is also some evidence that a similar proportion of the halogenated organic matter produced by chlorination can be considered non-volatile[3]. Methods for the characterisation of non-volatile organics in treated water are therefore urgently needed to provide information on the bulk of the organic matter in water which is not amenable to GCMS. The application of short-term bio-assays, such as the Ames test, to treated waters is emphasising the need for analytical methods for identifying non-volatile organics since many of the mutagens detected are apparently not identifiable by GCMS[4,5].

There are many problems associated with the anlysis of non-volatile organics. Most are due to the trace amounts usually present and the wide range of compound types and molecular weights encountered. The various methods used for anlysis of such compounds are reported under the general headings of isolation, separation and identification. Two aspects of iden-tification of non-volatiles not covered by this review are the use of specific detectors in high-performance liquid chromatography[6] and the use of combined liquid chromatography-mass spectrometry[7] since these are the subjects of separate reviews presented at this meeting. One further approach not reviewed is the study of byproducts formed by chemical degradation of naturally occurring organic compounds. This method for identifying the range of organic compounds likely to be present in potable waters has been recently reviewed[8].

2. ISOLATION TECHNIQUES

Organic compounds are generally present in water in trace amounts, at the microgram per litre level or less, and as part of a complex mixture. Thus, before separation and identification can be attempted, compounds usually must be isolated from a very dilute aqueous solution. A wide variety of techniques have been used for this purpose and these have recently been reviewed[9].

Table 1 lists some of the more commonly used techniques for the isolation of non-volatile organics. Further discussion will be limited to the techniques listed in Table 1, but other methods reported include vacuum distillation[10], ultrafiltration[11], ion-exchange resins[12] and adsorption on C_{18} bonded silica gel[13,14].

Table 1. Methods for isolating non-volatile organics from water

Method	Comments
Freeze-drying	Low temperature process and low contamination but disadvantages of also concentrating inorganics and is slow and comparatively expensive. Concentrates polar and non-polar compounds.
Reverse-osmosis	Large volumes can be processed but problems from contamination and the concentration of inorganics. Concentrates polar and non-polar compounds.
Activated carbon adsorption	Large volumes can be processed but only limited recovery of adsorbed organics and elution of artefacts from the adsorbent. Can be used for a wide polarity range but better for concentrating non-polar compounds.
XAD resin adsorption	Large volumes can be processed but only a limited range of compounds are adsorbed. Suitable for concentration of non-polar to moderately polar compounds.

2.1. Freeze-drying

In this process the water sample is frozen and the water removed by sublimation under vacuum. The more volatile organic constituents are lost. Although the recovery of organic matter (estimated by total organic carbon measurement) in the freeze dried solids can be very high with this technique, recovery of organics after extraction of the solids, which is composed largely of inorganic material is more limited. Crathorne et al[15,16] and Watts et al.[17] have shown that a wide range of organic compounds can be extracted from drinking water using this technique. Although recoveries obtained for particular compound types were found to be solvent dependent, methanol provided highest overall recovery[17]. The inorganic content of methanol extracts was found to present several problems when they were analysed, in particular when using field-desorption mass spectrometry[17]. A desalting process was found to be necessary and a method using ion-pair high-performance liquid chromatography is now in use at WRC. Freeze-drying has also been used to prepare concentrates from waste water samples for chemical analysis and biological testing[18]. Commercial freeze-dryers are available which are capable of processing several litres of water per day. However this can only be done in a single batch hence the processing of large volumes of water is extremely slow.

2.2. Reverse-osmosis

Reverse-osmosis involves passing water through a membrane with a very small pore size which exludes large molecules, thereby forming a concentrate of these constituents. The membrane is generally made of either cellulose acetate or nylon. The advantage of this technique is that it is continuous and hence large volumes of water can be processed. Some commercial units can process several hundred litres of water per day. Disadvantages are that membranes can release contaminants into the sample and may also adsorb constituents from the sample(19,20). Many inorganic compounds are also concentrated using this technique. One further drawback is that reverse-osmosis produces a concentrated aqueous solution of organics and the compounds of interest have to be extracted from this concentrate using a different technique. Kopfler et al.[21] have described a scheme whereby organics were recovered from tap water using a combination of reverse-osomosis, dialysis and solvent extraction. With this scheme approximately 30-40% of organics were recovered from tap water samples. Further work by Kopfler et al.[19] showed that approximately 85% of the total organic carbon from a drinking water sample was rejected using a cellulose acetate membrane. Reverse-osmosis in combination with freeze-drying, XAD resin adsorption and activated carbon adsorption was used by Deinzer et al.[20] to extract organics from drinking water samples. The percentage of organic material recovered, based on total organic carbon measurement, varied between 29 and 60%. A combination of reverse-osmosis and XAD resin adsorption was used by Loper et al.[22] for the preparation of extracts for mutagenicity testing. Although it was estimated that only 40% of the total organic carbon was recovered, appreciable mutagenicity was found in the samples.

2.3. Activated carbon adsorption

In this process the water sample is passed through a column of activated carbon which adsorbs the organic compounds and these are subsequently extracted from the carbon with a suitable solvent. The technique has been extensively used to prepare organic extracts from water for many years and activated carbon columns have been designed which are capable of processing several thousands of gallons[23]. A major disadvantage of the process is that recovery of organic material is incomplete. Normal recovery procedure involves drying the carbon and then recovering adsorbed material by sequential Soxhlet extraction with chloroform and ethanol[24]. Experiments with a series of chlorinated pesticides added to water showed that their adsorption varied between 90-100% but that recovery from the resin was in the range 70-85%[25]. Adsorption and recovery efficiency of a number of organic compounds including chlorinated phenols and geosmin was found to depend on the range of organic compounds present in the sample, the humic and fulvic acid content and the pH of the sample[26].

2.4. XAD resin adsorption

The adsorption of organic compounds from aqueous solution onto XAD resins has been used most extensively for the study of volatile organic compounds. The commonly used resins are XAD-2 and -4, both of which are styrene-divinylbenzene copolymers and XAD-8 which is an acrylic ester polymer. XAD-2 (and -4) have a high affinity for non-polar compounds[27] whereas XAD-8 has been found to be more efficient in the recovery of polar compounds such as humic acid[28]. XAD-2 and XAD-8 have been used to isolate non-volatile organics prior to organohalogen measurement[3]. In this work Glaze et al. found that natural organics do not adsorb as efficiently onto XAD resins after chlorination of the sample. However with careful pH adjustment and control of flow rate, it has been estimated that about 50% of the organics in an average water sample can be concentrated using XAD-8[29,30]. Adsorbed organics are recovered from the resin by elution with a suitable solvent, generally diethyl ether, methanol or acetone, although it has been shown that not all adsorbed organics can be recovered from the resin. For example it has been shown that only 70% of adsorbed fulvic acid and 80% of adsorbed humic acid can be recovered from XAD-2 resin[30]. A further disadvantage is that XAD resins require extensive clean-up before use[31] and each batch prepared must be evaluated to ensure that resin contaminants are negligible.

2.5. Summary of methods

Due to the wide variety of chemical compounds presents in water, no single extraction technique is capable of isolating all organics from the water sample. A satisfactory comparison of the isolation methods in use is impossible due to the wide range of water samples examined and the variety of methods used to analyse the organic extracts. However, it would appear from the literature that reverse-osmosis and freeze-drying result in extracts containing the widest range of organic compounds, although for ease of operation, particularly when processing large volumes of water, a column concentration technique may be preferred.

A comparison of the range of organics extracted using XAD-2 resin and freeze-drying can be made from some recent work carried out at WRC. Figure 1 shows reversed-phase HPLC separation of organics isolated using freeze-drying and XAD-2 resin adsorption. A wide range of compounds is shown to be present in both extracts. The freeze-dried sample (Fig. 1a) contains a large number of compounds of high polarity (i.e. early-eluting components) whereas the XAD extract (Fig. 1b) produces a chromatogram virtually free of early-eluting compounds. This indicates that many of the polar compounds present in water are not efficiently adsorbed onto XAD-2. A chromatogram from the separation of the XAD extract using normal-phase chromatography is shown in Fig. 2. This type of chromatography is more suitable for the separation of non-polar compounds and the presence of a wide variety of this type of compound is indicated in the XAD extract.

3. SEPARATION METHODS

The ideal technique for separation of non-volatile organics should have a high resolving power and sensitivity of detection for separated compounds. In addition, there should be no need for volatilisation or derivatisation before analysis and the technique should not chemically

alter the organics during separation. High-performance liquid chromatography (HPLC) fits these ideals best and is the method of choice for
separation of non-volatile organics.

HPLC utilises several chromatographic methods and most of these have
been applied to the separation of non-volatile organics isolated from water.

3.1. Reversed-phase chromatography

Reversed-phase chromatography employs a non-polar stationary phase
(i.e. the column packing) and a polar mobile phase. Thus highest retention
occurs with the most polar eluent (i.e. water) and retention is decreased
by addition of a less polar eluent (e.g. methanol).

Work at WRC and elsewhere has shown that reversed-phase HPLC can
provide good separation of organics extracted from water (see for example,
Fig. 1). Details of this work have been previously reported[15,16,17].

In reversed-phase HPLC, compounds are eluted in an order based essen-
tially on polarity. The presence of a wide variety of organics is illustrated ranging from highly polar compounds eluted with mainly water as
eluent to non-polar compounds eluted with 90% methanol.

Reversed-phase HPLC has also been used for the separation of non-
volatile organics extracted from drinking water to provide fractions for
mutagenicity testing[32] and as part of a general separation scheme for
characterisation of natural organics in both raw and drinking water[33].
Organics extracted from drinking water using either activated carbon adsorption or reverse-osmosis/XAD resin adsorption were separated on reversed-
phase HPLC using a water to acetonitrile gradient. This separation has
been used as part of a fractionation scheme described by Tabor et al.[34]
and by Loper[5] for the identification of mutagenic compounds. Reversed-
phase separation of XAD resin extracts of treated waters was also used by
Cheh and Carson as part of a method for the detection of mutagenic
compounds[35].

3.2. Normal-phase chromatography

In normal-phase chromatography a polar stationary phase is used with a
non-polar eluent. Polar compounds are most strongly retained and are
eluted by increasing the polarity of the eluent.

Separations of organics using normal-phase HPLC have been reported far
less frequently than those using reversed-phase systems. However, normal-
phase columns can be used to obtain good separation of organics isolated
from water (see Fig. 2). This type of chromatography is most suited to the
analysis of moderately polar or non-polar compounds as illustrated by the
good separation obtained for organics extracted from water by adsorption on
XAD-2. Normal-phase HPLC on silica was used by Shinohara et al.[36] to
separate neutral organics in an XAD-resin extract of a drinking water
sample. Results showed the presence of a complex mixture of organic
compounds which was confirmed by the subsequent analysis of HPLC fractions
by field-desorption mass spectrometry.

3.3. Size-exclusion chromatography

Size-exclusion chromatography (SEC) is a technique which separates
molecules according to molecular size. This separation takes place by
exchange of the solute molecules between the mobile phase and the stagnant
liquid within the pores of the packing.

SEC has shown that the molecular weights of organics in water can
cover a wide range. Several types of SEC column have been used, including
Sephadex (Pharmacia, Sweden) and TSK gels (Toya Soda Company, Japan) for
the comparison of organics in groundwater, surface water and waste water
samples before and after treatment[37,38]. This work showed that although
an increase in lower molecular weight compounds was generally observed

after treatment, particularly when using ozone, a large proportion of the high molecular weight material remained unaffected. TSK gels have also been used for the analysis of humic acids[39]. Glaze et al. have used a controlled porosity glass bead column for molecular weight characterisation of organics in water before and after treatment with chlorine[3], ozone, and ozone with u.v. irradiation[33,40]. Glaze observed a general reduction in molecular weight after ozone treatment, but chlorination was found to have little effect on the molecular weight range in the sample.

One drawback in the use of SEC to separate organics according to size results from the competition between exclusion processes and non-exclusion processes such as adsorption. This is a particular problem for an unknown mixture containing compounds of widely different type and polarity since it becomes virtually impossible to calibrate the column accurately[8].

3.4. Ion-exchange chromatography

Separation takes place in ion-exchange chromatography because of competition between ions in the eluent and in the solute for oppositely charged sites on the stationary resin. Thus if the solute bonds strongly to the resin it will be well retained whereas a weak bond will result in rapid displacement by eluent ions and quick elution.

The use of ion-exchange chromatography for the separation of organic compounds extracted from untreated and chlorinated water has been reported by Jolley and co-workers in several papers[41,42]. Although long elution times were required, very good separations of a wide range of polar and charged organic compounds were obtained, leading to the identification of over 40 organics. An ion-exchange system has also been used for the separation of organics from ozonated waters prior to identification and mutagenicity testing[43]. A general separation scheme devised by Glaze et al.[33] for the characterisation of organics extracted from raw and treated water included the use of weak anion-exchange chromatography with pH gradient elution. Three fractions were obtained from the separation of an extract of an ozonated surface water, one of which was shown to contain a mixture of phenols and another a mixture of benzene carboylic acids.

3.5. Ion-pair chromatography

In ion-pair chromatography a suitable counter-ion is added to the mobile phase or stationary phase in order to promote 'salt formation' (i.e. the formation of ion-pairs) with the acidic or basic solute. The 'ion-pairs' formed alter the retention behaviour of the solute thus allowing optimisation of a separation by adjustment of the type and concentration of counter ion and the pH.

Ion-pair chromatography is now used for the separation of a wide variety of organic compounds. However, it is mainly of use for the analysis of specific compound types when the column and mobile phase can be carefully adjusted to enable separation of the compounds of interest. For the separation of complex mixtures containing many compound types, ion-pair chromatography is of only limited use.

3.6. Summary of the use of HPLC for separation of non-volatile organics

A range of chromatographic methods have been used for the separation of organics extracted from water and each technique can provide a good separation of a certain range of compounds. To some extent the techniques are complementary. Reversed-phase chromatography can be used for separation of the widest range of compounds and is particularly useful for high and medium polarity compounds. Normal-phase chromatography can provide efficient separation of non-polar organics while size-exclusion chromatography is best used for providing information on the high molecular weight compounds present in water. Ion-exchange and ion-pair chromatography can be used for the separation of ionisable compounds. Although HPLC shows inherent advantages for the separation of non-volatile organics it also suffers from two serious limitations. There is no sensitive universal detector for HPLC comparable with the flame ionisation detector for GC. Several highly sensitive detectors are available but these are also highly selective and inappropriate for the detection of a wide range of compounds. Detection is normally carried out by u.v. absorption and is therefore limited to those compounds containing a u.v. chromophore. Secondly, since non-volatile organics isolated from water usually contain a large number of compounds of diverse chemical type, even high-efficiency HPLC columns cannot provide separation into individual components[8].

4. IDENTIFICATION

Any technique that is used to provide structural information on non-volatile organic compounds in extracts or after separation by HPLC, must be capable of handling very low quantities and complex mixtures.

It has been shown that even when examining HPLC fractions which apparently consist of single component (as observed by u.v. detection) there is invariably a mixture of compounds present[8]. Thus most of the classical methods of identification e.g. nuclear magnetic resonance or infra-red spectroscopy are unsuitable and some type of mass spectrometry must be used. Some structural information may be obtained using specific HPLC detectors, for example, electrochemical or post-column reaction detectors. However, these will probably at best only indicate compound type rather than identity. The use of specific detectors for HPLC is the subject of another paper at this meeting[6].

Directly coupled LCMS offers some promise for identifying non-volatile organics although many problems still need to be solved. A detailed review of LCMS is the subject of a paper at this meeting[7]. Therefore this paper only considers mass spectrometry used off-line for identification of non-volatile organics and in particular outlines the technique of field-desorption mass spectrometry. A more detailed review of modern mass spectrometric techniques applied to the analysis of organics in water is provided at this meeting by Cornu[44].

Conventional electron impact mass spectrometry using a direct insertion probe offers a considerable extension to the types of compound that can be handled by GCMS. However, electron impact suffers from two limitations; the compound of interest has to be volatilised prior to ionisation and the molecular ion is frequently of low intensity with respect to fragment ions. The first limitation can be overcome to some extent by the use of rapid heating and 'in-beam' techniques, which have been recently reviewed[45]. The second limitation, however, is difficult to overcome and results in extremely complex, overlapping mass spectra in the case of mixtures.

Ideally then, for mass spectrometry of mixtures of non-volatiles, a method is required that can ionise compounds in the solid state with transfer of very little excess energy. This would provide a mass spectrum consisting of intense molecular ions with few, if any, interfering fragment ions. Several ionisation methods which can achieve this are available, for example direct chemcial ionisation (DCI)[45], plasma desorption (PD)[46], laser desorption (LD)[47], field desorption (FD)[48] and, most recently, fast atom bombardment (FAB)[49]. All of these techniques provide intense molecular ions and only limited fragmentation for certain types of non-volatile organics and are, to some extent, complementary. At the present time FDMS has received more extensive application than any of the other techniques and appears to be applicable to the widest range of compound types[50]. Surprisingly, the use of FDMS for the identification of non-volatile organics separated by HPLC has received only limited study. However it has been used by Schulten and Stöber[51] for analysis of herbicides and their by-products in surface waters and by Watts et al.[8] and Shinohara et al.[36,52] for analysis of organics in drinking water.

While this approach has contributed to the identification of only few compounds so far, for example polyvinylacetates have been identified in drinking water samples[52], it has clearly shown that a complex mixture of non-volatile organics is generally present in potable water.

An example from work at WRC of a FDMS of a reversed-phase HPLC fraction of a drinking water extract is shown in Fig. 3. This represents one of the more complex fractions obtained but illustrates the wide range of compounds present, since the majority of ions can be considered to be molecular ions under the FDMS conditions used.

Halogenated compounds can be detected in drinking water extracts as is illustrated by FDMS of another HPLC fraction (Fig. 4). This shows several ions with isotope ratios indicative of compounds containing chlorine and/or bromine. Identification of these compounds presents many problems and the identifications reported to date[15] generally constitute cases where HPLC fractions have contained only one or two major compounds.

A limitation of FDMS for the identification of non-volatiles is the lack of structural information provided. Accurate mass measurement of molecular ions can provide empirical formulae, but a number of structures will still be possible. Recently, methods for obtaining structural data from selected ions, namely collisional activation (CA)[53] coupled with either linked scanning at a fixed B/E ratio[54] or mass analysed ion kinetic energy spectrometry (MIKES)[55,56] have become more widely available. Useful fragmentation information similar to that obtained from electron impact mass spectrometry can be obtained for any ion (of sufficient intensity) in a mass spectrum, which makes the method advantageous for the analysis of mixtures. Reports of the combined use of FDMS and CA have only recently appeared, but the technique will be particularly valuable for the identification of non-volatile organics[57]. At WRC, FDCA is used as part of a scheme for identification of non-volatiles using mass spectrometry. This scheme is illustrated in Fig. 5.

Firstly, low resolution FDMS is obtained on a fraction to provide nominal mass data for the organics present and information on the emitter current at which they desorb. Then accurate mass data and hence empirical formulae are obtained on each FDMS peak of interest by using suitable reference compounds. Either a calibration table is created by the datasytem which then automatically performs the accurate mass determination or the peak is manually matched against a reference ion and its accurate mass calculated. Structural information is then obtained for suitable compounds by the use of collisional activation and linked scanning at a fixed B/E ratio. This technique produces the equivalent of a normal electron impact mass spectrum for any molecular ion of sufficient intensity generated by FDMS.

An example of a FDCA mass spectrum obtained by linked scanning is shown in Fig. 6. The compound used to produce this spectrum was tetrabutyl ammonium chloride and the FDCA spectrum was obtained on the tetrabutyl ammonium ion, m/z 242. The structure of this ion is clearly indicated by the elimination of methane (m/z 226), ethane (m/z 212), propane (m/z 198) and butane (m/z 184). The formation of the tributyl ammonium ion (m/z 185) and similar eliminations from this (e.g. m/z 169 – elimination of methane) provide further structural information. The combination of this data with the empirical formula information would allow relatively easy identification of this compound.

An example of the type of information that can be obtained from an extract is provided by the FDCA spectrum (Fig. 7) of a non-volatile organic compound isolated from drinking water. The major loss from the molecular ion (m/z 638.6) is 15 a.m.u. indicative of a methyl group. The sequential losses of m/z 28 suggests a quinonoid moiety and this is in agreement with the empirical formula obtained from accurate mass measurement.

Application of this approach is still in its early stages, but it is possible to comment on some of the problems that have arisen. FDMS is a relatively sensitive technique, requiring only a few nanograms of compound, but produces low intensity fluctuating ion beams as compared to EIMS. In order to carry out collisional activation and linked scanning on an FD ion beam, a minimum of ∿0.5 μg of each compound must be deposited on the emitter, although this sample requirement can be reduced by the use of computer acquisition and averaging techniques. This means that there is a relatively high (for mass spectrometry) minimum sample requirement and also, if there is a complex mixture, it may only be possible to examine one or two of the compounds present before exhausting all of the sample. The presence of large amounts of inorganics in extracts of water (this is a particular problem with methanol extracts of freeze-dried residues) and in some derived HPLC fractions can interfere with FDMS in several ways; (i) they limit the absolute amount of a fraction that can be deposited on a FD emitter, (ii) they can give rise to high molecular weight cluster ions with polar organic molecules[58] and (iii) they can yield, by formation of cluster ions, FDMS spectra with ions at similar molecular weights to compounds of interest, which can complicate interpretation of the mass spectrum[48]. However the problems due to inorganic constituents have been alleviated by including a desalting stage and improved FDMS handling techniques.

Although only a few identifications have been made using this approach, it has provided structural information on non-volatile organics present in a complex mixture that is not available from any other technique. Further, it is possible to obtain complete structural identification with these techniques, an important consideration when there is such a paucity of mass spectral reference data for non-volatiles. Regardless of these problems, FDMS with collisional activation and linked scanning is the most promising technique currently available for the identification of non-volatile organics isolated from water samples.

5. DISCUSSION

Non-volatile organic compounds account for most of the organic matter present in both raw and drinking water. Many of the industrial, agricultural and pharmaceutical compounds in widespread use are non-volatile and undoubtedly some proportion of them find their way into the aquatic environment. Studies of the organic by-products of chlorination and ozonation of water have shown that a significant proportion of these are non-volatile. The increasing application of short-term bioassay tests to drinking water has shown that non-volatile organics could be responsible for some of the mutagenic response.

Until recently there have been no really suitable techniques available for the isolation, separation and characterisation of non-volatile organics from water. Consequently, little data is available on the types and amounts of non-volatile organics present in drinking water.

Several extraction techniques are capable of isolating non-volatile organics from water. However, it is clear that no one technique is capable of isolating all organics from the water sample. In order to achieve this it will probably be necessary to use several techniques.

High-performance liquid chromatography is the most appropriate separation technique available for non-volatiles. Good separations of non-volatiles have been obtained using reversed-phase, normal-phase, size-exclusion and ion-exchange chromatography. However even with the use of highly efficient columns it is still not possible to separate a complex extract from water into individual components. Some progress in this area can be expected in the next few years, possibly through the development of capillary or open-tubular columns capable of generating extremely high plate numbers. Alternatively, if enough sample were available then some form of column switching technique could be developed from which fractions containing only one major component may be obtained.

The development of a sensitive universal detector for HPLC would be on aid to identification since it would indicate the major components present and enable straightforward fractionation of an extract. Suitable specific element or functional group detectors (for example a specific halogen detector), if available, could also provide useful structural information.

Recently developed mass spectrometric techniques provide powerful tools for the identification of non-volatile organics. In particular the approach using FDCA mass spectrometry in combination with HPLC separation appears to offer great promise.

6. CONCLUSIONS

The development of methods for the identification of non-volatile organics is an urgent and important task. For example, an accurate assessment of the health risk associated with the long-term exposure to organics in water cannot be made until more is known about the non-volatile compounds present. This review has described techniques which are capable of isolating, separating and identifying non-volatile organics in water. The limited application of such techniques to date has demonstrated the presence in water of a wide range of non-volatile compounds, some halogenated, and virtually all unidentified. The identification of these compounds is proceeding slowly and one of the reasons for this is that few laboratories are actively involved. A more widespread application of these new analytical techniques is essential to speed progress in the identification of non-volatile organics in water.

ACKNOWLEDGEMENTS

The work carried out at WRC is funded by the Department of the Environment. Permission from the Department of the Environment and the Water Research Centre to publish this work is gratefully acknowledged.

REFERENCES

1. JOLLEY, R.L., BRUNGS, W.A. and CUMMING, R.B., Eds, 'Water Chlorination: Environmental Impact and Health Effects', Vol. 3, Ann Arbor Science, Michigan, 1980.
2. KEITH, L.H., Ed., 'Identification and analysis of organic pollutants in water', Ann Arbor Science, Michigan, 1976.
3. GLAZE, W.H., SALEH, F.Y. and KINSTLEY, W. In 'Water Chlorination: Environmental Impact and Health Effects', Vol. 3, Ann Arbor Science, Michigan, 99, 1980.
4. FALLON, R.D. and FLIERMANS, C.B. Chemosphere, 9, 385, 1980.
5. LOPER, J.C. Mutat. Res., 76, 241, 1980.
6. POPPE, H. Presented at the 2nd European Symposium on 'Analysis of Organic Micropollutants in Water', Killarney, 1981, this volume, p. 141.
7. LEVSEN, K. Presented at the 2nd European Symposium on 'Analysis of Organic Micropollutants in Water', Killarney, 1981, this volume, p. 149.
8. WATTS, C.D., CRATHORNE, B., FIELDING, M. and KILLOPS, S.D. Fundamental Appl. Toxicol. In press. Presented at symposium on 'Health effects of drinking water, disinfectants and disinfectant by-products', Cincinnati, Ohio, April 21-24th, 1981.
9. JOLLEY, R.L. Environ. Sci. Technol., 15, 8, 874, 1981.
10. JOLLEY, R.L., KATZ, S., MROCHEK, J.E., PITT, W.W. and RAINEY, W.T. Chemtech, 5, 312, 1975.
11. MILANOVITCH, F.P., IRELAND, R.R. and WILSON, D.W. Environ. Lett., 8, 4, 337, 1975.
12. BAIRD, R.B., GUTE, J., JACKS, C., JENKINS, R., NIESESS, L., SCHEYBELER, B., VAN SLUIS, R. and YANKO, W. In 'Water Chlorination: Environmental Impact and Health Effects', Vol. 3, Ann Arbor Science, Michigan, 925, 1980.
13. GRAHAM, J.A. and GARRISON, A.W. In 'Advances in the Analysis and Identification of Organic Pollutants in Water', Keith, L.H., Ed., Ann Arbor Science, Michigan, In press.
14. WAGGOTT, A. and CONNOR, K.J. In 'Proceedings of the 1st European Symposium on Analysis of Organic Micropollutants in Water', Berlin, 1979.

15. CRATHORNE, B., WATTS, C.D. and FIELDING, M. J. Chromatogr., *185*, 671, 1979.
16. CRATHORNE, B., WATTS, C.D. and FIELDING, M. In 'Proceedings of the 1st European Symposium on Analysis of Organic Micropollutants in Water', Berlin, 1979. In press.
17. WATTS, C.D., CRATHORNE, B., CRANE, R.I. and FIELDING, M. In 'Advances in the Analysis and Identification of Organic Pollutants in Water', Keith, L.H., Ed., Ann Arbor Science, Michigan, In press.
18. JOLLEY, R.L., LEE, N.E., LEWIS, L.R., MASHNI, C.I., PITT, W.W., THOMPSON, J.E. and CUMMING, R.B. In 'Advances in the Analysis and Identification of Organic Pollutants in Water', Keith, L.H., Ed., Ann Arbor Science, Michigan, In press.
19. KOPFLER, F.C., COLEMAN, W.E., MELTON, R.G., TARDIFF, R.G., LYNCH, S.C. and SMITH, J.K. Ann. N.Y. Acad. Sci., *298*, 20, 1977.
20. DEINZER, M., MELTON, R. and MITCHELL, D. Wat. Res., *9*, 799, 1975.
21. KOPFLER, F.C., MELTON, R.G., MULLONEY, J.L. and TARDIFF, R.G. In 'Fate of Pollutants in the Air and Environment', Suffet, I.H., Ed., J. Wiley-Interscience, New York, Vol. 2, 419, 1975.
22. LOPER, J.C., LANG, D.R. and SMITH, C.C. In 'Water Chlorination: Environmental Impact and Health Effects', Vol. 2, Ann Arbor Science, Michigan, 433, 1978.
23. MIDDLETON, F.M., PETTIT, H.H. and ROSEN, A.A. In 'Proceedings of the 17th Industrial Waste Conference', Purdue Univ., Engr. Ext. Ser., *112*, 454, 1962.
24. KOPFLER, F.C. In 'Application of Short Term Bioassays in the Analysis of Complex Environmental Mixtures', Sandhu, S.S., Ed., Plenum Press, New York, Vol. 2, 1981.
25. ROSEN, A.A. Analyt. Chem., *31*, 1729, 1959.
26. SNOEYINK, V.L., McCREARY, J.J. and MURIN, C.J. EPA-600/2-77-223, 1977.
27. JUNK, G.A., RICHARD, J.J., GRIESER, M.D., WITIAK, D., WITIAK, J.L., ARGUELLO, M.D., VICK, R., SVEC, H.J., FRITZ, J.S. and CALDER, G.V. J. Chromatogr., *99*, 745, 1974.
28. THURMAN, E.M., MALCOLM, R.L. and AIKEN, G.R. Analyt. Chem., *50*, 13, 1836, 1978.
29. MALCOLM, R.L., THURMAN, E.M. and AIKEN, G.R. Presented at symposium on 'Concentrating Organics for Toxicity Testing', Houston, Texas. In Preprints of Papers, ACS Environ. Chem. Div., Vol. 20, 1, 107, 1980.
30. AIKEN, G.R., THURMAN, E.M. and MALCOLM, R.L. Analyt. Chem., *51*, 11, 1799, 1979.
31. JAMES, H.A., STEEL, C.P. and WILSON, I. J. Chromatogr., *208*, 89, 1981.
32. CUMMING, R.B., LEE, N.E., LEWIS, L.R., THOMPSON, J.E. and JOLLY, R.L. In 'Water Chlorination: Environmental Impact and Health Effects', Vol.3, Ann Arbor Science, Michigan, 881, 1980.
33. GLAZE, W.H., JONES, P.C. and SALEH, F.Y. In 'Advances in the Analysis and Identification of Organic Pollutants in Water', Keith, L.H., Ed., Ann Arbor Science, Michigan, In press.
34. TABOR, M.W.. LOPER, J.C. and BARONE, K. In 'Water Chlorination: Environmental Impact and Health Effects', Vol. 3, Ann Arbor Science, Michigan, 899, 1980.
35. CHEH, A.M. and CARSON, R.E. Analyt. Chem., *53*, 7, 1001, 1981.
36. SHINOHARA, R., KOGA, M., KIDO, A., ETO, S., HART, T. and AKIYAMA, T. Preprints of Papers, ACS Environ. Chem. Div., Vol. 19, 1, 151, 1979.
37. GLOOR, R., LEIDNER, H., WUHRMANN, K. and FLEISCHMANN, Th. Wat Res., *15*, 457, 1981.
38. GLOOR, R. and LEIDNER, H. Presented at the VTH International Symposium on Column Liquid Chromatography, Avignon, May 11-15, 1981. To be published in J. Chromatogr.
39. SAITO, Y. and HAYANO, S. J. Chromatogr., *177*, 390, 1979.
40. GLAZE, W.H., PEYTON, G.R., SALEH, F.Y. and HUANG, F. Int. J. Environ. Analyt. Chem., *7*, 143, 1979.

41. PITT, JR, W.W., JOLLEY, R.L. and KATZ, S. In 'Identification and Analysis of Organic Pollutants in Water', Keith, L.H., Ed., Ann Arbor Science, Michigan, 215, 1976.

42. JOLLEY, R.L., JONES, JR, G., PITT, JR, W.W. and THOMPSON, J.E. In 'Identification and Analysis of Organic Pollutants in Water', Keith, L.H., Ed., Ann Arbor Science, Michigan, 233, 1976.

43. JOLLEY, R.L. and CUMMING, R.B. Ozone Sci. Eng., *1*, 31, 1979.

44. CORNU, A. Presented at the 2nd European Symposium on 'Analysis of Organic Micropollutants in Water', Killarney, 1981, this volume, p. 242.

45. COTTER, R. Analyt. Chem., *52*, 1589A, 1980.

46. McNEAL, C.J., MacFARLANE, R.D. and THURSTON, E.L. Analyt. Chem., *51*, 2036, 1979.

47. POSTHUMUS, M.A., KISTEMAKER, P.G., MEUZELAAR, H.L.C. and TEN NOEVER DE BRAUW, M.C. Analyt. Chem., *50*, 985, 1978.

48. BECKEY, H.D. Principles of Field Ionisation and Field Desorption Mass Spectrometry. Pergamon Press, London, 1977.

49. BARBER, M., BORDOLI, R.S., SEDGWICK, R.D. and TYLER, R.N. Nature, *293*, 270, 1981.

50. SCHULTEN, H.R. Int. J. Mass Spectrom. Ion Phys., *32*, 97, 1979.

51. SCHULTEN, H.R. and STÖBER, I. Fresen. Z. Analyt. Chem., *293*, 370, 1978.

52. SHINOHARA, R. Environ. Int., *4*, 1, 31, 1980.

53. McLAFFERTY, F.W. Phil. Trans. Roy. Soc. Lond., *293A*, 93, 1979.

54. BEYNON, J.H. and BOYD, R.K. Org. Mass Spectrom., *12*, 163, 1977.

55. BOZORGZADEH, M.H., MORGAN, R.P. and BEYNON, J.H. Analyst, *103*, 613, 1978.

56. KONDRAT, R.W. and COOKS, R.G. Analyt. Chem., *50*, 81A, 1978.

57. STRAUB, K.M. and BURLINGAME, A.L. Advances in Mass Spectrometry, 8, In press.

58. REICHSTEINER, JR, C.E., YOUNGLESS, T.L., BURSEY, M.M. and BUCK, R.P. Int. J. Mass Spectrom. Ion Phys., *28*, 401, 1978.

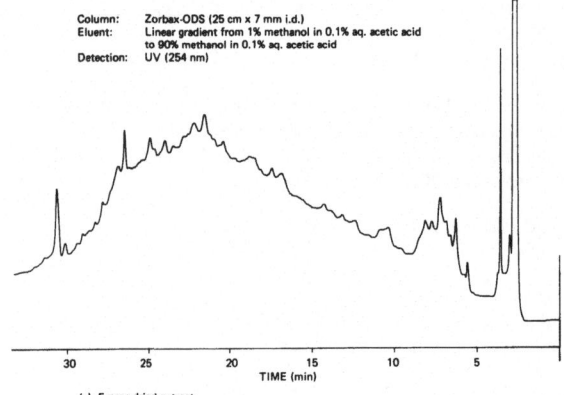

Column: Zorbax-ODS (25 cm x 7 mm i.d.)
Eluent: Linear gradient from 1% methanol in 0.1% aq. acetic acid to 90% methanol in 0.1% aq. acetic acid
Detection: UV (254 nm)

(a) Freeze-dried extract

Fig. 1 :

Reversed-phase HPLC chromatograms of (a) a methanol extract of a freeze-dried drinking water sample and (b) a diethyl ether/XAD-2 resin extract of drinking water from the same site

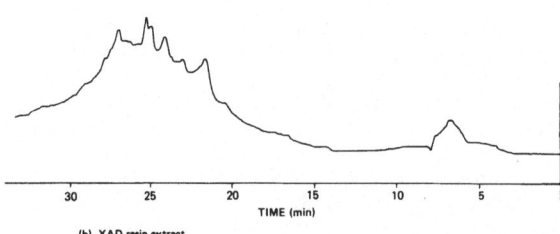

(b) XAD resin extract

Column: Spherisorb-CN (25 cm x 7 mm i.d.)
Eluent: Linear gradient from 1% isopropanol in n-hexane to 50% isopropanol in n-hexane ie
Detection: UV (254 nm)

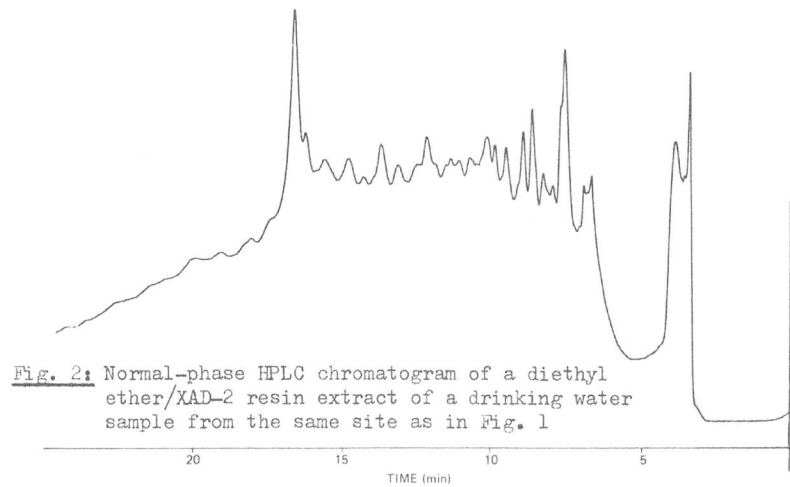

Fig. 2: Normal-phase HPLC chromatogram of a diethyl ether/XAD-2 resin extract of a drinking water sample from the same site as in Fig. 1

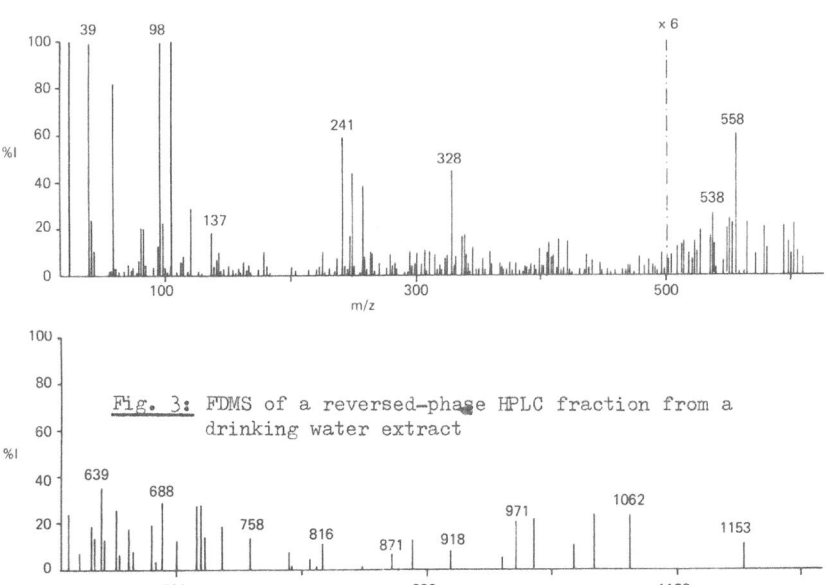

Fig. 3: FDMS of a reversed-phase HPLC fraction from a drinking water extract

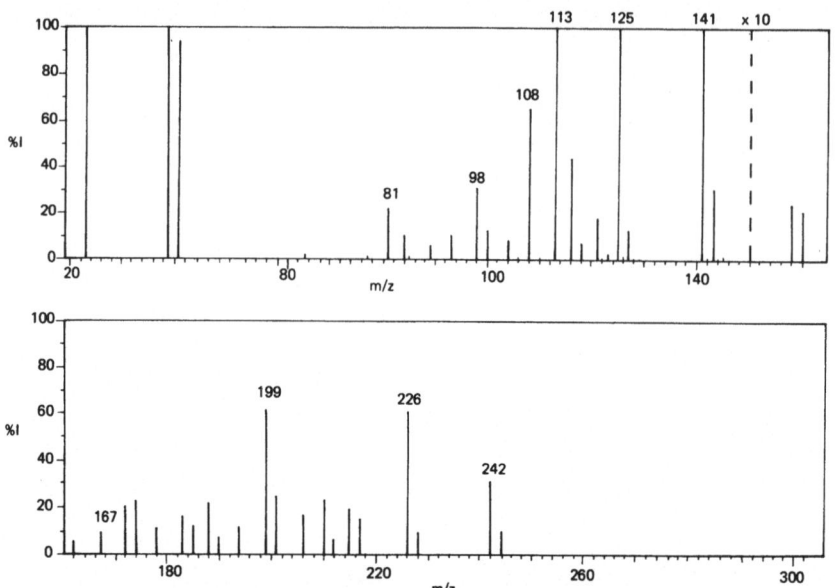

Fig. 4: FDMS of a reversed-phase HPLC fraction from a drinking water extract indicating halogen containing compounds

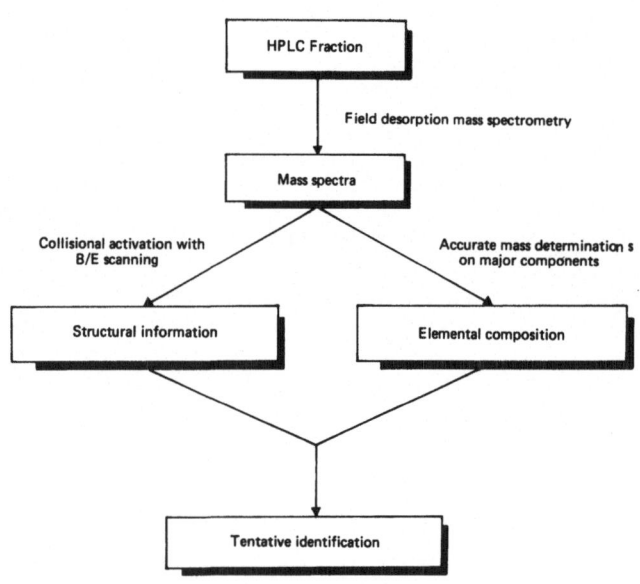

Fig. 5: Scheme for mass spectrometric analysis

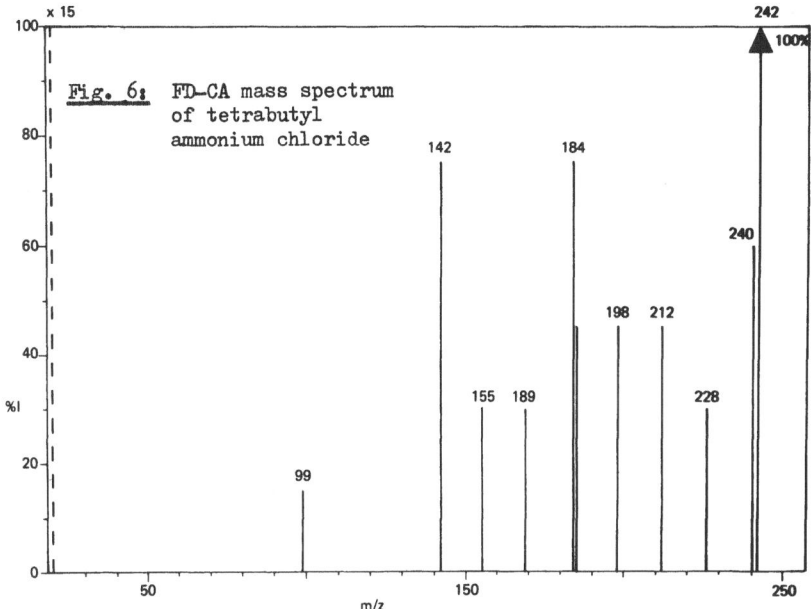

Fig. 6: FD-CA mass spectrum of tetrabutyl ammonium chloride

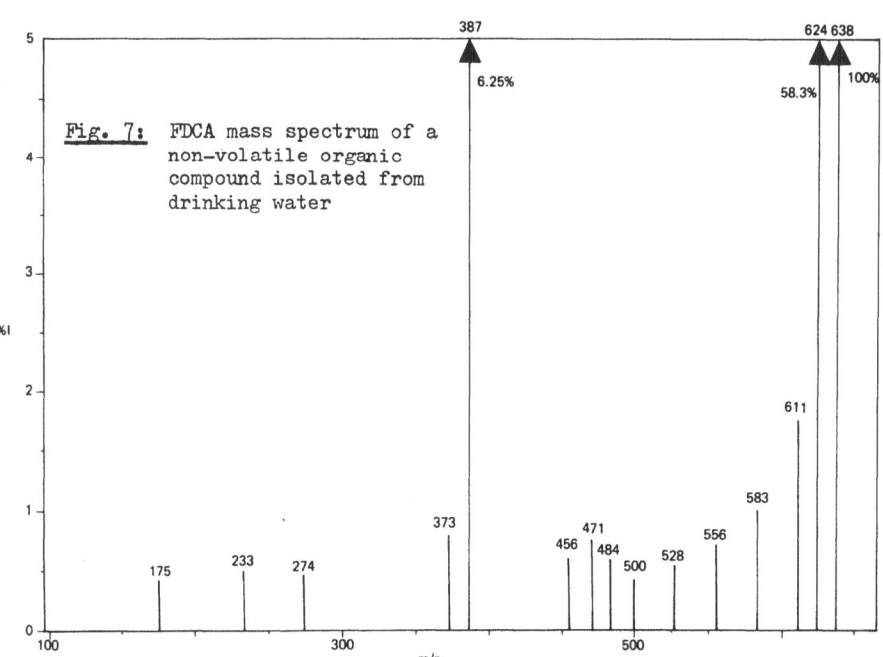

Fig. 7: FDCA mass spectrum of a non-volatile organic compound isolated from drinking water

ROUTINE HPLC OF POLYNUCLEAR AROMATIC HYDROCARBONS

C. O'DONNELL

Water Resources Division, An Foras Forbartha,
St. Martin's House, Waterloo Road,
Dublin 4, Ireland.

Summary

This paper describes the HPLC condition used for routine screening
of water extracts for PAH. The effect of temperature on the
separation is discussed and the use of a novel step programme is
described.

1. INTRODUCTION

Analysis of PAH in environmental samples is generally done by reverse-
phase HPLC on octadecyl silica columns using aqueous methanol or
acetonitrile as the mobile phase. As more sophisticated instruments are
developed the tendency seems to be away from the earlier methods using
isocratic conditions with methanol-water combinations, to the use of
solvent programming, usually with acetonitrile-water solvent combinations.

For routine screening of extracts containing PAH at about 0.01 µg/ml,
an isocratic method was chosen to avoid operator time being spent on
reversal of gradients and subsequent equilibration of the column prior to
the next injection. This also had the advantage of single pump operation,
leaving gradient elution equipment free for other work, which is a factor
to be considered in view of the relatively long run times involved.

2. EXPERIMENTAL

Reagents Methanol was distilled in glass and gave low fluorescence
under the conditions used. Distilled water was purified by ion-exchange
resin, charcoal filtration and membrane filtration in a Milli-Q apparatus
(Millipore).

Fluoranthene, benzo(a)pyrene and benzo(g,h,i)perylene were purchased
from Koch-Light Laboratories. Benzo(b)fluoranthene, benzo(k)fluoranthene
and indeno(1,2,3-c,d)pyrene were "Nanogen" 10 µg/ml solutions in toluene.
Solutions were prepared by dissolving crystalline standards in chloroform
and by dilution with methanol to give a working standard of 0.01 µg/ml.
This standard was stable for several months at room temperature.

Apparatus Solvent delivery was by a Waters pump model 6000A.
Samples were introduced via a Rheodyne valve, model 7120. The column was
15 cm x 4.6 mm I.D. packed with LiChrosorb RP-18, 5 µ particle size,
protected by a guard column 5 cm long packed with ODS-Sil-X-2 (35 µ size
pellicular). Detection was by a Perkin Elmer LC-1000 fluorescence detector
with excitation at 360 $\pm$ 15 nm using a wide-band interference filter and
emission > 430 nm using a cut-off filter.

3. DISCUSSION

Flow rate was kept at 0.8 ml/min for this work, giving a column head
pressure of c.1400 psi (100 bar). The six W.H.O standard PAH are easily
separated into three groups on HPLC. Fluoranthene elutes rapidly, followed
by a group made up of benzo(b) - and benzo(k)fluoranthene with benzo(a)-
pyrene, in that order. The last two compounds to elute are benzo(g,h,i)-
perylene and indeno(1,2,3-c,d)pyrene, which are the most difficult to
separate, and in fact these two compounds influence the conditions
required for the complete run.

Using pure methanol the mixture gave only four discrete peaks, and these could be resolved adequately with 91% methanol in about 50 minutes (fig.1).

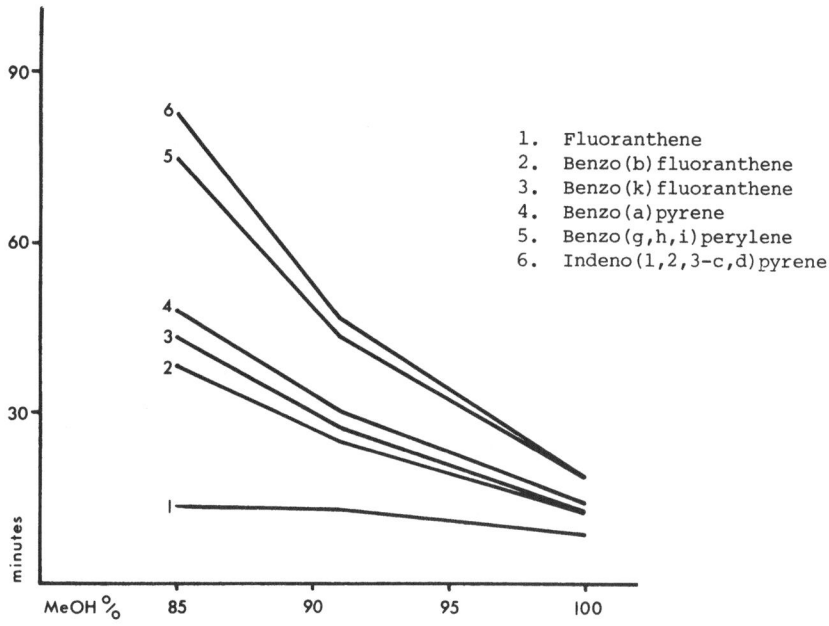

1. Fluoranthene
2. Benzo(b)fluoranthene
3. Benzo(k)fluoranthene
4. Benzo(a)pyrene
5. Benzo(g,h,i)perylene
6. Indeno(1,2,3-c,d)pyrene

Fig. 1.   Change in retention time with solvent composition at 16.5°C

The retention times are quite sensitive to temperature changes and it was found that the resolution of the last two compounds was also dependent on the temperature (fig. 2).

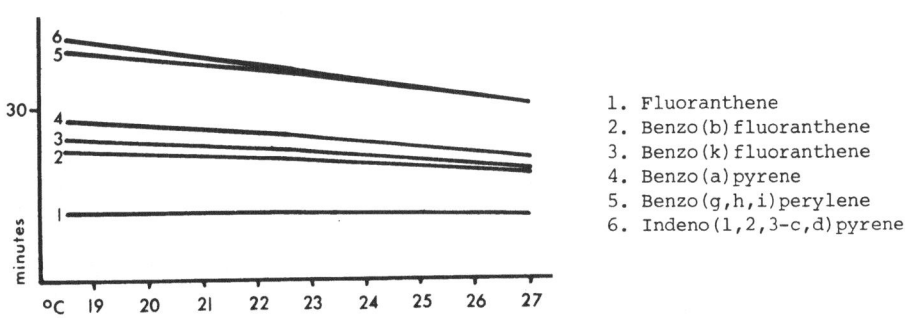

1. Fluoranthene
2. Benzo(b)fluoranthene
3. Benzo(k)fluoranthene
4. Benzo(a)pyrene
5. Benzo(g,h,i)perylene
6. Indeno(1,2,3-c,d)pyrene

Fig. 2.   Change in retention time with temperature in 91% MeOH

It can be seen from fig. 2 that over the range 18.5 to 27°C the resolution of benzo(g,h,i)perylene from indeno(1,2,3-c,d)pyrene decreases, becoming unsatisfactory above 21°C. Using 85% methanol as the mobile phase, it was found that the rise in temperature from 16.5 to 26.5°C had almost destroyed the resolution of benzo(g,h,i)perylene and indeno(1,2,3-c,-d)pyrene while the run time was still over 50 minutes (fig. 3). The column temperature was conveniently controlled at 17 to 19°C by immersion in a cold waterbath. This varied by less than 1°C over a working day.

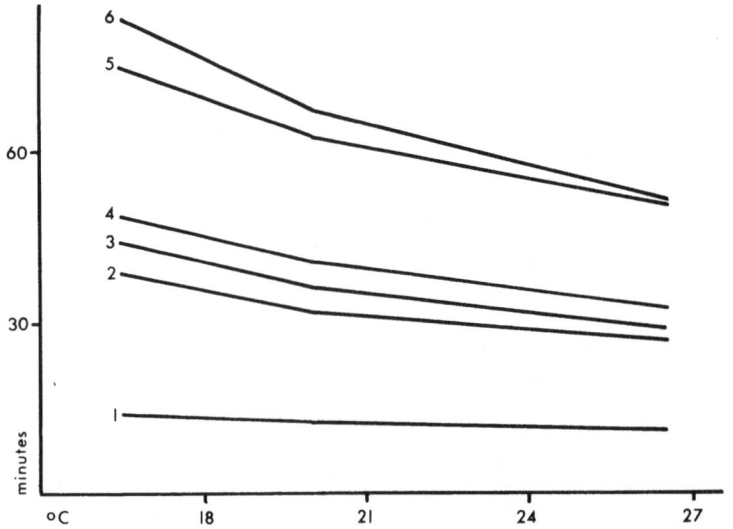

Fig. 3.   Change in retention time with temperature in 85% MeOH

Under these conditions an injection of 100 μl of the solution containing 0.01 μg of each PAH per ml gave an easily measured response (fig. 4). It is interesting to note that the temperature of the column is critical, even though this parameter is rarely mentioned in the literature for HPLC work done at "ambient" temperatures. It seems likely that diffusion at raised temperatures outweighs the advantage of faster equilibration between stationary and mobile phases.

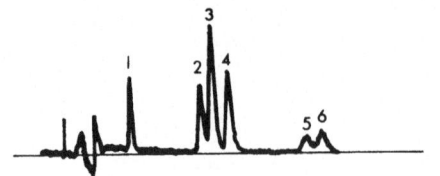

1.   Fluoranthene
2.   Benzo(b)fluoranthene
3.   Benzo(k)fluoranthene
4.   Benzo(a)pyrene
5.   Benzo(g,h,i)perylene
6.   Indeno(1,2,3-c,d)pyrene

Fig. 4.   Separation of PAH standards at 17°C in 91% MeOH at 0.8 ml/min

The samples to be analysed were concentrated on XAD-2 resin, and generally contained little material eluting after the indeno-pyrene, but had a large quantity of early-eluting material which tailed towards the fluoranthene peak. In the absence of a solvent programme, a large (1.53 ml) sample loop was installed on the Rheodyne valve. This was flushed with water prior to each injection and the 100 μl aliquot was then injected.

On switching the loop to the "inject" position the sample entered the column first, followed by an exact quantity of water giving a reproducible step gradient which helped to concentrate the sample at the head of the column. This step gradient lasted for almost two minutes at the flow of 0.8 ml/min and required no equilibration of the column between runs.

Build-up of less mobile material on the column was removed by periodic flushing with pure methanol or by repeated injection of dimethyl sulphoxide (1.5 ml). Immobile material was physically removed by cleaning out the guard column.

The system described provides a convenient reproducible analysis of PAH in the 0.01 µg/ml range with a run time of approximately 55 minutes and operator time of less than one minute.

EVALUATION OF A MASS DETECTOR FOR HPLC DETERMINATION OF ORGANIC
COMPOUNDS IN WATER

K.J. CONNOR and A. WAGGOTT
Water Research Centre, Stevenage Laboratory, Elder Way,
Stevenage, Herts. SG1 1TH

Summary

According to its manufacturers, the ACS 750/14 Mass Detector is
totally mass responsive for compounds of molecular weight exceeding 250, and could therefore fill an outstanding requirement in
the field of HPLC detection of non-volatile organic compounds.
The instrument has been assessed in an HPLC system developed to
concentrate and resolve organics from aqueous samples in the
0.1-1.0 µg/l range. Its behaviour during both isocratic and
gradient elution has been studied, as well as its performance
with various water-miscible solvents and HPLC columns. Preliminary work has shown that it may prove to be a valuable asset
for HPLC detection when placed in series after a UV absorbance/
fluorescence monitor.

1. INTRODUCTION

One of the disadvantages of using UV absorbance and/or fluorescence
detection in HPLC is the vast difference in response experienced for
different organic compounds. This necessitates calibration of such detectors for each organic compound examined to determine concentration. In
water pollution analysis, a sample chromatogram will usually contain peaks
which are unidentified and calibration will then be impossible. The
addition of a detector which is "totally mass responsive" and which has no
molecular structure dependence would therefore be very useful.
The ACS 750/14 Mass Detector (produced by Applied Chromatography
Systems Ltd, Luton, Bedfordshire, England) has recently been introduced
and, according to its manufacturers, is "totally mass-responsive"; can detect
compounds from molecular weight 250 upwards; has excellent linearity and is
unaffected by changes in ambient temperature. Its main application to date
is in the analysis of carbohydrates for which there is no satisfactory
method of detection using UV absorbance, because these compounds have an
absorbance < 200 nm[1]. The detection limit for most compounds is claimed to
be in the 500 ng range when using this instrument.

2. DESCRIPTION OF THE MASS DETECTOR

A diagrammatic representation of the instrument is given in Fig. 1.
The operation of the mass detector can be split into three stages:
nebulisation, evaporation and detection.
    (a) Nebulisation
The solvent stream is pumped into the atomiser in the top of the heated
evaporation column. Primary carrier gas assists the atomisation of the sol-
vent stream.
    (b) Evaporation
The solvent stream passes at high speed down the evaporation column.
The volatile solvent is flash-evaporated and eluted solutes, less volatile

than the solvent, remain as a fine cloud of particles to be carried through the evaporation column and past the light source. A secondary carrier gas stream enters at the bottom of the evaporation column.

(c) Detection

Light from a lamp is collimated and passed through the instrument at right angles to the direction of carrier gas flow. A light trap, located opposite the light source, eliminates internal reflections inside the instrument body. When only pure solvent vapour passes through the light path, the amount of scattered light reaching the photomultiplier does not cause detectable change in the recorder baseline. If, however, solute is present, a particle cloud passes through the light path and the light scattered by this intrusion is detected by the photomultiplier located at an 120° angle to the incident light beam. For maximum sensitivity and minimum noise, the detection system is modulated. The amplified signal from the photomultiplier was found to be proportional to concentration.

A more detailed account of the theoretical considerations of this type of system is given by Charlesworth[2].

3. INSTRUMENT INSTALLATION

The instrument was installed downstream of a CECIL CE2012 variable wavelength UV monitor, set at 254 nm. The liquid pump used was a Varian LC 5000 connected via the appropriate packed column. All interconnecting tube was of either stainless steel or polytetrafluoroethylene and of 0.25 mm i.d.

Solvents used were either acetonitrile or methanol (HPLC grade, Rathburns Ltd, Peeblesshire, Scotland) and double-distilled water. Because of the flammable nature of the organic solvents, it was considered necess-ary to use nitrogen (oxygen-free) as the carrier gas.

4. OPERATION OF THE INSTRUMENT

Generally the instrument proved difficult to handle; the reasons for this were several:

(a) The users manual left much to be desired, giving newcomers to the Mass Detector few basic operating instructions. The manual lacked accuracy both in describing the elementary theory of the instrument and in describing the instrument itself and this caused a number of problems in the current work. For example, the secondary carrier gas is sup-posed to rise up between the gauge heater column and the outer evaporation column tube, thus acting as a make-up gas for the primary gas from the atomiser. Because of the geometry of the actual instrument supplied (which is given accurately in Fig. 1), the secondary carrier gas rises only as far as the light path tube and is then forced back towards the exhaust by pri-mary carrier gas issuing from the base of the gauge heater column. This may possibly cause turbulence within the light path tube and affect base-line stability (see later).

(b) The temperature controller on the instrument, a variable resis-tance connected to the gauge heater column, was infinitely variable and uncalibrated. The only temperature read-out was from the detector unit it-self which was in the instrument base. The temperature within the evap-oration column was much higher than the meter value because of the tem-perature gradient existing between heater column and detector.

In practice the temperature required considerable time to stabilise (1-1½ hours) and setting exactly the same temperature between one day and the next proved very difficult and time-consuming. This is a crucial point considering that detector response is dependent upon detector temperature.

Though the instrument was supposed to operate between ambient tem-perature and 200°C, it proved difficult to elevate the temperature above 120°C on most occasions, even though the evaporation column was eventually lagged with asbestos rope surrounded by a glass-fibre bandage.

(c) The carrier gas flow to the system was recommended to be 4-6 litres per minute at 20 psi (~140 kPa). This meant that a large volume of nitrogen mixed with either methanol or acetonitrile vapour (1 ml or more per minute as liquid) would be escaping from the instrument exhaust and it was considered unwise to let this exhaust mixture enter the laboratory atmosphere. Venting was therefore another difficulty. It was decided to vent through a 1" diameter tube, attached to the exhaust, to the atmosphere outside the laboratory. Alternatively, the instrument might have been set-up in a fume cupboard.

Because of the non-existence of a make-up gas facility within the instrument (see section 4(a)), it was necessary to increase the total carrier gas flow to a pressure of ca. 30 psi (215 kPa). This meant that nitrogen was being consumed at a rate of one cylinder per full working day.

(d) Because of the high carrier gas flows required as outlined in 4(c) the instrument needed to be reasonably gas-tight. Unfortunately, it turned out in practice that this was not so. The atomiser insert which fitted into the top of the evaporation column leaked gas both between the insert and the evaporation column and also where the solvent stream tube entered the atomiser unit. The leakage was reduced by using PTFE tape and a 1/16" pipe fitting on the affected instrument parts. The modifications here described were assessed by injection of several (approximately 10) 10 µl portions of a 1000 mg/l aqueous solution of acid blue dye 3 (CI No. 42051) into the working mass detector through a Rheodyne valve (6-port; 2 position) encompassing a 10 µl loop and an HPLC column. Blue deposits which were apparent both on the atomiser and at the junction of the atom-iser insert and the evaporation column before sealing were no longer apparent after the modification.

(e) As a consequence of inspecting the instrument for solid deposition in 4(d), it became apparent that some of the acid blue dye was depositing at the base of the gauge column heater and therefore was not reaching the lamp beam for light scattering detection. After several weeks use it was discovered that minor corrosion of the gauge heater column within the evap-oration column was occurring. Blue deposits were also observed in the base of the detector unit adjacent to the exhaust.

(f) A final point of consequence for HPLC analysis is that, because of their nature, buffers and ion-pairing reagents commonly used in conjunc-tion with eluting solvents in HPLC cannot be used in this system because they will give a response in a light scattering detector, masking real sample peaks.

5. APPLICATIONS IN WATER ANALYSIS

Having made the modifications to the instrument described in section 4(d), its applications in the field of micropollutant analysis could be examined.

The mass detector was assessed in a High Performance Liquid Chromato-graphic system used in the reversed-phase mode. The only solvent systems used during the assessment were either acetonitrile-water or methanol-water, in varying concentrations. According to the manufacturers, other solvent systems using non-water miscible solvents may also be employed for mass detection but this paper has not included these.

(a) Detector response to different compounds

To assess the total mass response of the detector, which the manu-facturers claimed was independent of the structure of each compound, a 10 µl loop was connected directly to the solvent inlet to the evaporation column and aliquots of various compounds at concentrations between 200 and 1000 mg/l, dissolved in methanol-water mix, were injected, in batches of five. The operating conditions and results are presented below in Table 1.

Table 1. Mass Detector Response to Various Organic Compounds

Solvent 90% methanol in water; Flow rate 1 ml per min
Detector temp. 60°C; Nitrogen carrier at 215 kPa
HV coarse 3 Gain x4   Chart Recorder range 0-1 V

Compound	µg per 10 µl	Peak height (mm)	Peak height corrected to 10 µg	Limit of detection (2:1 signal:noise)
Acid Blue 1	10	84	84	480 ng/10 µl
Acid Blue 3	10	105	105	380 ng/10 µl
Acid Blue 9	10	104	104	380 ng/10 µl
Hostalux PRT	10	160	160	250 ng/10 µl
Blankophor REU-P	10	170	170	240 ng/10 µl
Tinopal UPM	10	130	130	310 ng/10 µl
Tinopal STP	10	132	132	300 ng/10 µl
Phenidone	10	123	123	330 ng/10 µl
Potato Starch	2	No response	–	–
D-Maltose	10	73	73	550 ng/10 µl
Fructose	2	13	65	570 ng/10 µl
Sucrose	2	14	70	570 ng/10 µl
Glucose	2	14	70	570 ng/10 µl
Tannic Acid	2	14	70	620 ng/10 µl
Malachite Green	2	17	85	470 ng/10 µl

Baseline noise ± 2 mm

The limit of detection under the conditions stated was in the 500 ng range
but there was obviously a variation in response depending upon the organic
compound.
    It appeared that response was dependent on molecular structure and/or
on molecular weight to some extent.
    (b) Variation of detector response with carrier gas and solvent flow
         rates
Nitrogen pressures below 170 kPa appeared, in this instrument, to
affect baseline stability. The instrument was used at a nitrogen pressure
of 215 kPa, though at this pressure small variations had little noticeable
affect.
    To determine the effect of changing the solvent flow on both response
and baseline stability and noise, the solvent flow was varied linearly
between 0.5 and 1.5 ml/min over a period of 20 minutes. There was no
noticeable affect on the baseline stability; and only the normal effect of
changing flow rate on peak shape, i.e. the faster the flow of solvent, the
narrower the peak width.
    (c) Variation of detector response with temperature
    (i) Acetonitrile-water
The response of the mass detector to changes in detector temperature
at different compositions of water and acetonitrile was determined using
Acid Blue 3 (CI No. 42051). The mass detector response was measured purely
as a peak height, and variations in peak height occurring as a result solely
of changing solvent composition were compensated for over the solvent composition range by dividing peak height given by the mass detector (PHMD)
and the corresponding peak height given and a UV absorbance monitor, set at
254 nm and at attenuation 1 (PHUV). It was assumed that the ratio
PHMD/PHUV would give a reflection of the temperature effect alone on mass
detector response over a range of acetonitrile-water compositions.
    The effect of temperature variation on response is shown graphically
in Fig. 2. It can be seen that the temperature for optimum response at any
solvent composition is between 80 and 90°C.

Table 2. below gives the corresponding baseline noise exhibited by the instrument at various temperatures and acetonitrile-water compositions.

Table 2. Baseline Noise, Acetonitrile-water System

% Acetonitrile	Baseline noise (mm) at different temperatures ($^{\circ}$C)			
	60°C	80°C	100°C	120°C
40	± 10	± 6	± 9	-
60	± 6	± 4.5	± 10	± 13
80	± 6	± 4	+ 6	± 9

Generally at temperatures above 100°C, detector response dropped off rapidly with the corresponding increase in baseline noise.

(ii) Methanol-water

The methanol-water solvent system proved more difficult to use than acetonitrile-water. Fig. 3. shows the temperature-detector response relationship for 90% methanol-water. 80°C was the temperature for maximum response but above this temperature baseline instability increased and response decreased more rapidly than with the acetonitrile-water system.

At compositions of 60% methanol-water and below, results were useless due to excessive baseline noise. The authors consider part of the reason for this might be hydrogen-bonding effects of the water-methanol system which affects the smooth vaporisation of the solvent stream within the evaporation column. Increasing detector temperature at low methanol concentrations increased baseline noise; at similarly low concentrations of acetonitrile increased temperature improved baseline noise.

(d) Reproducibility of Response

Day-to-day reproducibility of response was difficult to obtain because of the problems with the temperature controller discussed in section 4(b). At any particular temperature and solvent composition, after allowing stabilisation for about 1 hour, reproducibility of consecutive sample injections was quite acceptable. Figure 4, for example, gives the reproducibility of 12 injections, compared with the UV response for the same injections. Working conditions are outlined on the chromatogram. The coefficient of variation for PHMD/PHUV was 5.54%.

(e) Detector response to variation of solvent-water composition

The mass detector response to variation of the acetonitrile-water composition of the solvent stream at a series of temperatures between 60°C and 110°C is given in Fig. 5. It can be seen that response decreases with decreasing acetonitrile composition, i.e. the detection limit of the mass detector increases with increasing amounts of water within the solvent stream. This decrease in response is even more apparent with methanol-water systems.

(f) Isocratic elutions

A mass detector chromatogram of an isocratic elution of a sewage sludge extract, together with a low-sensitivity UV absorbance chromatogram for comparison, is shown in Fig. 6. The composition is 90% methanol-water. Here the conditions for response and baseline stability are good. At lower percentages of organic solvent, the baseline stability degenerates rapidly.

(g) Gradient elutions

The effect of a gradient elution on the mass detector response can be demonstrated with the aid of Fig. 7. A 2% per minute gradient was run from 100% methanol to 100% water. At 80%, 60% and 40% methanol, 10 µl aliquots of acid blue dye 3 were injected and compared with the injection at 100% methanol.

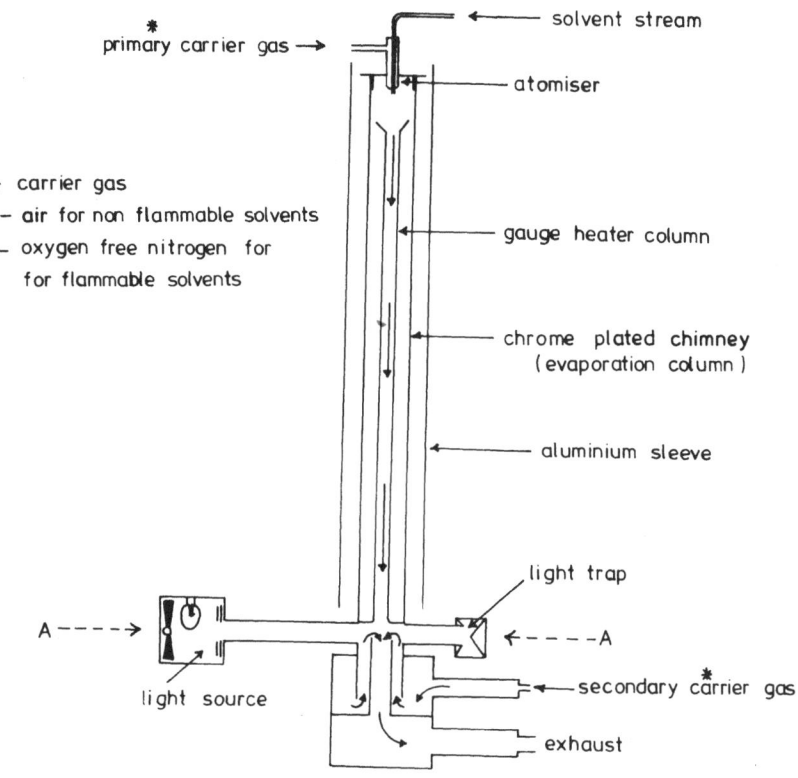

* primary carrier gas →

← solvent stream

— atomiser

* carrier gas
 — air for non flammable solvents
 — oxygen free nitrogen for
   for flammable solvents

— gauge heater column

— chrome plated chimney
  (evaporation column)

— aluminium sleeve

light trap

A - - - →

← - - - -A

light source

← secondary carrier gas

exhaust

(a) Schematic diagram of mass detector

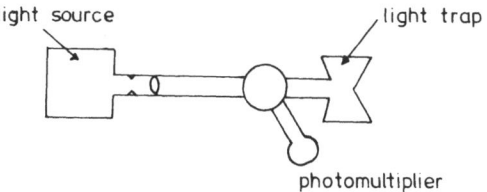

light source

light trap

photomultiplier

(b) Cross section through A –A

Fig.1   ACS   750 / 14   mass detector   (schematic)

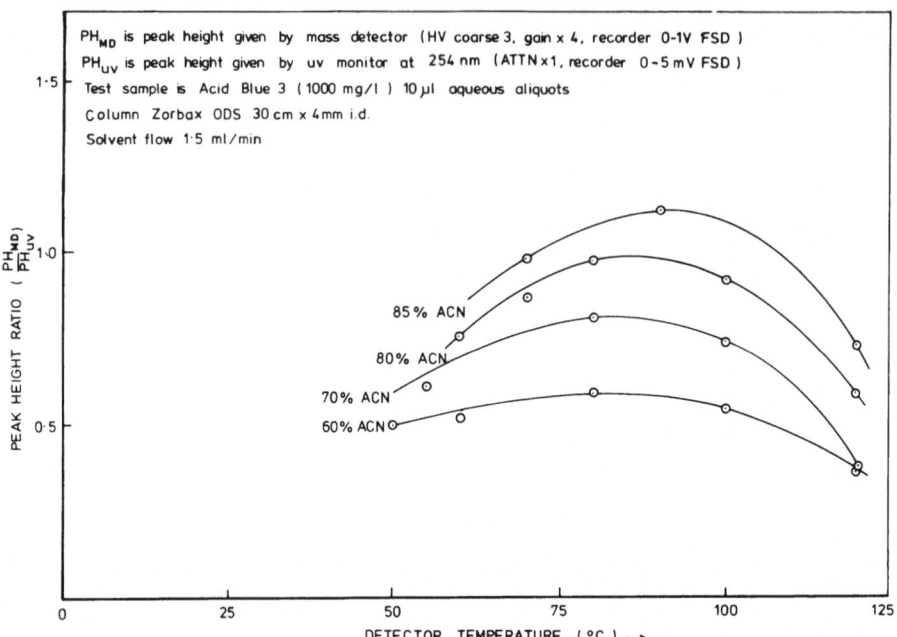

Fig. 2   Effect of detector temperature on response (acetonitrile–water)

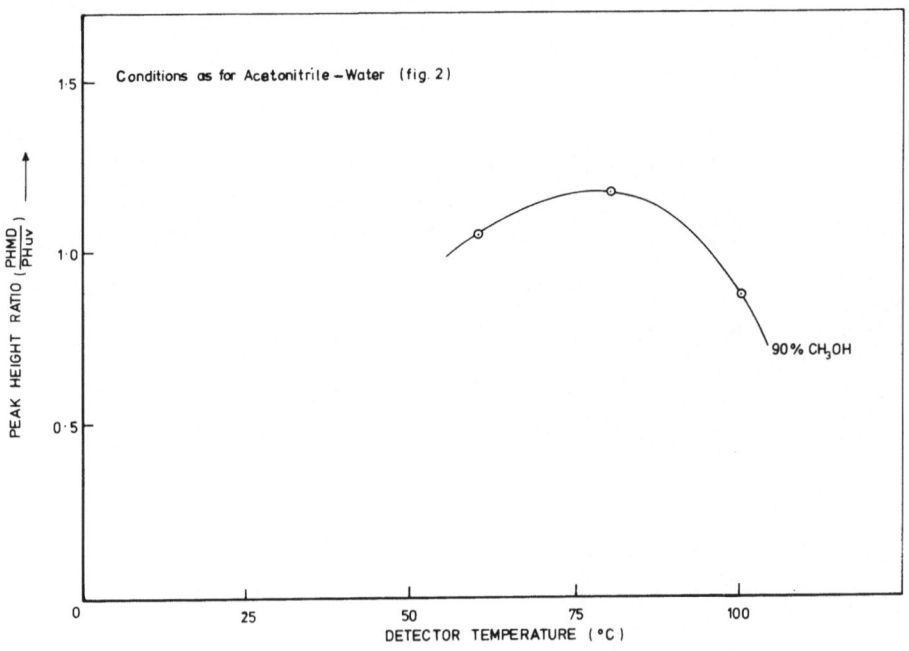

Fig. 3   Effect of detector temperature on response (Methanol–Water)

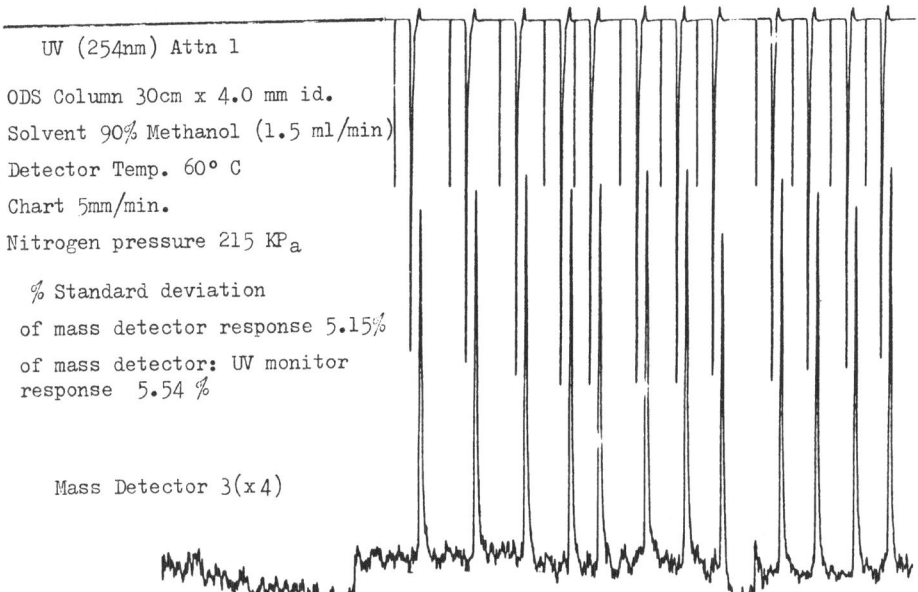

UV (254nm) Attn 1

ODS Column 30cm x 4.0 mm id.

Solvent 90% Methanol (1.5 ml/min)

Detector Temp. 60° C

Chart 5mm/min.

Nitrogen pressure 215 KP$_a$

% Standard deviation
of mass detector response 5.15%

of mass detector: UV monitor
response   5.54 %

Mass Detector 3(x4)

Fig. 4   Reproducibility of consequtive injections on mass detector

Test compound : ACID BLUE 3 (1000 mg/l) 10 μl injection

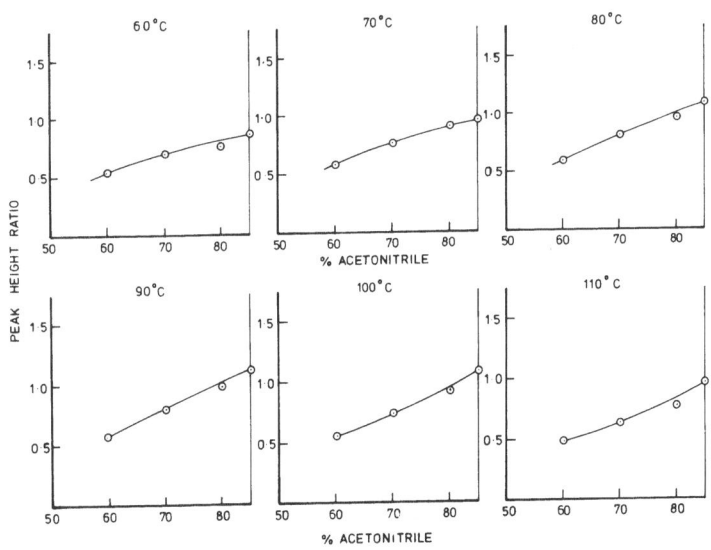

Fig. 5   Affect of Acetonitrile-water composition on response

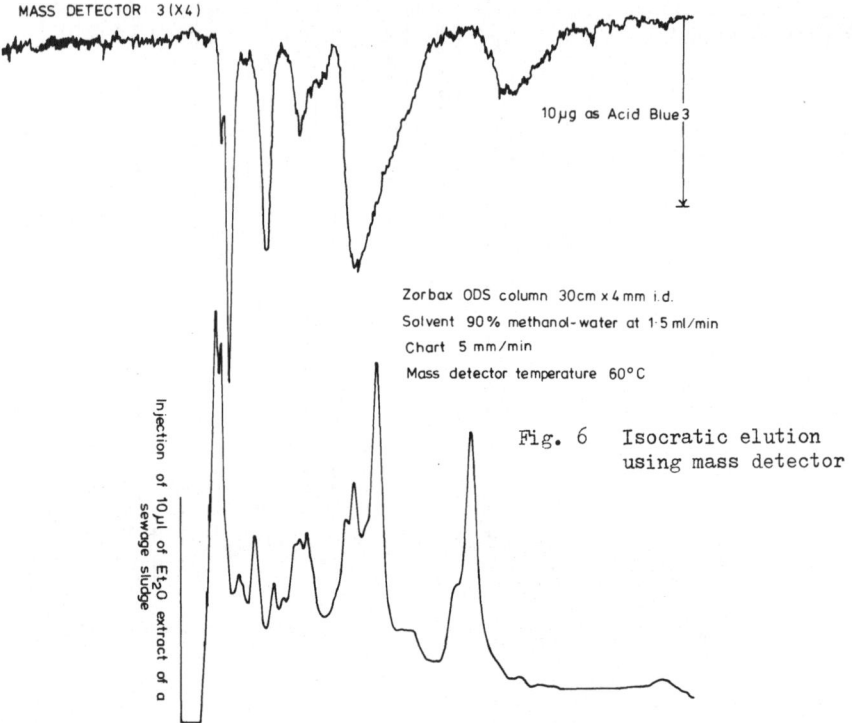

MASS DETECTOR 3 (X4)

10 µg as Acid Blue 3

Zorbax ODS column 30 cm x 4 mm i.d.
Solvent 90% methanol-water at 1·5 ml/min
Chart 5 mm/min
Mass detector temperature 60°C

Fig. 6   Isocratic elution
using mass detector

Injection of 10 µl of Et₂O extract of a sewage sludge

UV DETECTOR AT 254 nm (ATTN x 1)

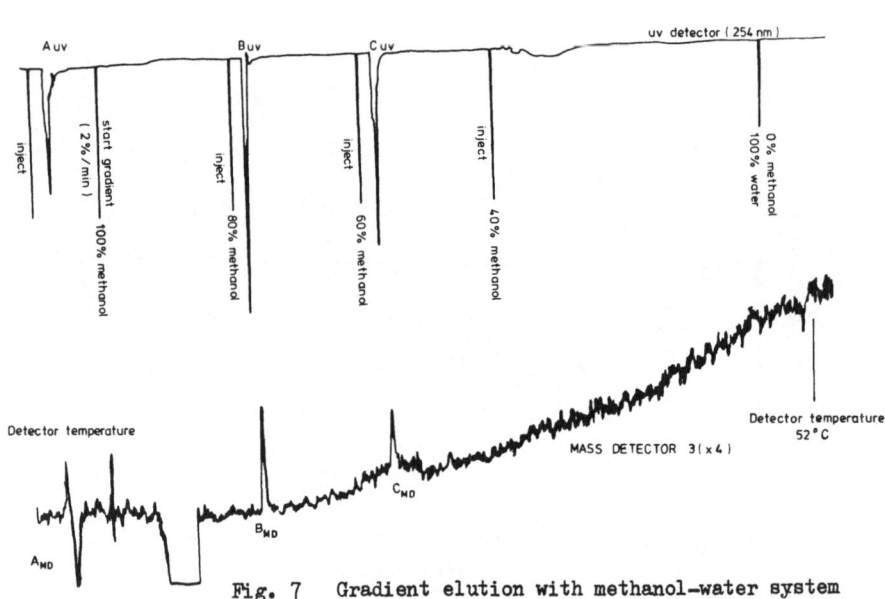

uv detector (254 nm)

A uv   B uv   C uv

inject
(2%/min)
start gradient
100% methanol

inject
80% methanol

inject
60% methanol

inject
40% methanol

0% methanol
100% water

Detector temperature

Detector temperature
52°C

MASS DETECTOR 3 (x 4)

A_MD   B_MD   C_MD

Fig. 7   Gradient elution with methanol-water system

% Methanol	Ratio of peak heights (= response)	
100	AMD/AUV	= 0.44
80	BMD/BUV	= 0.39
60	CMD/CUV	= 0.27
40	-	-

From the gradient elution it could be seen that:

(1) Response of the mass detector to acid blue dye decreases with increasing water composition i.e. sensitivity decreases.

(2) The baseline of the mass detector chromatogram becomes more unstable as elution proceeds.

(3) The detector temperature drops from $80^{\circ}C$ to $52^{\circ}C$ during the elution program. In fact after stabilisation at 0% methanol the temperature had dropped below $50^{\circ}C$.

Point (3) is to be expected because methanol and water vapours have different thermal conductivities and latent heats of evaporation. Therefore changing the methanol-water composition will also change the heat transfer from the evaporation column to the detector unit.

6. CONCLUSIONS

(a) Practical limitations on use

From what has been discussed above, it can be seen that the mass detector examined here has a limited potential for gradient elutions in HPLC systems.

In isocratic elutions and at compositions of organic solvent 70% and above, it lives up to the manufacturers claim of a 500 ng detection limit but even this is far removed from detection limits obtainable with other HPLC detectors such as UV fluorescence and absorbance.

The instrument examined is not "totally mass responsive" (as shown in section 5(a)) and in view of the handling difficulties such as temperature stabilisation as well as safely venting away the toxic exhausts, the authors are unable to recommend this instrument as a practically viable instrument for most applications in water pollution analysis.

(b) Future design

However, the idea of the instrument is basically a very good one. This instrument is the first production model and can only improve. Certain of its design characteristics are open to change and if these are changes for the better, a similar detector may yet find a role in pollution analysis.

REFERENCES

1. MACRAE, R., and DICK, J. Analysis of carbohydrates using the mass detector. J. Chromatogr. 210 (1981), 138-145.
2. CHARLESWORTH, J. M. Evaporative Analyser as a Mass Detector for Liquid Chromatography. Anal. Chem. 50 (11): 1414-1420 (9178).

DETERMINATION OF POLYCYCLIC AROMATIC HYDROCARBONS (PAH's) AT THE LOW ng/l LEVEL IN THE BIESBOSCH WATER STORAGE RESERVOIRS (NETH.) FOR THE STUDY OF THE DEGRADATION OF CHEMICALS IN SURFACE WATERS

N. VAN DEN HOED and Ms M.T.H. HALMANS
KONINKLIJKE/SHELL-LABORATORIUM, AMSTERDAM
(Shell Research B.V.)
and
J.S. DITS
N.V. WATERWINNINGBEDRIJF BRABANTSE BIESBOSCH

Summary

The concentrations of six polycyclic aromatic hydrocarbons (PAH's) in water of the storage reservoirs of the Dutch Biesbosch have been determined since December 1980 to study photochemical degradation processes in large aquatic ecosystems. The analytical procedure developed comprises sample concentration with SEP-PAK^{TM} cartridges followed by reversed phase HPLC with fluorescence detection. The procedure has been optimized to allow PAH detection down to concentrations of 0.1 ng/l. Results obtained so far are being used to validate the EXAMS program, a mathematical model that predicts the enviromental fate of toxic chemicals.

1. INTRODUCTION

EXAMS (= EXposure Analysis Modelling System) is a computer program released by U.S. Environmental Protection Agency. It can be used to predict concentrations of chemicals in aquatic ecosystems from relevant data on the chemicals, the environment and the load of these chemicals in the environment.

Since EXAMS is based mainly on well-controlled laboratory investigations there is a need to evaluate its validity for natural ecosystems. As has been shown recently (1), the water storage reservoirs "GYSTER", "HONDERD EN DERTIG", and "PETRUSPLAAT" in the Dutch "Brabantse Biesbosch" provide a unique opportunity for such an evaluation.

A collaborative program was set up by Koninklijke/Shell-Laboratorium, Amsterdam (KSLA) and N.V. Waterwinningbedrijf Brabantse Biesbosch to follow the fate of several selected trace organics – ambient in the river water feeding these reservoirs – on their way through the basins, which are connected in the above-mentioned order. In the first part of this program some of the ubiquitous polycyclic aromatic hydrocarbons (PAH's) were chosen as models for chemicals that are supposed to be removed primarily by photo-degradation. In view of the large amount of samples that had to be analysed and the generally low (ng/l) concentration levels of the PAH's in these reservoirs, the existing analytical procedures comprising multiple solvent extraction, clean-up, volume reduction and gas chromatography were inadequate for our purposes. By adopting the procedure outlined below we have succeeded in developing a rapid and very sensitive method for the determination of PAH's in water at the ng/l level.

2. EXPERIMENTAL
Sampling

Water samples were taken from the effluents of the reservoirs every two weeks. These samples were pumped in situ through SEP-PAK^{TM} disposable adsorption cartridges (supplier Waters Ass. Inc/Milford Massachusetts) to adsorb organic compounds of low and medium polarity. Loss of PAH's (by adsorption on surfaces of sample container, pump and tubing) was prevented by the addition of isopropyl alcohol (IPA) to the sample (2). Cartridges loaded with the isolated material were transported to KSLA for further analyses. The stagnant water was sucked off and desorption of the organics effected by percolating tetrahydrofurane (THF) through the cartridge. The eluate was then analysed for PAH's by reversed-phase HPLC with fluorescence detection.

As representative PAH compounds fluoranthene, benzo(b)fluoranthene, benzo(k)fluoranthene, benzo(a)-pyrene, benzo(g,h,i)perilene and ideno(1,2,3-c,d)pyrene were selected for this investigation.

Analytical procedure:

Water samples containing 15 % v/v of IPA, spiked with the six PAH's at the 100 ng/l level, were worked up according to the method outlined above, with a second cartridge connected in series to the first one to check for the possible breakthrough of PAH's. During desorption THF was collected in fractions of 1 ml. These fractions were analysed separately to determine the elution behaviour of the PAH's.

Recoveries and concentration factors

The mean recoveries for the six PAH's applying different sample volumes are given in Table I. The results indicate that the critical sample volume is close to 2 l. For this particular volume Table II gives more detailed information. The third column of this table shows the concentration factors (ratio between concentrations in THF and water/IPA phase, respectively) obtained for the first ml of THF that was eluted from the cartridge during desorption. On an average this factor amounts to 1200. In view of the required sensitivity this value should preferably be about 10 000. This could easily be realised by means of volume reduction.

In practice the desired concentration factor was attained as follows. The first 3 ml of THF that were eluted during desorption were collected and dried by percolation through a small column packed with anhydrous sodium sulfate. After washing this column with another two bed volumes of THF the combined fractions were reduced to about 100 μl by passing a gentle stream of nitrogen over them. The mass of the residual solution was determined gravimetrically.

The resultant recoveries and concentration factors are presented in Table III. It shows a mean concentration factor of 14200, which is attained at an average recovery of 60 %. This means that the detection limits of the various PAH's will be of the order of 0.1 to 0.5 ng/l.

Measurements based on the procedure outlined above have been performed since December 1980.

TABLE I

MEAN RECOVERIES OF SIX PAH's
FOR VARIOUS SAMPLE VOLUMES

Sample volume, l	Mean recovery, %	
	1st adsorption cartridge	2nd adsorption cartridge
1	84	<1
2	73	6
5	27	n.a.*

* not analysed

TABLE II

RECOVERIES AND CONCENTRATION FACTORS OF SIX INDIVIDUAL PAH's OBTAINED
FROM THE FIRST ADSORPTION CARTRIDGE FOR A 2-LITRE SAMPLE VOLUME

PAH	Recovery, %		Concentration factor
	1st ml THF	2nd ml THF	1st ml THF
Fluoranthene	78	5	1390
Benzo(b)fluoranthene	74	6	1330
Benzo(k)fluoranthene	73	6	1300
Benzo(a)pyrene	70	6	1250
Benzo(g,h,i)perilene	58	<1	1000
Indeno(1,2,3-c,d)pyrene	59	<1	1090

TABLE III

RECOVERY AND CONCENTRATION FACTORS FOR 2-LITRE SAMPLE
FOR PROCEDURE INCLUSIVE VOLUME REDUCTION

	Recovery, %	Concentration factor
Fluoranthene	78	18500
Benzo(b)fluoranthene	70	16700
Benzo(k)fluoranthene	68	16300
Benzo(a)pyrene	43	10200
Benzo(g,h,i)perilene	50	12000
Indeno(1,2,3-c,d)pyrene	48	11500

TABLE IV

CONCENTRATION OF PAH's (ng/l) IN EFFLUENT WATER
OF BRABANTSE BIESBOSCH STORAGE RESERVOIRS

	Sampling date					
	1-4-1981			13-4-1981		
Reservoir / PAH	De Gijster	Honderd en Dertig	Petrus-plaat	De Gijster	Honderd en Dertig	Petrus-plaat
Fluoranthene	3.0	5.1	6.7	3.5	6.0	2.9
Benzo(b)fluoranthene	1.4	1.1	0.8	1.5	1.1	0.3
Benzo(k)fluoranthene	0.8	0.6	0.2	0.7	0.5	<0.1
Benzo(a)pyrene	1.0	0.7	0.3	1.0	0.8	0.2
Benzo(g,h,i)perilene	2	1.5	1	2	1.5	1
Indeno(1,2,3-c,d)pyrene	1.2	0.9	0.5	1.3	1.0	0.3

HPLC conditions: column 250 mm, 4 mm ID Lichrosorb RP-18; mobile phase acetonitrile/water 85/15, 1 ml/min; detection Perkin Elmer 3000 fluorescence detector, excitation 360 nm, emission 460 nm; sample volume 10 μl.

3. DISCUSSION OF RESULTS

In principle the detection limits that can be attained by the procedure described are low enough to measure the expected low PAH concentrations in the storage reservoirs quantitatively. Whether or not the low detection limits will be reached in practice will largely depend on the blank values associated with this procedure. For this reason blank determinations have been performed simultaneously with all determinations since December 1980. These blank runs showed a dramatic increase on January 26th, which could be traced back to a relatively high PAH contamination level of one particular IPA batch. Once a batch of a sufficiently good quality had been identified large quantities of that batch were ordered to guarantee constant performance over a long period of time.

For all routine measurements two adsorption cartridges connected in series (to check for breakthrough) were used. At random some of these back-up cartridges were analysed for PAH's. Up till now no significant PAH concentrations have been found in them.

Sampling, PAH adsorption on SEP-PAK^{TM}, transport and sample preparation for a full series of six samples including blanks and checks for breakthrough could be performed within 48 hours, the HPLC analyses requiring another 48 hours.

Some typical results are presented in Table IV. From this table it can be concluded that the quality of the water clearly improves on its way downstream the reservoir system with respect to all PAH's analysed except for fluoranthene. The observed rates of degradation are now being compared against the rates predicted by the EXAMS program.

4. CONCLUSION

This study has shown that adsorption columns such as the SEP-PAK^{TM} cartridges can be used efficiently to isolate and concentrate PAH's from water samples. The simplicity of the method allows the isolation to be carried out in the field thereby decreasing the risks of deterioration of samples by degradation processes or contaminations in the period between the moments of sampling and instrumental analysis. In combination with HPLC separation and fluorescence detection the analytical procedure is further simplified as no clean-up is necessary. Besides, the high sensitivity of fluorescence detection for PAH's allows the determination of these compounds at the low parts-per-trillion level.

In summary, it can be concluded that a very sensitive, convenient and fast method for the determination of ng/l levels of PAH's in surface water has been developed.

REFERENCES

1. C.J.M. Wolff, N. van den Hoed, H.B. van der Heijde, G. Oskam, J.S. Dits and L.W.C.A. van Breemen; H2O 14 (1981) 278 (In Dutch; English translation available on request).
2. V.F. Eisenbeiß, H. Hein, R. Jöster and G. Naundorf; Chemie-Technik 6 (1977) 227.

ASSESSMENT OF A MOVING BELT TYPE HPLC-MS INTERFACE
WITH RESPECT TO ITS USE IN ORGANIC WATER POLLUTION ANALYSIS

H. SCHAUENBURG, H. SCHLITT AND H. KNÖPPEL

JOINT RESEARCH CENTER ISPRA
COMMISSION OF THE EUROPEAN COMMUNITIES

Summary

The Finnigan moving belt type HPLC-MS interface has been used for
the analysis of organic compounds extracted from surface water
samples. One of the major reasons of scarce results was a low
sensitivity due to high chemical background.

A series of experiments with test mixtures have been made in order
to assess the amount of sample required for detection and characterization of a compound as a function of different experimental parameters and to assess the source of the background signals.

Some measures taken to reduce background and their consequences
are described.

1. INTRODUCTION

A major obstacle to the survey analysis of (unknown) non-volatile
compounds is the lack of a detector for HPLC which meets the three following requirements :

- SENSITIVITY   - NON-SELECTIVITY   - SPECIFICITY.

Unfortunately there are considerable obstacles to the use of a mass
spectrometer as on-line HPLC detector in analogy to its use in GC-MS.
These obstacles result from a comparison of the carrier gas used in capillary GC and the solvent used in HPLC :

	CAPILLARY GC	REVERSED PHASE HPLC
CARRIER/SOLVENT	inert gas	polar liquid
FLOW RATE	1 - 6 ml/min	0.5 - 2 ml/min (liquid) corresponding to 620 - 2480 ml/min water vapour 210 - 850 ml/min acetonitrile vapour 280 - 1110 ml/min methanol vapour
BOILING POINT	- 195° C	$\geq$ 50 ° C

The large amount of solvent emerging from a HPLC column cannot be admitted
to a mass spectrometer. An approach to overcome this difficulty is the socalled 'MOVING BELT' HPLC-MS interface developed by Finnigan Instruments.

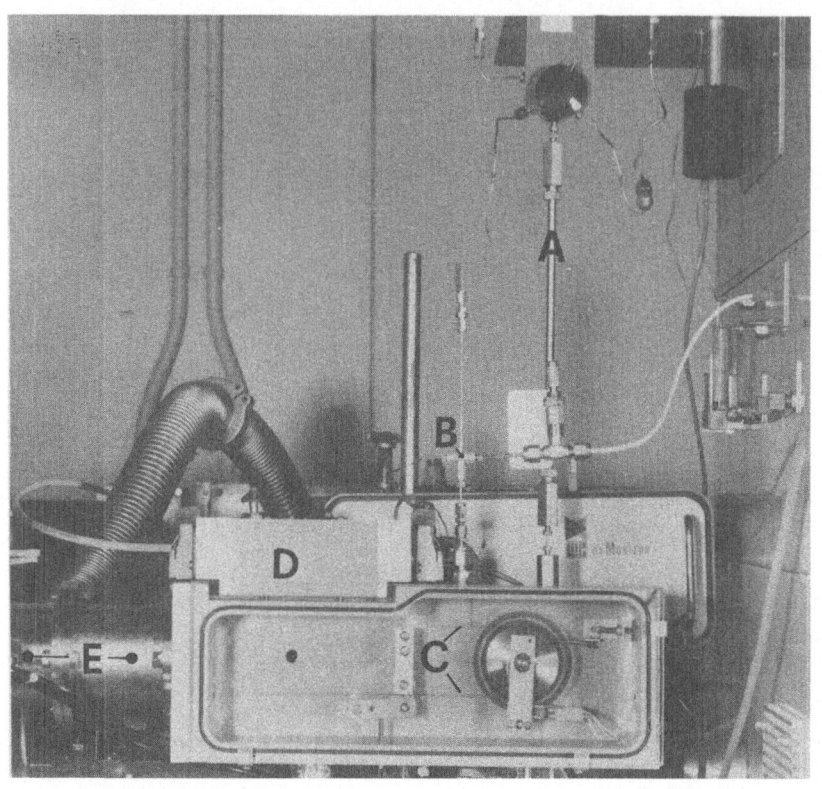

Fig. 1 - FINNIGAN MOVING BELT TYPE HPLC—MS INTERFACE

Part of the solvent leaving the HPLC column **A** is split to the interface
by means of a no-dead volume valve **B**. In the sample load chamber **C**
the effluent is deposited on a moving belt **C**. The solvent is evaporated
under a infrared heater **D** and in two differential pumping chambers **E**.
Solutes are vaporized near to the MS ion source. Electron impact (EI) and
chemical ionization (CI) can be used.

Belt materials are :

	ADVANTAGES	DISADVANTAGES
KAPTON Ⓡ	• good mechanical properties (low friction) • compounds desorb at ~ 50°C lower temperature (see, however, below!)	• high MS background • only limited amount of effluent (water) accepted
STAINLESS STEEL	• low MS background • higher amounts of effluent accepted	• mechanical difficulties (high friction) • compounds desorb at higher temperatures only

Maximum admitted effluent flows for Kapton belts are:
    water: 0.05 - 0.08 ml/min; acetonitrile: 0.2 - 0.3 ml/min;
    n-hexane: ~ 1 ml/min

2. RESULTS

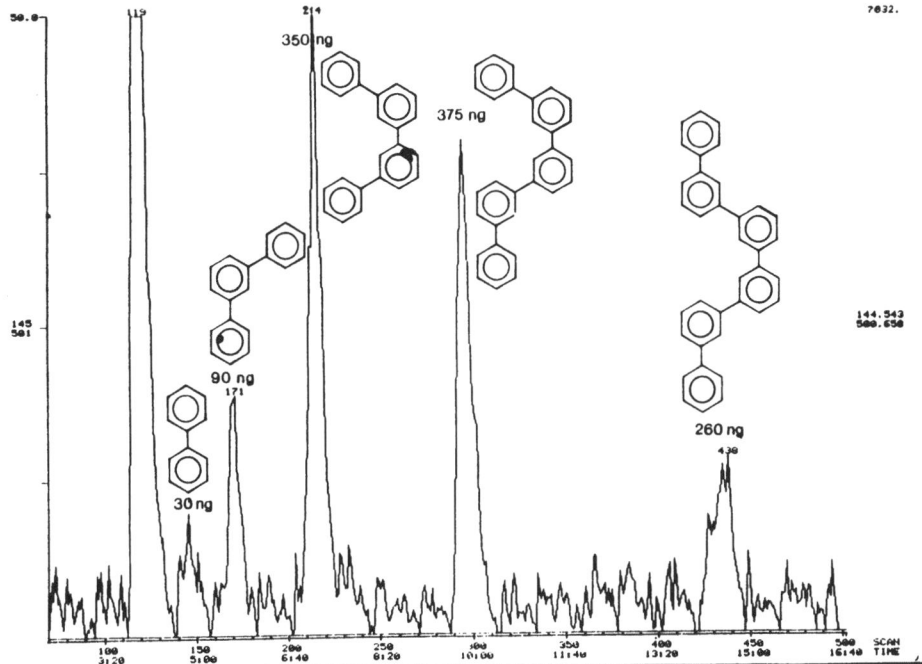

INFLUENCE OF SOLUTE VOLATILITY ON THE SENSITIVITY OF THE HPLC-MS COMBINATION

EXPLANATION

● HPLC-MS CHROMATOGRAM OF A POLYPHENYL MIXTURE. EQUAL AMOUNTS OF THE INDICATED POLY-
PHENYLS HAVE BEEN INJECTED.

● INDICATED AMOUNTS HAVE BEEN DETECTED.

RESULTS

● ONLY COMPOUNDS OF A LIMITED VOLATILITY RANGE ARE TRANSMITTED QUANTITATIVELY TO THE
MASS SPECTROMETER.

● THE USEFUL RANGE MAY BE ADJUSTED WITHIN RELATIVELY NARROW LIMITS ADJUSTING THE
SOLVENT AND THE SOLUTE VAPORIZER TEMPERATURES.

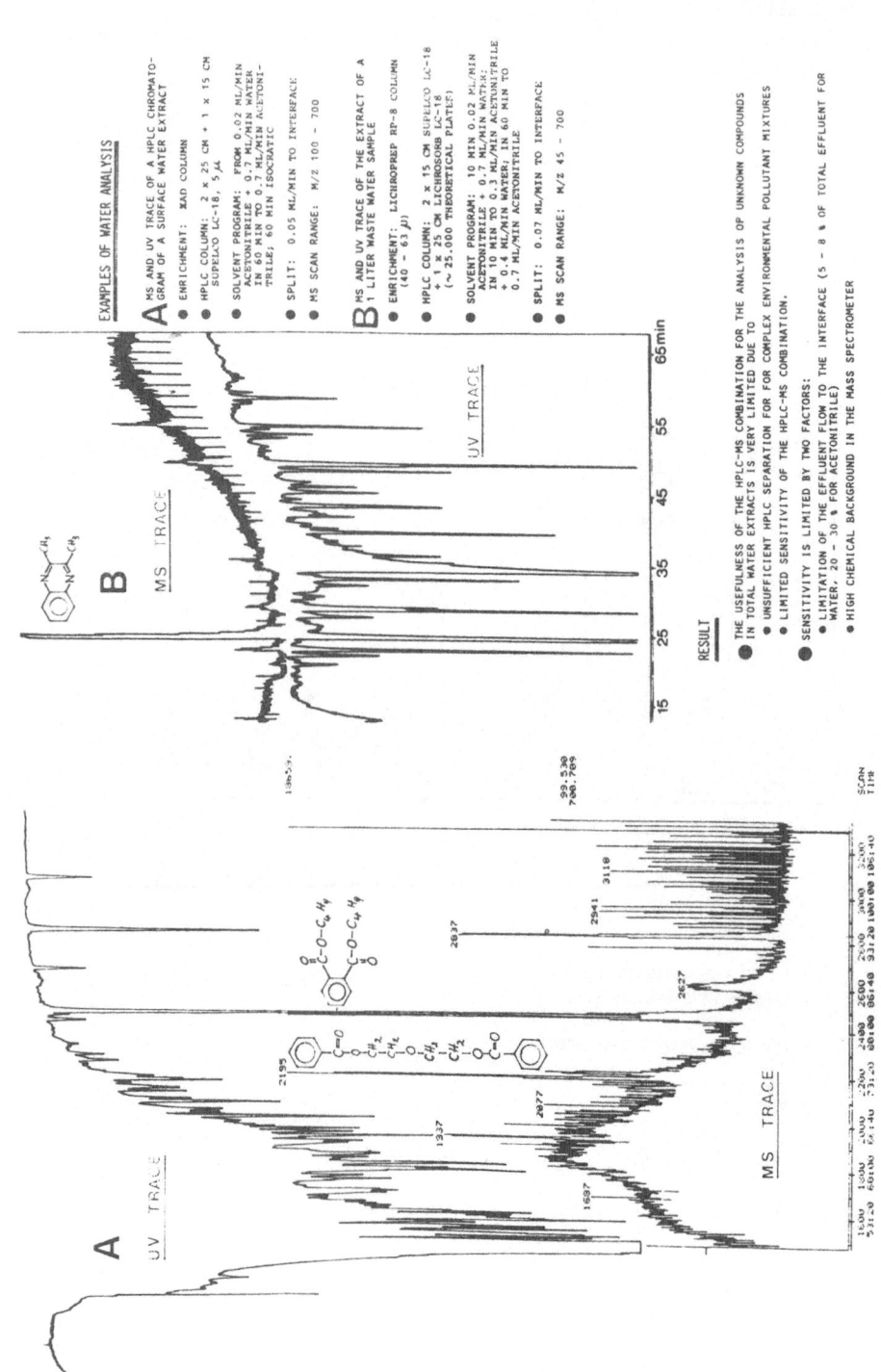

EXAMPLES OF WATER ANALYSIS

A MS AND UV TRACE OF A HPLC CHROMATO-
GRAM OF A SURFACE WATER EXTRACT

- ENRICHMENT: XAD COLUMN

- HPLC COLUMN: 2 x 25 CM + 1 x 15 CM
  SUPELCO LC-18, 5 µ

- SOLVENT PROGRAM: FROM 0.02 ML/MIN
  ACETONITRILE + 0.7 ML/MIN WATER
  IN 60 MIN TO 0.7 ML/MIN ACETONI-
  TRILE; 60 MIN ISOCRATIC

- SPLIT: 0.05 ML/MIN TO INTERFACE

- MS SCAN RANGE: M/Z 100 - 700

B MS AND UV TRACE OF THE EXTRACT OF A
1 LITER WASTE WATER SAMPLE

- ENRICHMENT: LICHROPREP RP-8 COLUMN
  (40 - 63 µ)

- HPLC COLUMN: 2 x 15 CM SUPELCO LC-18
  + 1 x 25 CM LICHROSORB LC-18
  (~ 25.000 THEORETICAL PLATES)

- SOLVENT PROGRAM: 10 MIN 0.02 ML/MIN
  ACETONITRILE + 0.7 ML/MIN WATER;
  IN 10 MIN TO 0.3 ML/MIN ACETONITRILE
  + 0.4 ML/MIN WATER; IN 60 MIN TO
  0.7 ML/MIN ACETONITRILE

- SPLIT: 0.07 ML/MIN TO INTERFACE

- MS SCAN RANGE: M/Z 45 - 700

RESULT

- THE USEFULNESS OF THE HPLC-MS COMBINATION FOR THE ANALYSIS OF UNKNOWN COMPOUNDS
  IN TOTAL WATER EXTRACTS IS VERY LIMITED DUE TO
  - UNSUFFICIENT HPLC SEPARATION FOR COMPLEX ENVIRONMENTAL POLLUTANT MIXTURES
  - LIMITED SENSITIVITY OF THE HPLC-MS COMBINATION.

- SENSITIVITY IS LIMITED BY TWO FACTORS:
  - LIMITATION OF THE EFFLUENT FLOW TO THE INTERFACE (5 - 8 % OF TOTAL EFFLUENT FOR
    WATER, 20 - 30 % FOR ACETONITRILE)
  - HIGH CHEMICAL BACKGROUND IN THE MASS SPECTROMETER

- 196 -

INFLUENCE OF SCANNED MASS RANGE ON SIGNAL/NOISE RATIO

- REVERSED PHASE LIQUID CHROMATOGRAM OF A PAH MIXTURE. MS DETECTION.
- 25 % OF HPLC EFFLUENT SPLIT TO THE INTERFACE. SUBSTANCE FLOW AT THE TOP OF PEAK A IS ∼ 1 NG/SEC. PEAK HALF WIDTH IS 24 SEC.
- THE DIFFERENT TRACES REPRESENT THE ION CURRENT SUM OF THE INDICATED MASS RANGES.

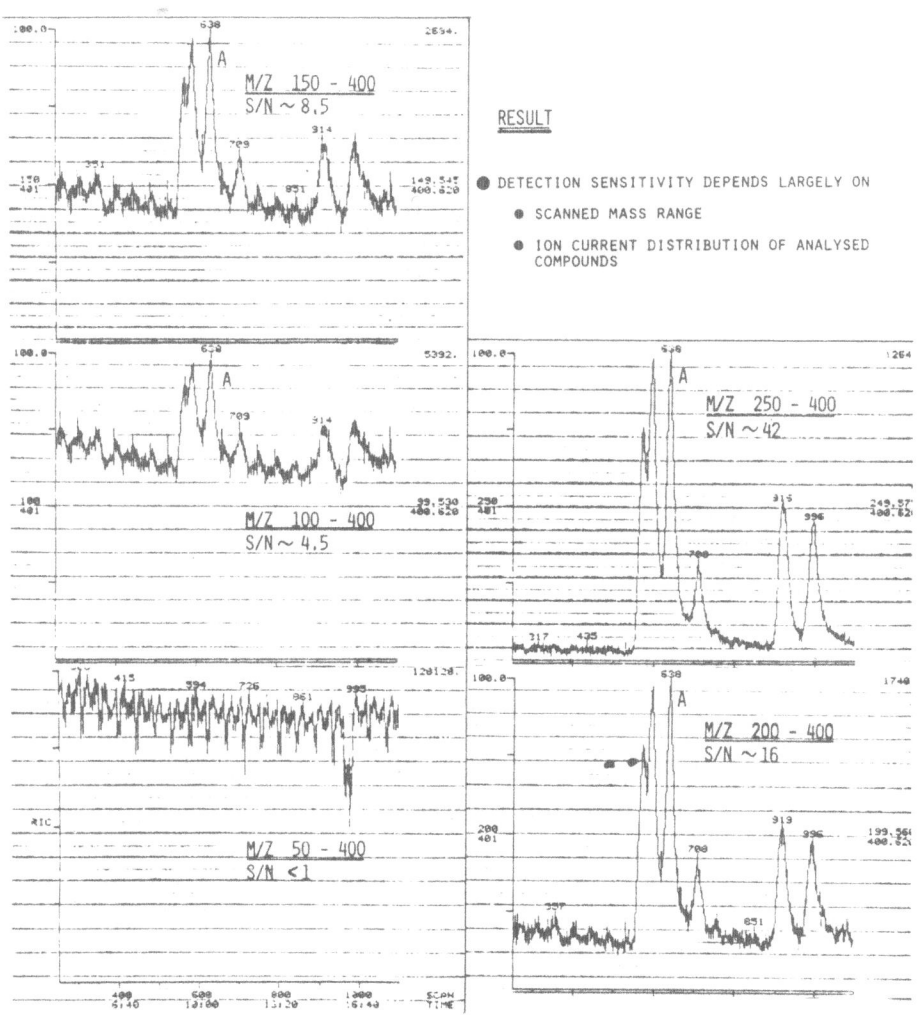

RESULT

- DETECTION SENSITIVITY DEPENDS LARGELY ON
  - SCANNED MASS RANGE
  - ION CURRENT DISTRIBUTION OF ANALYSED COMPOUNDS

COMPARISON OF ELECTRON IMPACT AND CHEMICAL IONIZATION

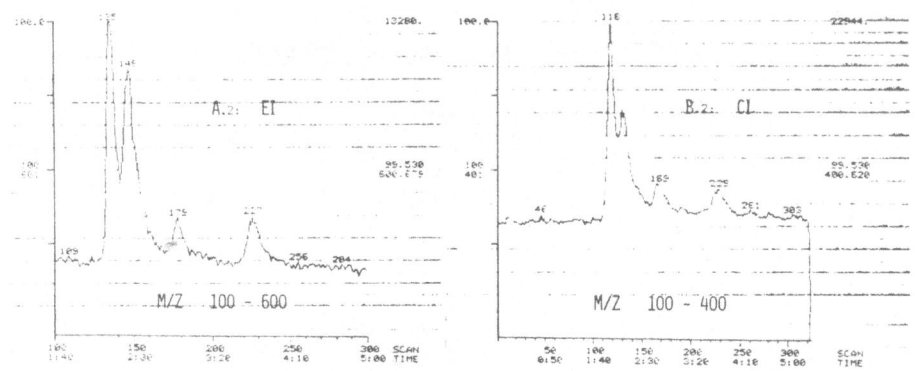

RECONSTRUCTED CHROMATOGRAMS OF A COLORANT MIXTURE OBTAINED BY EI AND CI MS DETECTION

<u>RESULT</u>

● USING METHANE CHEMICAL IONIZATION BOTH, SIGNAL AND BACKGROUND IONS ARE SHIFTED TO
HIGHER MASSES WITH NO SIGNIFICANT CHANGE OF THE SIGNAL / NOISE RATIO.

TENTATIVES TO REDUCE BACKGROUND

● Substitution of the vacuum lines connecting the differential pumping
chambers with rotary pumps: stainless steel instead of plastic tubing.
● Insertion of molecular sieve traps in the pumping lines.
<u>RESULT:</u> only small effect (at high background level).

● Substitution of ambient air pumped through the sample load chamber by
a clean gas (Argon).
<u>RESULT:</u> low effect at high background levels, noticeable effect at
low background level.

● Removal o rubber seal of the sample load chamber cover plate.
<u>RESULT:</u> Important background improvement.

CONCLUSION : KAPTON BELT MATERIAL HAS A KEY FUNCTION

● Starting work with a new belt results in high background for many
working hours. Simultaneously the belt can be easily wetted and solutes
vapourize at ∼ 50°C lower temperatures than with stainless steel belts.

In these conditions background increases markedly if the sample load
chamber is evacuated (closed).

● After extensive use of a belt background decreases considerably
(10 - 100 times). Simultaneously it becomes difficult to wet the belt
and the evaporation temperature of solutes increases.
In these conditions background decreases if the sample load chamber
is evacuated.

SESSION IV - MASS-SPECTROMETRY

Chairman: R. FERRAND, CERCHAR, France

Review paper :

- Apport de la spectrométrie de masse avec collisions (CID/MS/
  MS) à l'étude des micropolluants organiques

APPORT DE LA SPECTROMETRIE DE MASSE AVEC COLLISIONS (CID/MS/MS)
A L'ETUDE DES MICROPOLLUANTS ORGANIQUES

A. CORNU - Service d'Etudes Analytiques
C.E.A. - CENG 85 X - 38041 GRENOBLE Cédex (France)

Summary

A review of the litterature of the MS-MS technique has been attempted to
elucidate if this technique can be as promising for the multicomponant
analysis of micropollutants, as it seems to be for structural chemistry
and biology.
MS/MS avoids the use of the chromatographic column, with its inherent
disadvantages, and could provide a more exhaustive method of detection
and identification.
Its main advantages are : possibility of using all known sources of
ionisation (specially for non volatile and labile compounds), possibility of special scans for systematic search of compounds according to
their functionality, elimination of troubles due to background compo-
nants, and products of so called chemical pollution of the samples.
The sensibility could be comparable or better than for GC-MS, if one uses
an apparatus specially constructed for analytical purposes, and not for
structural studies.

1. INTRODUCTION

L'analyse des micropolluants organiques, et plus généralement de tous
les mélanges complexes de produits organiques, se fait principalement par
couplage GC-MS.
    Des progrès constants ont été accomplis dans cette technique et malgré
tout il reste certaines limitations, en particulier celles inhérentes à
l'emploi du chromatographe :
    - nécessité de volatiliser l'échantillon et de bloquer les sites
actifs par dérivatisation.
    - saignement de colonne qui ajoute à la complexité des spectres.
    - élution d'un grand nombre d'hydrocarbures non séparés qui ajoutent
encore une forêt de pics dans le spectre de masse, masquant les pics des
constituants mineurs intéressants.
    - résidus dus aux impuretés des solvants et réactifs utilisés pour la
préparation de l'échantillon car, même avec la méthode GC-MS, on n'introduit que très rarement un échantillon brut dans l'appareil.
    Tout ceci constitue ce que l'on appelle pollution chimique de l'échan-
tillon et il serait bien préférable de disposer d'une méthode permettant de
s'affranchir de ces servitudes, grandes consommatrices de main d'oeuvre.
    L'introduction récente du couplage LC-MS n'échappe pas à ces critiques
avec la grande quantité de solvant mise en cause par rapport à la quantité
d'éluat utile.

L'avantage indéniable de la méthode GC-MS ou LC-MS est qu'elle est sensiblement panoramique, c'est-à-dire détecte à peu près toutes les molécules présentes sous réserve qu'elles soient ionisées dans la source sans qu'il soit nécessaire, à priori, de connaître la nature des produits à rechercher.

Une autre méthode est en train de se développer très rapidement sous le terme général de MS/MS, dans laquelle l'échantillon global est introduit, sans préparation, dans un premier spectromètre de masse dont on extrait un type d'ion défini par la masse m_1 : cet ion, après passage dans une chambre de collision où il acquiert un supplément d'énergie interne, se fragmente selon un processus unimoléculaire, et le spectre de masse correspondant est analysé par le second spectromètre. (Figure n° 1)

Ce spectre, assez voisin des spectres habituels obtenus par impact électronique, permet une identification quasi certaine de l'espèce ionique m_1, triée par l'appareil à partir de l'ensemble de tous les ions produits par l'échantillon total, qui n'a subi aucune séparation préalable.

2. NATURE DES IONS PRODUITS PAR COLLISION

Depuis longtemps, divers laboratoires (BEYNON, Mac LAFFERTY...) ont étudié les décompositions spontanées (en 10^{-4} - 10^{-6} s) produisant des ions métastables et le parti que l'on pouvait en tirer, surtout pour les études fondamentales de structure des ions fragments, isomérisation, énergies de liaison, libération d'énergie lors des ruptures...

Mac LAFFERTY (16) a fait remarquer que l'étude du spectre des métastables pouvait permettre d'identifier une molécule diluée dans un mélange mais les pics métastables ne sont observables que s'ils ont des durées de vie situées dans une fourchette assez étroite dépendant de l'instrument utilisé, et sont peu abondants.

Un progrès important est intervenu en utilisant une chambre de collision qui ajoute aux décompositions spontanées un processus de décomposition induite par collision, qui est alors général pour tous les types d'ions incidents et produit de nombreux fragments intenses qui, par chance, sont à peu près les mêmes que ceux produits par impact électronique ; ils redonnent ainsi des spectres familiers. (Fig. n° 2 et 3)

Cette méthode devient alors analytique, c'est-à-dire intéresse non plus seulement les fondamentalistes pour des études de structures, mais les analystes pour l'identification qualitative des constituants d'un mélange et la mesure quantitative des teneurs.

3. PHENOMENES DE BASE

Prenons l'exemple d'un spectromètre à double focalisation à géométrie inversée (c'est-à-dire électro-aimant en tête). (Fig. n° 3)

La source peut être n'importe quelle source utilisée en spectrométrie de masse et non plus celles réservées au couplage GC-MS.

Selon la nature des produits on pourra utiliser l'impact électronique (EI) à 70 eV ou en basse tension, l'ionisation chimique (CI), la désorption chimique (D/CI), la désorption par champ (FD) ou l'ionisation par champ (FI), la sonde à pointe d'or, le bombardement ionique (SIMS) ou le bombardement par jet de particules neutres (FAB).

L'échantillon global est introduit dans la source, sans séparation préalable, résidu d'évaporation ou de lyophilisation déposé à l'extrémité d'une sonde d'introduction directe, dépôt d'une infime couche obtenue par évaporation d'une goutte d'eau sur une plaquette métallique (pour SIMS et FAB). Pour les produits aisément vaporisables, on peut utiliser un réservoir d'introduction, soit même un chromatographe servant de dispositif d'introduction.

Pour les produits plus lourds on peut, par chauffage de l'échantillon placé sur la sonde, effectuer une évaporation progressive, voire même une pyrolyse ménagée.

Quelle que soit la source, elle produit un grand nombre d'ions différents dus à tous les composants du mélange complexe. Les sources à ionisation douce ont l'avantage de ne donner presque que des ions moléculaires ou quasi moléculaires avec peu de fragments et sont alors mieux adaptés à l'analyse des polluants par la méthode MS/MS.

Les ions sont accélérés par la tension V. On peut extraire soit les ions positifs soit les ions négatifs.

Le premier séparateur est l'électro-aimant produisant l'induction magnétique B. Il sort un faisceau d'ions de masse m_1 (dispersion en mv). Au-delà de la fente de sortie, le faisceau entre dans une chambre de collision dans laquelle on peut soit faire le vide, soit introduire un gaz inerte (N_2, A, He) sous une pression faible (quelques 10^{-3} torr) mais finement réglable.

Nota : Le potentiel de cette chambre peut être réglé entre 0 et $\pm V'$ ($< V$) si l'on désire distinguer les fragments induits par collision (CID) des fragments spontanés (métastables).

a) Spectre de fragmentation induite par collision : $m_1^+ \rightarrow m_2^+ + m_3$

Dans cette chambre de collision s'il y a un gaz, les ions m_1 se fragmentent et le faisceau sortant est analysé par un second séparateur qui est souvent l'analyseur électrostatique du spectromètre ; dans ce cas on effectue un balayage de E, ce qui donne ainsi un spectre en énergie (dispersion en mv^2) ; ceci revient à un spectre de masse des fragments issus de m_1 mais la dispersion énergétique $\Delta V/V$ propre à ces ions fragments est importante de sorte que la résolution est assez médiocre (inférieure à la résolution unitaire au-delà de la masse 100). (Fig. n° 3)

Cette dispersion énergétique contient d'ailleurs des informations sur l'énergie libérée lors de la rupture des liaisons.

Pour améliorer la résolution, on peut soit post-accélérer le faisceau pour réduire $\Delta V/V$, soit placer un spectromètre à double focalisation complet avec un autre électro-aimant ; on a ainsi un spectromètre à trois secteurs, mais cette solution est évidemment plus coûteuse. (13)

Pour l'analyse d'un mélange de produits inconnus, on choisira de préférence comme ions m_1 des ions moléculaires ou quasi moléculaires qui se fragmenteront dans la chambre de collision.

Cette fragmentation, due à une augmentation d'énergie interne suivie d'une décomposition unimoléculaire par clivage simple donnera des fragments m_2 dont les intensités relatives, constantes sont caractéristiques de la structure de l'ion et reproduisent sensiblement le spectre habituel obtenu par impact électronique. On observe parfois plus de fragments que par impact électronique car de nouveaux clivages peuvent apparaître fournissant des détails supplémentaires pour aider à distinguer entre isomères possibles.(16) (voir Fig. n° 3).

Pour chaque ion m_1 successivement on obtiendra ainsi un spectre de collision qui ressemble au spectre familier par impact électronique de la molécule m_1, ce qui permet l'identification progressive des constituants du mélange global par comparaison avec les banques de données.

Il faut signaler immédiatement qu'il existe une autre configuration instrumentale permettant d'effectuer des spectres de collision : le système à triple quadripole.

Pour le cas qui nous intéresse, le premier quadripole trie la masse m_1, le second, alimenté seulement en tension HF, sert de chambre de collision et le troisième enregistre le spectre des fragments m_2, sans subir l'élargissement des raies dû aux libérations d'énergie de rupture.(22) (Fig. n° 4)

b) Observation en fonction du temps d'une ou plusieurs fragmentations spécifiques (single or multiple reaction monitoring) (8)

L'échantillon introduit dans la source est chauffé progressivement. On enregistre, en fonction du temps, l'intensité du pic m_2 provenant de la fragmentation par collision de m_1.

Ceci doit caractériser le produit recherché grâce à la réaction spécifique :
$$m_1^+ \rightarrow m_2^+ + m_3$$

On mesure ainsi sélectivement l'abondance de ce produit, noyé dans une matrice complexe pendant qu'il s'évapore progressivement jusqu'à épuisement. L'aire comprise sous le tracé donne une mesure quantitative de la quantité de ce produit initialement contenue dans l'échantillon, sans subir la perturbation des nombreux autres constituants, si la réaction est bien choisie. (voir Figure n° 6).

On peut simultanément faire un balayage comparatif sur la réaction de fragmentation d'un produit traceur, marqué aux isotopes stables, quantitativement introduit au préalable dans l'échantillon, ce qui accroît la précision du dosage.

c) Utilisation des ions négatifs

En utilisant, par exemple, une source à ionisation chimique en mode négatif (par exemple avec OH- ou MeO-) on obtient les fragments principaux par collision, qui apportent de précieux renseignements pour l'identification des molécules ionisés dans la source. (7) (4) (2).
Par exemple :
$(M - H)^- \rightarrow (M - H - CO_2)^-$ pour les acides carboxyliques.
Perte de NO et de NO_2 pour les dinitrophénols (Figures n° 7 et 8).
Fragmentation d'ions de type (M + Cl) avec attachement d'un chlore pour les sucres, par exemple avec le glucose :
$$M + Cl^- \rightarrow (M + Cl)^-_{m/e = 215-} \rightarrow (M - H)^-_{m/e = 179-} + HCl$$

d) Réaction d'acquisition de charge (charge striping) (1) (10)
$$m^+ + gaz \rightarrow m^{++} + gaz + e^-$$
Cette réaction permet de détecter dans un mélange les composés donnant aisément des ions doublement chargés, comme les hétérocycles azotés ou les aromatiques polycycliques.

Un balayage du spectre, en réglant la tension de l'analyseur à E/2 au lieu de E, avec un appareil à deux secteurs, donne directement le spectre des espèces moléculaires concernées avec leurs proportions relatives.

e) Inversion de charge

La réaction d'inversion de charge peut se produire dans la cellule de collision, par exemple à partir d'un ion négatif qui devient positif : (17)
$$[M]^- + N \rightarrow M^+ + N + 2e$$
comme pour le nitrobenzène : $[C_6H_5NO_2]^- \rightarrow [C_6H_5NO_2]^+$ qui, par collision avec l'hélium redonne son spectre habituel de fragmentation (comme par impact électronique), alors que le spectre complet d'ions négatifs serait touffu et non aisément reconnaissable. (Figure n° 9).

4. MODES DE BALAYAGES PROPRES A LA TECHNIQUE MS-MS

Nous avons vu qu'en triant l'ion m_1 par le premier séparateur et en lui faisant subir une collision, on utilise le second séparateur pour obtenir son spectre de dissociation, c'est-à-dire tous les fragments m_2 issus de m_1, ceci par balayage de E avec l'appareil à deux secteurs.

Il existe d'autres modes de balayge intéressants pour l'analyse des mélanges complexes.

a) <u>Balayage à neutre constant</u> dans la réaction $m_1^+ \rightarrow m_2^+ + m_3$

On combine les paramètres instrumentaux de façon à enregistrer tous les ions m_1 tels que m_3 = constante. m_3 est un fragment neutre expulsé lors de la fragmentation de m_1 en m_2. (22) (24)

Par exemple, les polychlorodiphényls donnent tous des réactions :

$\phi - \phi \ Cl_n^+ \rightarrow \phi - \phi \ Cl_{n-1}^+ + Cl$

$\phi - \phi \ Cl_n^+ \rightarrow \phi - \phi \ Cl_{n-2}^+ + Cl_2$

de sorte qu'en effectuant un balayage à neutre constant m_3 = 35 ou m_3 = 70, on enregistre le profil de tous les ions moléculaires de diphénylchlorés contenus dans le mélange.

Ce mode de balayage demande le pilotage du spectromètre par ordinateur, mais il est plus aisé à réaliser avec l'appareil à triple quadripole.

b) <u>Recherche des ions précurseurs d'un fragment donné</u> :

Ce mode de balayage consiste à retrouver tous les ions moléculaires donnant un fragment observé dans le mélange ; par exemple pour retrouver tous les phtalates on recherchera les précurseurs des fragments 104 et 76 observés dans le mélange, ou encore les différents alkylphénols ou chloro-alkylphénols donnant le fragment C_7H_7O (m/e = 107). (Figure n°10).(18) (21)

5. INSTRUMENTATION

Les configurations les plus utilisées sont les suivantes :

a) <u>Secteur magnétique - Secteur électrostatique</u> (soit double focalisa-tion à géométrie inverse). C'est l'appareil actuellement le plus répandu pour la technique MS/MS. Son avantage est une grande souplesse d'utilisation avec les divers modes de balayages combinés et la possibilité de mesurer les énergies de ruptures de liaisons. L'inconvénient est la faible résolution du spectre final (on n'obtient pas la séparation unitaire au-dessus de la masse 100), précisément à cause de la sensibilité de l'analyseur électrostatique à la dispersion énergétique des ions (l'ESA disperse selon mv^2).(Fig. n° 3)

En somme cet appareil est bien mieux adapté aux études de structure qu'à l'analyse.

b) <u>Appareil à plus de deux secteurs</u> : spectromètres tandem. Ces appa-reils ont pour but d'obtenir un spectre final à haute résolution. Evidemment on gagne en finesse de détection pour des mélanges de molécules de masses très proches. On perd forcément en sensibilité et surtout le coût de l'ensem ble croît avec la complexité du système. C'est aussi un appareil de recher-che structurale fine, généralement complété par des détecteurs intermédi-aires disposés sur le parcours du faisceau ionique pour mesurer la cascade d'événements qu'il subit dans les diverses régions sans champ du système et en retirer des informations complémentaires surtout sur les structures des fragments et les forces de liaison, en somme des études fondamentales sur la chimie des ions. (Fig. n° 11)

Chose intéressante pour l'analyse, ces appareils sont actuellement les seuls à pouvoir traiter les grosses molécules dans la gamme de 1000 à 3000 et plus. En effet, les sources FD, D/CI et FAB sont susceptibles de donner les ions lourds, mais l'étude de leur fragmentation pour leur identification demande des appareils à haut pouvoir séparateur, en particulier des aimants à très haut champ (> 2 Teslas).

Les travaux du Professeur Mac LAFFERTY, dans ce domaine, explorent(13) toutes les possibilités de cette coûteuse technique avec des combinaisons judicieuses. Le Professeur BOERBOOM a également construit un instrument très sophistiqué dans le même but. (9)

c) Système à triple quadripole : Un autre mode de montage MS/MS consiste à associer en série trois quadripoles : (Fig. n° 12)
- le premier fonctionne en séparateur de la masse m_1.
- le second a la particularité de n'être alimenté qu'en tension FH, sans la composante continue, de sorte qu'il ne filtre pas les masses mais sert de cellule de collision à basse énergie, avec un excellent rendement.
- le troisième quadripole, alimenté normalement, analyse les ions produits par collision et en donne le spectre avec une bonne résolution car il est insensible à la dispersion énergétique des ions (séparation en m/z).

Ce système s'avère le plus efficace pour l'analyse parce que, complété par un bon système d'informatique pour le traitement des signaux, il permet d'effectuer correctement :
. l'enregistrement classique du spectre du mélange, sans gaz de collision et avec les spectromètres II et III alimentés en RF seul.
. l'enregistrement des ions issus d'un ion m_1 trié par le spectromètre I, obtenus soit par décomposition spontanée (métastables) dans le spectromètre II s'il n'y a pas de gaz ajouté, soit par collision s'il y en a, et balayés par le spectromètre III qui en donne un spectre bien résolu.
. l'enregistrement de tous les ions parents d'un fragment m_2 trié par le spectromètre III, alors que l'on balaie avec le spectromètre I, le spectromètre II servant de chambre de collision.
. le balayage à neutre constante est également facile à obtenir en balayant simultanément les spectromètres I et III et en imposant, par l'ordinateur, une différence de masse définie et constante.

Le spectromètre quadripolaire atteint facilement une résolution de 1000 à 1500 et, si le balayage est réalisé en mode "résolution constante", le rendement du filtre quadripolaire est maintenu sur tout le domaine de masse de sorte que jusque vers la masse 1000 au moins le spectre de l'appareil quadripolaire est très voisin de ceux donnés par un appareil à secteur magnétique.

Le gros avantage, pour l'analyste, du quadripole est qu'il est insensible à la dispersion énergétique des ions engendrés par collisions et que, par conséquent, la résolution des spectres de collision est semblable à celles des spectres habituels, ce qui simplifie la comparaison avec les données de référence pour l'identification.

Le système à triple quadripole a aussi un grand avantage pour l'analyste, c'est sa grande sensibilité. La source produit environ 1 ion utile pour 2.10^{-5} molécules. La transmission des spectromètres I et III est 10 à 15 %. Le filtre intermédiaire alimenté en RF seul a une transmission de plus de 60 % et l'efficacité de collision, vu l'allongement considérable des trajectoires ioniques et la basse énergie des particules, est proche de 100 %, si l'on ajuste bien la pression du gaz neutre (quelques 10^{-4} torr à quelques 10^{-3}, mais pas trop pour maintenir raisonnables les pertes par dispersion de trajectoires, plus importantes qu'avec des ions incidents à haute énergie). Ainsi la sensibilité de l'ensemble est de quelques centaines de femtogrammes, c'est-à-dire semblable à celle d'un spectromètre quadripolaire et pratiquement supérieure à celle d'un GC/MS grâce à l'élimination de la pollution chimique.

d) Système à double quadripole : Il est voisin du précédent, mais la cellule centrale est une cellule de collision simple ou une cellule dans laquelle un champ alternatif allonge les trajectoires ioniques, mais pas à proprement parler un quadripole. Les résultats sont assez analogues à ceux obtenus par le triple quadripole. (5)

6. UTILISATION D'UN SPECTROMETRE DE MASSE CONVENTIONNEL A DOUBLE FOCALISATION (type E.B)

Dans ce cas on peut seulement étudier les réactions qui se produisent dans la première région sans champ du spectromètre, c'est-à-dire entre la source et l'ESA, où l'on doit placer la cellule de collision (sinon on n'observera que les métastables se décomposant dans cet intervalle).

Il faut alors, grâce aux alimentations électriques pilotées par ordinateur, réaliser des balayages concertés :
. le balayage avec B/E = constante permet d'enregistrer tous les ions fragments produits par l'ion m_1 du réglage initial,
. le balayage avec B^2/E = constante permet d'enregistrer les précurseurs de l'ion fragment m_2 affiché pour le réglage initial,
. le balayage en tension d'accélération V permet aussi de retrouver les précurseurs d'un ion défini m_2,
. le balayage avec $E|1-(E/E_0)|^{\frac{1}{2}}/E = Cte = B_3/E_0$ permet l'enregistrement des ions moléculaires donnant un fragment neutre constant, dont la masse m_3 correspond au champ B_3 dans le réglage initial.

Ces possibilités sont réelles mais bien moins aisées qu'avec les appareils à géométrie inversée ou triple quadripole, à cause des limitations inéluctables :
. grande abondance d'ions différents dans le premier espace libre de champ, avec possibilité d'artefacts dus à des ions de masse voisine.
. limitation des domaines de balayage par la limitation des grandeurs de V et B.

7. APPLICATION AUX MICROPOLLUANTS ORGANIQUES

En résumé, les modes d'utilisation les plus utiles pour l'étude des micropolluants contenus dans un extrait complexe sont, à mon avis :
- à partir des masses des ions moléculaires, le balayage des spectres de dissociation donnant pour chacun d'eux un spectre reconnaissable.
- à partir des principaux fragments, le balayage des ions parents ou précurseurs susceptibles de produire le fragment observé.
- le balayage à neutre constant pour enregistrer les pics moléculaires de familles entières de composés, caractérisées par les pertes du fragment neutre choisi (Cl, NO, O_2...).
- le balayage, en fonction du temps et de l'élévation de la température de la source, du signal produit par une ou plusieurs réactions spécifiques $m_1^+ \rightarrow m_2^+ + m_3$ permettent de détecter des polluants bien définis dans un échantillon brut.

L'extension de ces procédés aux ions négatifs accroît la sensibilité pour certains types de composés, tout en abaissant l'émission ionique des autres.

8. CONCLUSION

Pour l'analyse des mélanges complexes, comme des micropolluants organiques, le système MS-MS peut se comparer au système GC-MS ou LC-MS de la façon suivante :

a) Sensibilité :

Elle est comparable pour les deux méthodes en utilisant les instruments les plus élaborés. Le type double ou triple quadripole semble le plus adapté à la détection de traces et probablement le plus sensible.

Il faut noter que la méthode MS/MS est plus avantageuse que la méthode GC-MS pour les produits lourds, soit qu'ils aient un long temps de rétention donc une moindre sensibilité chromatographique à cause de l'élargissement des pics, soit qu'ils soient en partie retenus par la colonne.

b) Spécificité :

De par son principe même, le système MS/MS est supérieur au système GC-MS en ce sens qu'il permet une meilleure élimination des effets parasites dus aux nombreux produits accompagnant les polluants recherchés ; comme la préparation de l'échantillon est fortement simplifiée, on évite l'ajout de pollution chimique, due aux traitements de dérivatisation et séparation par solvants.

c) Exhaustivité :

Il faut distinguer deux aspects :
. le système GC-MS est supérieur en ce sens qu'il permet l'analyse de tous les composants présents à la sortie du chromatographe, attendus ou non, sans qu'il soit nécessaire de faire aucune hypothèse sur la nature des polluants à rechercher ; mais cette analyse est limitée aux polluants qui peuvent traverser la colonne, soit dans leur état initial soit après dérivatisation, et pour les traces se trouve gênée par les polluants majoritaires ou diffus.
. inversement, le système MS/MS permet d'utiliser tous les types de source connus pour l'ionisation de l'échantillon global, et en particulier des composants lourds, labiles ou polaires. On peut utiliser non seulement EI et CI mais aussi, pour les produits peu volatils, la désorption chimique (DCI), la désorption par champ (FD), le bombardement par des ions ou par des neutres (FAB), etc..., ce qui élargit notablement le champ des polluants détectables.

d) Commodité de l'analyse :

Si l'on se fixe, à priori, certains types de polluants à rechercher (polluants prioritaires, par exemple), le système MS/MS permet de les faire apparaître successivement par familles entières grâce aux divers modes de balayages concertés (linked scans) et assistés par ordinateur incorporé. Ceci constitue une originalité de la technique MS/MS qui pourrait déboucher sur des modes opératoires définis et faciles à mettre en oeuvre, assurant une recherche rigoureuse des polluants présents pour chaque famille soumise à investigation.
On gagnera ainsi en sûreté de détection.

e) Rapidité :

La méthode MS/MS n'est plus liée aux temps de rétention chromatographiques qui sont, eux, incompressibles et souvent longs (jusqu'à 1 à 2 h).
Au contraire, la méthode MS/MS, assistée par ordinateur permet une analyse quasi instantanée à condition de bien définir au départ les types de polluants à rechercher. Grâce à cette rapidité, on peut utiliser sur une même prise d'échantillon plusieurs modes d'ionisation et de balayage, par exemple :
. enregistrement panoramique des pics m_1 avec source CI ou FD,
. enregistrement par nombre de carbone des précurseurs de certains m_2 significatifs,
. enregistrement des familles donnant des neutres m_3 significatifs,
. etc...

REFERENCES

(1) T. AST, C.J. PORTER, C.J. PROCTOR and J.H. BEYNON - Charge stripping reactions in mass spectrometry : ionisation energies of some mono- and disubstituted benzene ions. Bulletin de la Sté Chimique BEOGRAD 46 (5) 135-151 (1981).

(2) J.E. SZULEJKO, I. HOWE and J.H. BEYNON - Collision-induced charge-inversion reactions of negative ions : formation of excited states. International J. of Mass Spectrometry and Ion Physics, 37, 27-34 (1981).

(3) C.J. PROCTOR, B. KRALJ, A.G. BRENTON and J.H. BEYNON - Studies of consecutive reactions of organic ions in a reversed geometry mass spectrometer. Organic Mass Spectrometry, vol. 15, n° 12, 619-631 (1980).

(4) C.V. BRADLEY, I. HOWE and J.H. BEYNON - Analysis of underivatised peptide mixtures by collision-induced dissociation of negative ions. J. of the Chemical Society, 561-564 (1980).

(5) K.L. BUSCH, T.L. KRUGER and R.G. COOKS - Charge exchange using a double quadrupole mass spectrometer. Analytica Chimica Acta, 119 (1980) 153-156

(6) G.L. GLISH and R.G. COOKS - Direct analysis of mixtures by double quadrupole mass spectrometry. Analytica Chimica Acta, 119 (1980) 145-148.

(7) D.F. HUNT, J. SHABANOWITZ and A.B. GIORDANI - Collision activated decompositions of negative ions in mixture analysis with e triple quadrupole mass spectrometer. Anal. Chem. 52 (1980) 386-390.

(8) R.W. KONDRAT, G.A. McCLUSKY and R.G. COOKS - Multiple reaction monitoring in mass spectrometry/mass spectrometry for direct analysis of complex mixtures. Anal. Chem. 50, n° 14 (1978) 2017-2021.

(9) G.J. LOUTER, A.J. BOERBOOM, P.F. STALMEIER, H.H. TUITHOF and J. KISTEMAKER - A tandem mass spectrometer for collision-induced dissociation. Inter. J. Mass Spectrometry and Ion Physics, 33 (1980) 335-347.

(10) A. MAQUESTIAU - Détection et identification de composés dans les mélanges par MS/MS. Communication personnelle.

(11) F.W. McLAFFERTY, A. HIROTA, M.P. BARBALAS and R.F. PEGUES - Energy independence of the collisional activation mass spectra of benzoyl ions. Inter. J. Mass Spectrometry and Ion Physics, 35 (1980) 299-303.

(12) F.W. McLAFFERTY and F.M. BOCKHOFF - Separation/identification system for complex mixtures using mass separation and mass spectral characterization. Anal. Chem., 50, (1978) 69-75.

(13) F.W. McLAFFERTY, P.J. TODD, D.C. McGILVERY and M.A. BALDWIN - High resolution tandem mass spectrometer (MS/MS) of increased sensitivity and mass range. J. American Chemical Society, 102-10 (1980) 3360-3363.

(14) F.W. McLAFFERTY - Tandem mass spectrometry (MS/MS).(à paraître dans Science).

(15) T. MATSUO and H. MATSUDA - Field desorption-collisional activation mass spectrometry with accumulated linked-scan technique for peptide structure elucidation. Anal. Chem. 53 (1981) 416-421.

(16) F.W. McLAFFERTY and T.A. BRYCE - Metastable-ion characteristics : characterization of isomeric molecules. Chem. Communications (1967).

(17) C.J. PORTER, J.H. BEYNON and T. AST - The modern mass spectrometer - a complete chemical laboratory. OMS, 16, n° 3 (1981) 101-114.

(18) D.H. RUSSELL, E.H. McBAY and T.R. MUELLER - Mixture analysis by high resolution mass spectrometry/mass spectrometry. International Laboratory, (avril 1980) 49-61.

(19) B. SHUSHAN and R.K. BOYD - Scans for preselected neutral fragment loss in double-focusing mass spectrometry. Anal. Chem. 53 (1981) 421-427.

(20) J.R.B. SLAYBACK and M.S. STORY - Chemical analysis problems yield to quadrupole MS/MS. Industrial Research & Development (Feb. 1981) 129-134.

(21) M.W. SIEGEL - Collision induced dissociation apparatus with quadrupole field collision cell for mass spectrometry/mass spectrometry. Anal. Chem. 52 (1980) 1790-1792.

(22) R.A. YOST and C.G. ENKE - Triple quadrupole mass spectrometry for direct mixture analysis and structure elucidation. Anal. Chem. 51, 12 (oct. 1979) 1251-1264.

(23) D. ZAKETT and R.G. COOKS and W.J. FIES - A double quadrupole for mas spectrometry/mass spectrometry. Analytica Chemica Acta, 119 (1980) 129-135.

(24) D. ZAKETT, P.H. HEMBERGER and R.C. COOKS - Functional group screening of complex mixtures with a double quadrupole mass spectrometer. Analytica Chimica Acta, 119 (1980) 149)152.

(25) R.G. COOKS - Collison spectroscopy - p. 374 Plenum Press 1978

Figure n° 1

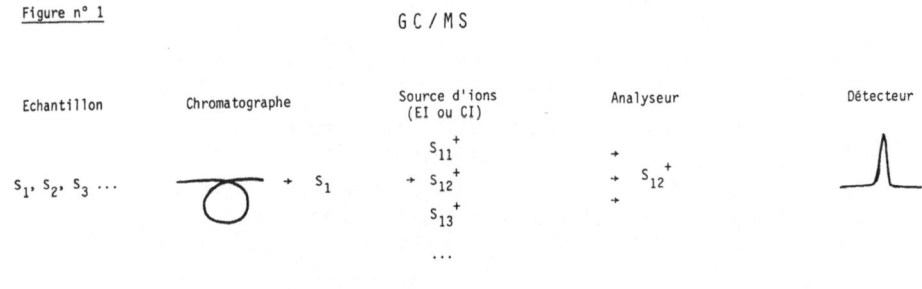

GC/MS

MS/MS

Fig. 2

Décomposition spontanée en 10^{-4} - 10^{-5} sec (métastable)

$$m_1^+ \rightarrow m_2^+ + m_3$$

Dissociation induite par collision

a) excitation sous le choc $\qquad m_1^+ + N \rightarrow m_1^{*+} + N$

b) dissociation unimoléculaire $\qquad m_1^{*+} \rightarrow m_2^+ + m_3$

Fig. 3

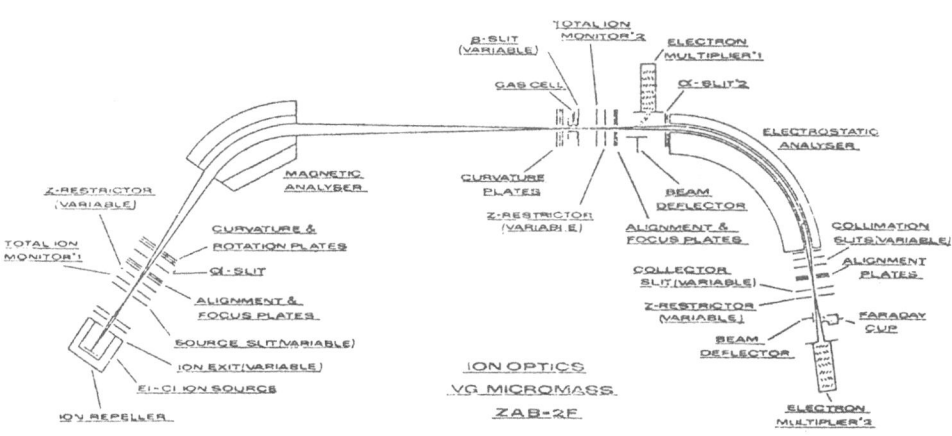

ION OPTICS
VG MICROMASS
ZAB-2F

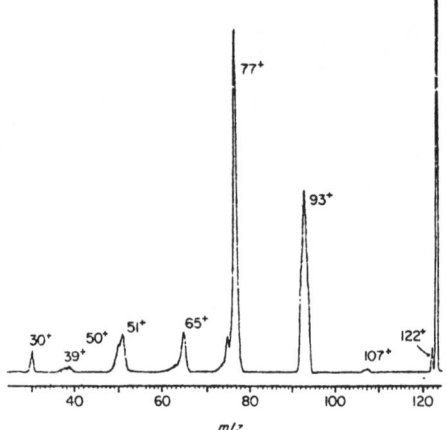

Fig. 4 – The positive collision induced MIKE spectrum of the molecular
ion of nitrobenzene

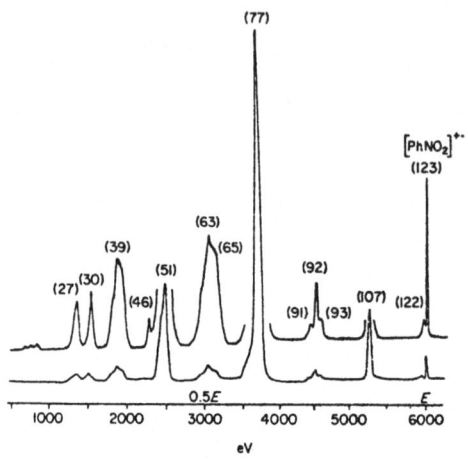

Fig. 9 – The MIKE spectrum of
$(C_6H_5NO_2)^+$ ions obtained by charge
inversion of $(C_6H_5NO_2)^-$ ions.

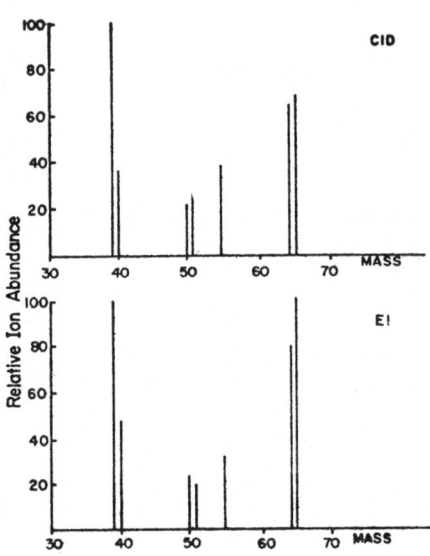

Fig. 5 – Comparison of major fragment ions in the electron impact
(EI) mass spectrum of phenol and the collision-induced dissociation
(CID) spectrum of the phenol molecular ion.

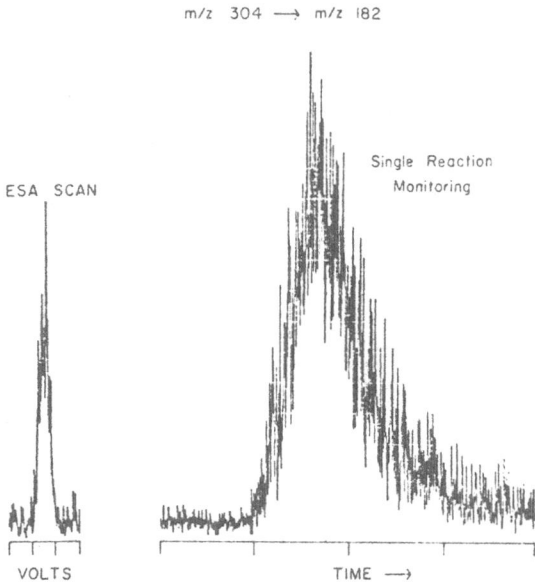

Cocaine (2 ng)

m/z 304 ⟶ m/z 182

ESA SCAN

Single Reaction
Monitoring

VOLTS

TIME ⟶

Comparison of responses observed for the reaction 304^+ ⟶ 182^+ for 2-ng cocaine samples by (left) scanning through the peak in the MIKE spectrum corresponding to this transition and (right) monitoring the transition as a function of time without scanning. The peak heights are approximately the same. The time scale is the same for both scans so that peak widths can be compared directly

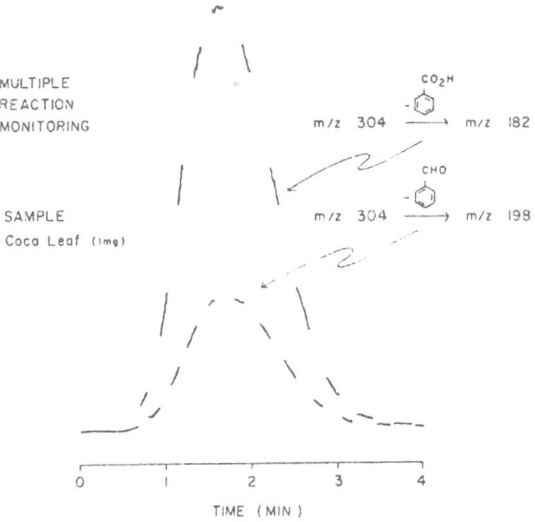

MULTIPLE
REACTION
MONITORING

CO_2H

m/z 304 ⟶ m/z 182

CHO

SAMPLE
Coca Leaf (1mg)

m/z 304 ⟶ m/z 198

0   1   2   3   4

TIME (MIN)

Fig. 6 – Multiple reaction monitoring (analog scan) for cocaine in coca leaf

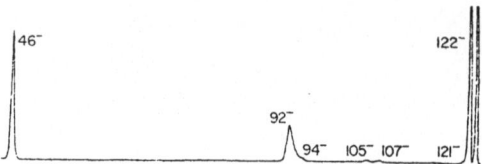

Fig. 7 – The negative collision induced MIKE spectrum of the molecular
ion of nitrobenzene (Ref. 17)

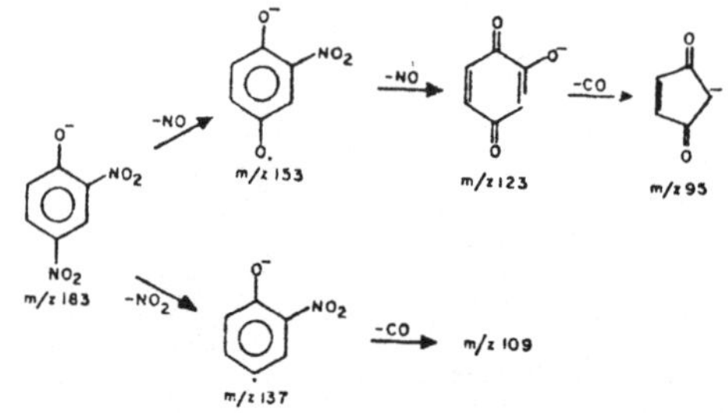

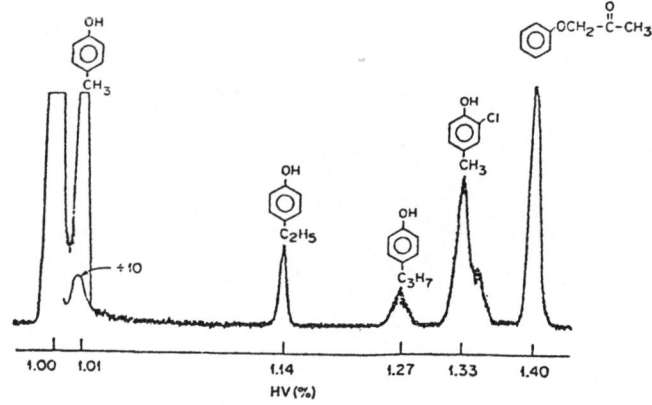

Fig. 8 – The CAD spectrum of the 2,4–dinitrophenol (13) M – 1⁻ion
shows the same two fragments plus several others. Plausible
structures and reaction pathways for the formation of these
ions are shown below (Ref. 7)

Fig. 9 placed beside fig. 4 for reason of comparison

Fig. 10 – Accelerating voltage scan obtained for the $C_1H_1O^+$ ion in
a mixture which contains alkyl phenols and realted
compounds (Ref. 18)

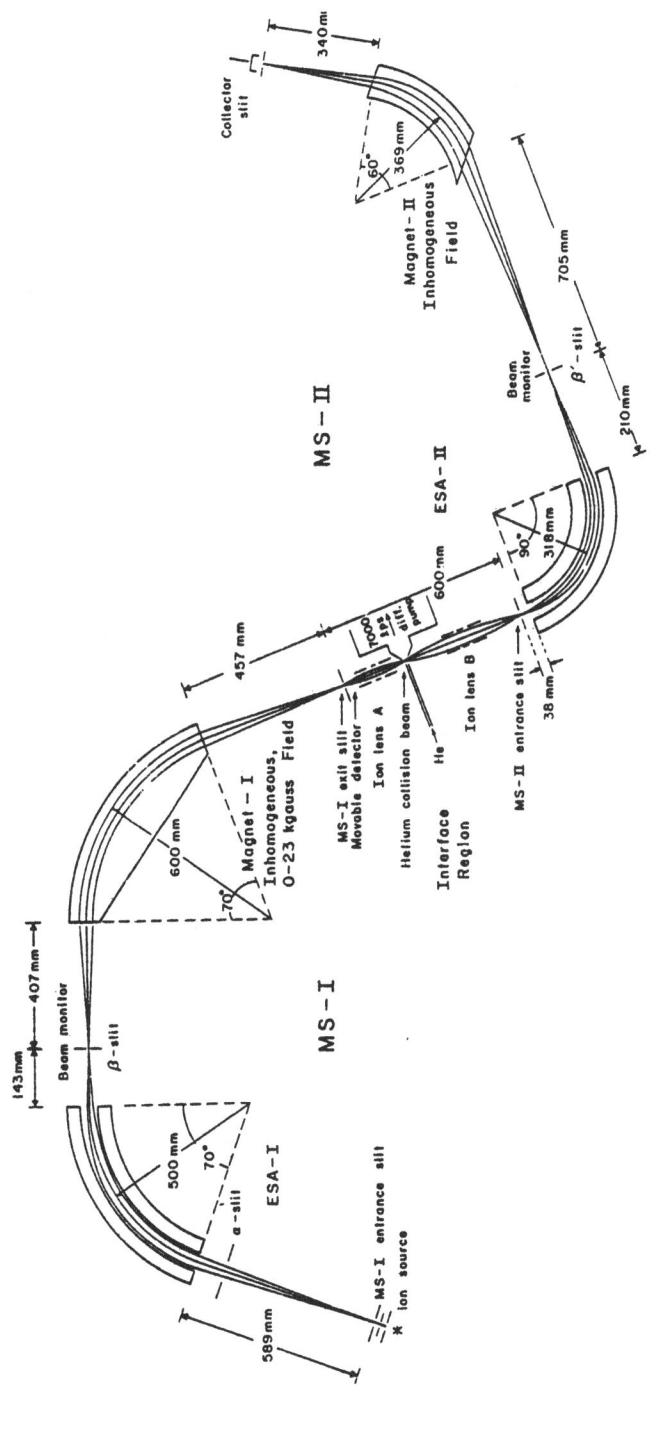

Fig. 11 – Hihg-resolution MS/MS instrument (Ref. 13)

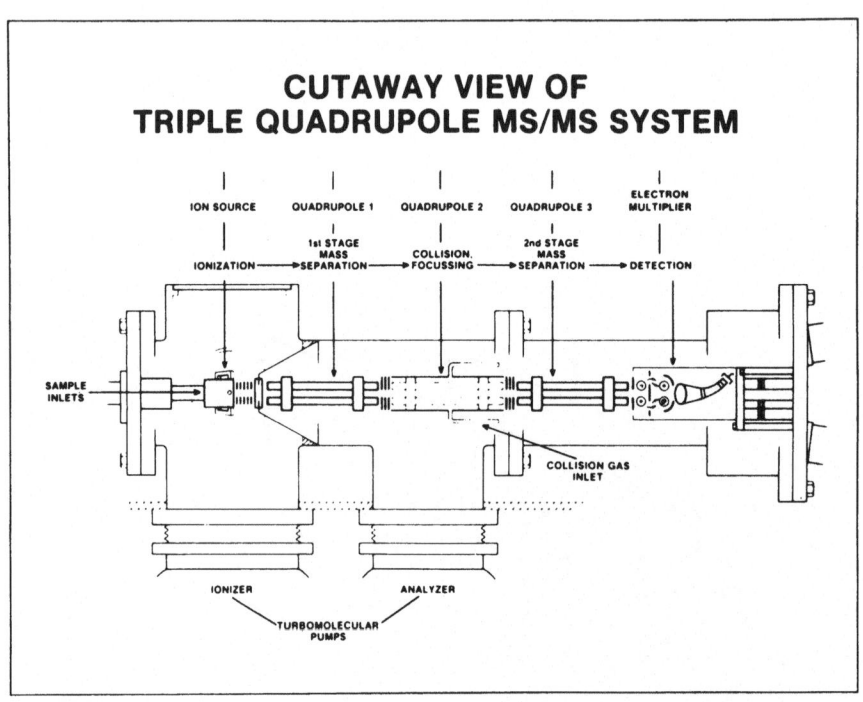

Fig. 12 — Triple quadrupole (montage Finnigan) (Ref. 20)

SESSION V - DATA PROCESSING

Chairman: P. GROLL, Kernforschungszentrum Karlsruhe,
Federal Republic of Germany

Review papers :

- Software systems for mass spectrometry - Remarks in view of
  GC/MS analysis of water pollutants

- Compilation of an inventory of organic pollutants in the
  aqueous environment

Short papers :

- Status of the computerized compilation of mass-spectra of
  organic pollutants

- Current status of the compilation of reference data

SOFTWARE SYSTEMS FOR MASS SPECTROMETRY

Remarks in View of GC/MS Analysis of Water Pollutants

D.Henneberg, MPI für Kohlenforschung Mülheim/Ruhr

Summary. Critical comments are made on the main points
of a computer aided GC/MS analysis of samples from
micro pollutants in water: Dynamic range of the measurement, selection of component spectra out of the
measured series, correction for background and overlap
and identification by library search.

By far the most software systems in use are of
commercial origin, only few developed independantly at
certain institutions. A data system with its hardware
and software acquires, manages, evaluates and stores
the data from the spectrometer, and more and more it
also controls parameters and functions of the measurement. The very close connection between spectrometer
and data system leads to very specialized hardware and
software, which therefore is normally not accessable
for the user. Furthermore, the extent of the information available in manuals is sometimes not sufficient
to understand important details of the function of the
programs. Therefore only some of the important
principles or algorithms of general functions mainly
involved in the field of the analysis of water
pollutants are discussed.

GC/MS contributes to the analysis of the volatile
organic part of the samples extracted from water. Two
different types of evaluation are used, (i) the
identification of the components of the mixture which
is mainly a qualitative analysis but may be extended to
a quantitative or at least half-quantitative
evaluation, and (ii) the exact quantitative
determination of selected priority pollutants.

The latter method is applied for current control in
series of analyses. For this purpose specialized
equipment has been developed which integrates all
functions of this task [1,2]: For a group of "compounds
of interest" or targets specific unique ions, response
factors and retention values are stored. A series of
spectra is measured and the targets are (i) located by
reverse search (see below) in a retention window, (ii)
a peak area is determined using specific ions selected
to be free of interference and (iii) the amount of the
component is calculated using the response factor and
the relation to an internal standard which was
determined in the same way as a target. All these steps
are performed automatically.

The other type of evaluation, the identification of
unknowns, can also be automized. Most software systems
allow the combination and execution of user defined
"procedures" which are assembled from commands or

operations existing in the system. We use a separate program especially designed for this purpose [3] including some additional algorithms needed for the automatic processing of spectra series from fractional evaporations.

The organic part of water pollutants is always a very complicated mixture, i.e. it contains many components, different classes of compounds and covers a large range of concentrations. Therefore even the high seperation efficiency of glass capillary columns is not sufficient to seperate all major components, not to mention the minor ones. Figure 1 gives an impression of the situation. It shows one fifth of a chromatogram of a fraction containing the aromatic and chlorinated hydro-carbons and oxygenated compounds. Therefore the analysis of such mixtures is mainly a problem of dynamic range of the measurement and of interference between unseparated components. Pure spectra of normal or even high quality are available only for a small percentage of the components so that identification as well as quantitation is complicated.

The following discussion therefore concentrates on

1. Dynamic range of the measurement
2. Selection of component spectra
3. Correction of background and overlap
4. Identification of components by library search

Dynamic range

If samples with a very high range of concentrations are to be analyzed in GC/MS measurements with capillary columns, the columns must often be overloaded for the major components. This broadens the corresponding peaks ("leading") with the consequence that the dynamic range of partial pressures in the effluent, seen by the mass spectrometer, is somewhat lower than the range of con-centrations in the sample. Nevertheless the resulting range covers at least two orders of magnitude and furthermore the pattern of each measured spectrum must contain a range of abundances between 10 and 1000 de-pending from the type of the spectrum and the required interpretation.

A dynamic range, i.e. the ratio between a highest and a lowest possible value, has to be considered for all constituents of the analyzing and measuring equipment: sample injection, column, ionization chamber, electron multiplier and amplifier.

The last link in the chain is the data system with its ranges for the digitization of the MS signal and for the peak detection from this signal. For an optimal performance of the whole system, which is essential for the occurring range of concentrations of the involved samples, all these ranges have to be properly adjusted to each other.

The dynamic range achieved by several data systems delivered some years ago is only about 1000. A range of 100.000 has been realized without changing any sensitivity parameter in the data system developed by us. Much higher values are claimed for systems which change the gain in the intervalls between spectra. Unfortunately, a reasonable definition as basis of an unambiguous procedure to determine the dynamic range is missing, so that the given numbers leave some doubts with regard to their practical utility. Sometimes, numbers are given which are not the range of measurable peaks but the range of D to A conversion or the range down to the noise of the electronic circuits.

Selection of component spectra

The first step in the evaluation of a spectra series is to find the best spectrum for each component. In the best spectrum, ideally, the mass chromatograms of all ions from the component run through a maximum. Therefore a search for a clustering or coincidence of maxima in mass chromatograms across the measured series will automatically locate the components.

If two overlapping components are separated at least partially, mass chromatograms may have their maxima in different, sometimes even in neighboring spectra. In many of these cases where the mass spectra allow to differentiate between two overlapping components, their presence cannot be derived from the total ion chromatogram.

Maximizing mass chromatograms in adjacent spectra, however, must not in any case result from different components. If two spectra are measured at comparable heights just before and after a peak maximum, the opposite pressure drop during the scan of these spectra leads to maximization of the high mass part in the first and the low mass part in the second spectrum (or reverse for downscan). An example is given in Figure 2.

Accidental maxima in mass chromatograms may arise by statistical fluctuations of constant or slowly changing partial pressures. Their influence can be suppressed if the algorithm determines only qualified maxima or looks for qualified clustering.

Many data systems plot the sum of maximizing ions instead of the entire total ion current as a so-called "enhanced chromatogram". Because of the mentioned ambiguity of the origin of maximizations the gain of information in these curves is very limited compared with the information from a critical inspection of the normal chromatogram.

Correction

The correction for background and overlapping components is urgently needed for library search methods, which are irritated by mixture spectra; but even with methods, which are not sensitive to impurities, a correction increases the probability and the certainty or the ease of an identification.

All data systems provide the functions needed for corrections carried out by the user, namely the partial subtraction of a spectrum in which the component to be eliminated is present in a higher percentage. Such corrections are time consuming because they need knowledge about the uniqueness of ions which in every particular case has to be established from mass chromatograms. For multicomponent mixtures which normally contain many interferences it is therefore desirable to have a procedure which operates automatically.

A very elegant principle to correct simultaneously for background and overlap was introduced by Biller and Biemann [4] and is used in most commercial data systems. The corresponding algorithm performs a "correction" in the following way: If a mass chromatogram runs through a maximum in the selected spectrum, the corresponding mass peak remains in the spectrum without change, if not, the peak is eliminated entirely. The mass chromatograms or peaks for m/z 103 and 111 in Figures 3 and 4, respectively, are examples for these two alternatives.

This very reasonable algorithm, however, can fail remarkably. A false elimination can happen if the ion is also present in the spectrum of an overlapping possibly isomeric component with an intensity sufficient to distort the maximum into a shoulder as for instance m/z 75 and 85 in Figures 3 and 4. Such a loss of information can complicate or even prevent an identification.

On the other hand the elimination of a background ion can be omitted because this ion has a maximum in the selected spectrum due to fluctuation as probably m/z 105 in Figures 3 and 4. A background ion will also remain in the corrected spectrum if a maximum is present due to a real but comparably small contribution from the component spectrum. If such an uneliminated background ion corresponds to a characteristic ion of a reference spectrum, an identification is complicated.

A correction as carried out by the principle of Biller and Biemann must provide precautions against the case of maximization in adjacent spectra caused by pressure changes. The original program avoids the resulting wrong correction by the addition of neighboring spectra across the whole spectra series prior to the determination of the mass chromatograms. But, as Figure 2 shows, an addition does not solve the problem in any case completely. Furthermore, if the adjacent maximizations are due to neighboring components, the addition deteriorates the chromatographic separation and the indication for the presence of two components is lost. Our program [3] is able to distinguish between the two cases by comparing the two spectra: If they are equal, as expected for one component, only the better one is kept, if they are not equal two neighboring components are positively indicated.

A more sophisticated program has been described [5] which performs intelligent corrections mainly for background, but it also eliminates non maximizing peaks. This program makes use of component peak profiles and therefore needs seven spectra across a peak, which in the case of the narrow peaks of modern capillary column chromatography can be measured only with a sacrifice in sensitivity and accuracy.

Our program determines for each selected component spectrum the best background spectrum and substracts this spectrum entirely. This program does not make corrections, but in the case of a very close neighbourhood of two selected spectra of different compounds it tries to use spectra which are more apart from the near neighbour and therefore perhaps less overlapped than the originally selected "best" spectrum. With respect to overlapping components this procedure is of course much less effective than the correction achieved with the Biller-Biemann algorithm. However, we avoid the described possible loss of information by wrong corrections. Furthermore, our search method is rather insensitive to the presence of interfering masses and well capable of identifying mixture components. At the moment we are developing a new method which makes a correction by subtraction of a suitable percentage of the complete spectrum of the overlapping component (lower spectrum in Figure 4).

It should be mentioned, however, that a search method capable of handling mixture spectra is needed anyhow because complete and therefore uncorrectable overlap of components happens occasionally even in columns with very high separation efficiency.

Identification

The first step of a computer aided identification is a library search with the optimally prepared spectrum of the unknown. The result of a library search is a hitlist with often only one, but in more advanced search methods several parameters which establish the order within the hitlist and characterize the comparison of the unknown with the respective reference spectrum numerically.

In principle, no objective numerical measure of similarity exists. Therefore, the numerical value for the degree of a match in the hitlist cannot indicate unambigously if a compound is truly identified or not. For the same reason the values achieved for an identification are sometimes lower than the values of best hits which are not an identification, a fact, which is almost neglected in the literature about library search. Therefore the result of a search has to be interpreted from the analyst, at best by inspection of the full spectra of unknown and reference but unfortunately most data systems have the full reference spectra not available!

This need for an individual interpretation of probably not all but many of the search results is the main hindrance for a complete automization of the computer aided analysis of mixtures by GC/MS.

The confirmation of a possible identification requires the reasonable explanation of the differences between the patterns of the unknown and the corresponding reference spectrum: The unknown may be contaminated or a mixture, or the differences may be due to temperature-, pressure- or mass discrimination effects. Furthermore, from our experience with large libraries containing spectra of different origin the reference spectrum may be suspect because its appearance is unusual in comparison to spectra of similar compounds or has no reliable intensity ratios because of a very bad dynamic range. After all in the case of an "identification" it must be considered also that isomers or compounds similar to the "identified" one (including those which are not in the library!) may have identical spectra.

For an assessment of a search result it is very important, that the analyst has broad experience with the behaviour of the method he is using, namely its response to certain types of spectra or degrees of contamination, distortion or admixture. Last not least the content of the library should be known roughly at least within the field of interest with regard to the presence or absence of certain compounds or compound classes as well as with regard to the quality of the spectra which is remarkably low in numerous cases (see for instance in [13]).

As an aid to become acquainted with the content of the library, we use a program which extracts references from the library according to heteroatom content, molecular weight (including limited areas or homologous series) and character strings within names in different logical combinations.

Two different search strategies, forward and reverse, can be used: Forward search is the comparison of a measured, corrected spectrum with all spectra of a library and aims at the identification of unknown compounds. Reverse search is the comparison of the measured spectra of a series (or only those in a retention window) with the library spectra of compounds of interest. The aim is to establish the presence and to locate these compounds in the chromatogram.

Sometimes the terms "forward search" and "reverse search" are used to characterize not the search method but the algorithm for the comparison of unknown and reference: If peaks or features which are present in the unknown but absent in the reference spectrum (the typical situation for mixtures or contaminations) are validated, the search is called "forward" or "the unknown is compared with the reference". If such peaks have no influence on the value of the match, the search is called "reverse" or "the reference is compared with the unknown".

This use of the same expression for different things is not only confusing, it may be misleading especially for the more sophisticated comparison algorithms. These consist of several terms which are mainly neutral (commutative) or forward or reverse (both non-commutative), and these terms of different type are weighted differently in the final match value. Types and weights of the used terms establish the essential differences between the methods.

Comparison of Search Methods

Often used and well tried search methods are the one described by Biemann and coworkers [6], the one used in the Finnigan/Incos data systems [7] and PBM [8,9] and STIRS [9,10] developed by McLafferty and coworkers. We use SISCOM [11], a method developed by us.

More detailed discussions of the aspects of library search are found in the literature [12,13,14]. A comparison of the efficiency of these different search methods in a strict sense is not possible mainly because the numbers which describe the degree of a match in the different systems are not comparable, but also because the libraries used in conjunction with the different methods differ more or less in their content.

Nevertheless we have carried out searches with the above mentioned methods using the same set of unknown spectra. This set contains measured spectra of pure compounds represented in the library and some not represented themselves but similar ones. Another group in the test set are spectra of mixtures or spectra with a pattern differing from the reference spectrum.

After all, the results of the different searches show that almost pure spectra without heavy distortions of their pattern relative to the available reference spectrum are easily identified by all methods.

The behaviour of the various methods against the typical complications is rather different. Table I gives a typical example for the retrieval of a mixture:

Table I Retrieval of an artificial 40/40/20 mixture
       (from[15])

A: Pyruvic Acid, Methylester, Oxim
B: Hexanoic Acid, Methylester
C: N-Methyladenin

	A	B	C
Biemann	–	–	–
Incos	(+)	–	+
PBM	+	(+)	+
SISCOM	+	+	+

(+): Retrieval in the residual spectrum
     after subtraction of the best match.

The Hertz-Biemann method is less suited for the
analysis of mixtures and strongly distorted spectra.
PBM and the Incos method can be used for mixtures, they
identify mostly at least one component and further
components after subtraction [15] of the best match
which enhances the match value for the remaining component. Both methods are sensitive to differing patterns.
PBM is oriented especially to identifications, similar
compounds are not predominant in the hitlist.

SISCOM often presents all major components of mixtures
without subtraction. Moreover, the hitlist of SISCOM
exhibits mainly similar compounds, sometimes even if
the spectral pattern looks very unsimilar. This feature
is very important for the analysis of compounds which
are not represented in the library. But the presen-
tation of many similar compounds also supports the
probability of an identification, which can be important if the obtainable match value is low because the
unknown and the reference spectrum differ too much.

STIRS is a method primarily devised to the
determination of substructures in pure spectra of compounds missing in the library. It is proposed to be
used if PBM was unable to identify the unknown.

The great success of SISCOM in the tested examples,
namely its reduced sensitivity against distortion of
abundances and its particular ability to recognize
structural similarities (see in [13]) is among others
due to the fact that the influence of abundances on the
resulting match value is reduced. SISCOM is the only of
the mentioned methods which valuates the presence of a
characteristic ion separately from its abundance and
with a higher weight.

The quality of the attainable information

During the rapidly changing partial pressures of the
components in the effluent of high resolution capillary
columns only short time is available to measure and

process the spectra. Since the typical half widths of peaks are in the range of few seconds and during this period at least two spectra are needed, cycle times between less than one up to three or four seconds may be desirable for spectra series. The choice of optimal parameters is very important because too short as well as too long cycle times reduce the quality of the attainable information.

Long cycle times lower the probability to measure a spectrum near the peak maximum, peak profiles are badly represented in reconstructed mass chromatograms and high chromatographic resolution cannot be used optimally. Long scan times increase the skew of spectra measured on a side of a peak, which may cause a heavy distortion of the pattern.

With short scan times on the other hand the influence of ion statistics is increasing. At the lower end of the intensity scale peaks originate from only few ions or even disappear. Therefore the sensitivity and the accuracy of the abundances is reduced. Moreover, short scan and cycle times limit the time available for the algorithms to process the measured data and do not allow to use sophisticated and possibly more efficient programs.

Modern equipment as powerful computers supported by microcomputers and fast scanning quadrupole and magnetic spectrometers can be operated with cycle times in the order of one second. But if these short times are really neccessary depends on the analytical problem. For an accurate quantitative analysis based on a spectra series with a method as mentioned above for priority pollutants, the determination of the peak areas requires exact profiles and therefore more than three spectra per peak. As a lower limit for a qualitative analysis, however, only two spectra per peak or a cycle time equal to a peak half width can be sufficient. For example, Figure 5 shows the region of the two peaks marked in Figure 1, measured with a cycle time (5 sec) equivalent to the peak half widths. With the exception of the 1% component all others are still retrieved.

In this context one should remind that often the chromatographic parameters can be chosen to meet the desired conditions with cycle times of two or more seconds.

Literature

1) W.E. Pereira and B.A. Hughes Journal American Water Works Assn. 72, 220 (1980)

2) R.E. Finnigan, D.W. Hoyt and D.E. Smith
   Environmental Science Technology 13, 534 (1979)

3) D. Henneberg, H. Damen, B. Weimann
   Adv. in Mass Spectrometry 7B, 975 (1979)

4) J.E. Biller and K. Biemann, Anal. Lett. 7 515 (1974)

5) R.G. Dromey, M.J. Stefik, T.C. Rindfleisch
   and A.M. Duffield, Anal. Chem. 48, 1368 (1976)

6) H.S. Hertz, R.A. Hites and K. Biemann Anal. Chem. 43,
   681 (1971)

7) S. Sokolow, I. Karnofsky and P. Guslafson,
   Finnigan Application Report 2, (1978)

8) F.W. Mc Lafferty, R.H. Hertel and R.D. Villwock
   Org. Mass Spectrom. 9, 690 (1974)

9) F.W. Mc Lafferty, B.L. Atwater, K.S. Haraki, K. Hosokawa,
   In Ki Mun and R. Venkataraghavan
   Adv. Mass Spectrom. 8B, 1564 (1980)

10) K.S. Haraki, R.Venkataraghavan and F.W. Mc Lafferty
    Anal. Chem. 53, 386 (1981)

11) H. Damen, D. Henneberg and B. Weimann
    Anal. Chim. Acta 103, 289 (1978)

12) F.W. Karasek and J. Michnowitz, Research and Development
    27, 38 (1976)

13) D. Henneberg, Adv. in Mass Spectrom. 8B, 1511 (1980)

14) F.W. Mc Lafferty and R. Venkataraghavan
    J. Chromatogr. Sc. 17, 24 (1979)

15) B.L. Atwater, R. Venkataraghavan and F.W. Mc Lafferty
    Anal. Chem. 51, 1945 (1979)

Section of Micro Pollutant Chromatogram
(17 Minutes from 90)

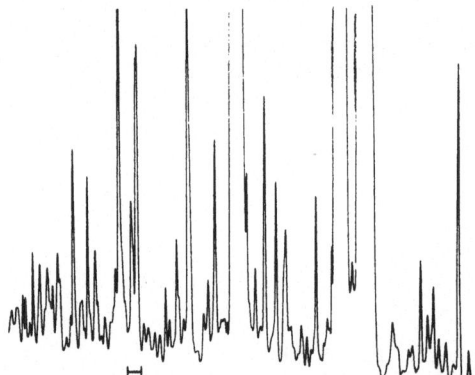

Figure 1

⊢ Two peaks with six components:

1. 2-Undecenal   (27%)
2. Nitro-Anisole   (39%)
3. Styrene, -Methyl Partial Structure (1%)
4. Biphenyl   (3%)
5. 1.1-Diethoxy Partial Structure   (16%)
6. Nitro-Phenetole   (15%)

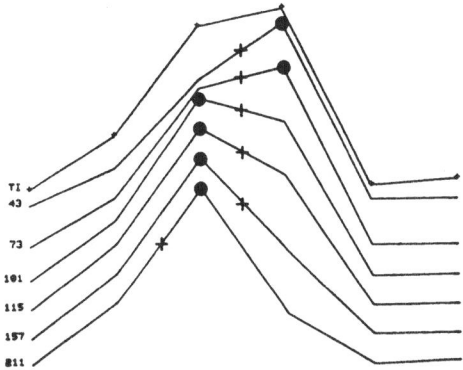

Maximization: Mass chromatograms of one component.

● Maxima in spectra measured before and after the peak maximum.

+ Maxima after addition of neighboring spectra.

Figure 2

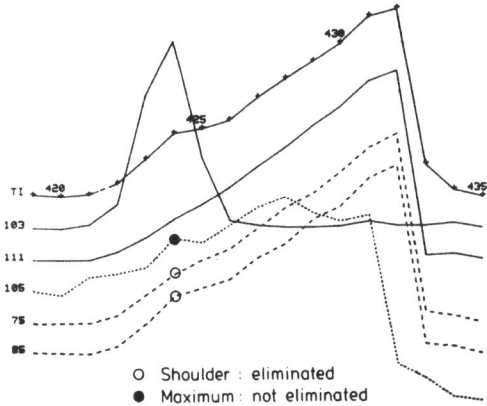

Correction:
Possible failure of the Bitter-Biemann Algorithm

O Shoulder : eliminated
● Maximum : not eliminated

Figure 3

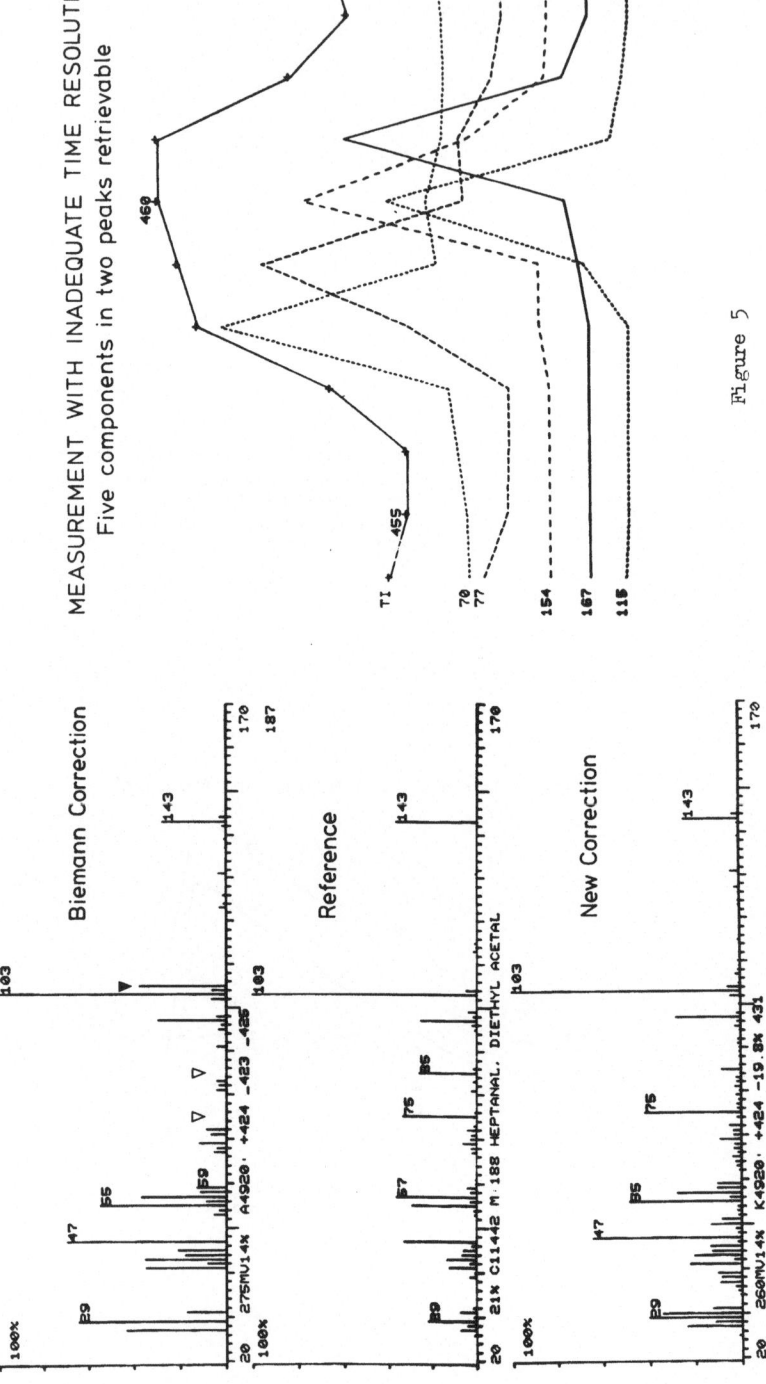

Results of different correction methods

Biemann Correction

Reference

New Correction

MEASUREMENT WITH INADEQUATE TIME RESOLUTION
Five components in two peaks retrievable

Figure 5

COMPILATION OF AN INVENTORY OF ORGANIC POLLUTANTS IN THE

AQUEOUS ENVIRONMENT

A. WAGGOTT and H.V. BRITCHER
Water Research Centre
Stevenage Laboratory
Elder Way
Stevenage
Herts SG1 1TH

Summary

An inventory of known pollutants in the aqueous environment has been
compiled at WRC. The work has progressed through several distinct
stages. Early editions of the inventory were produced by manual
means before the data was eventually entered onto a computer. This
has streamlined production of hard-copies which continues to be the
main priority. The problem of computer storage and handling of this
data has recently been re-examined. As a result a new system has
been commissioned based on software which orders and selects data
on the basis of chemical structure. This system will be developed
in order to promote greater flexibility in searching and retrieval
of the data as well as to allow more efficient data compilation. It
is eventually hoped that the data will be more freely accessible
by its entry into a large computer network.

1. INTRODUCTION

The Water Research Centre was asked by the Department of the
Environment of the UK to undertake the collation of an inventory of organic
pollutants which had been identified in the aqueous environment as part of
the UK contribution to the EEC Concerted Action Group dealing with organic
micro-pollutants in water (COST Project 64b). The work started in 1972
and has continued steadily ever since.
    Originally all data were processed and handled manually to produce
hard-copy versions in a tabulated format. The listings were published
under the title of "A Comprehensive List of Organic Pollutants which have
been Identified in Various Surface Waters, Effluent Discharges, Aquatic
Animals and Plants, and Bottom Sediments". The production of hard copy
versions continues to be the main priority although the data base has
been handled by a computer since 1976. Three editions of the data base
have now been published. Eventually it is hoped that the information
will be more freely available via a computer network.

2. AIMS AND INTENTIONS

The inventory is intended to be a useful source of reference for the
water user (water undertakings, industry, etc) and the research analyst
in the field of water pollution control. It provides listings of organic
compounds actually identified in waters, sediments, etc, of various types
and from various sources. The data can be and have been applied in many
ways eg to obtain precise information about organic pollution in a specific
location, to build up a picture of the overall pollution to be expected in
a certain type of geographical area, and to identify possible organic

pollutants arising from specific industrial activity.

3. INFORMATION INCLUDED

Organic pollutants identified in the aqueous environment are sub-classified into thirty-one substance classification groups (Table 1) and ordered alphabetically by compound name. Compounds may appear in one or more groups with the exception of pesticides and herbicides, optical brighteners, and surfactants. These three groups are also the only ones which are classified by a physical rather than a chemical property. Under each organic compound identification there are listed data on the type of sample, the location where it was taken, its concentration in the sample, the laboratory submitting the data, the date of sampling, the method of analysis or estimation, and a literature reference number for access to an extensive bibliographic section. If required space is also provided for additional information, eg to explain high or low concentration values. The format of a typical page of data is shown in Table 2.

Table 1.   Substance group classification scheme for CICLOPS and no. of organic compounds in each group (Nov 1981)

No	Substance group classification	No. in group
1.	Polynuclear aromatic hydrocarbons and benzene	63
2.	Alkyl substituted polynuclear aromatic hydrocarbons	70
3.	Aliphatic amines and derivatives	41
4.	Aromatic amines and derivatives	129
5.	Cyanides and azo compounds	33
6.	Nitro and nitroso compounds	70
7.	Organo phosphorus compounds	29
8.	Pesticides and herbicides	97
9.	Aliphatic organo halogens excl. pesticides and herbicides	223
10.	Aromatic organo halogens excl. pesticides and herbicides	365
11.	Organo metallics	13
12.	Mercaptans and miscellaneous sulphur compounds	114
13.	Phenols	231
14.	Quinones	13
15.	Heterocyclics	231
16.	Surfactants	9
17.	Optical brighteners	15
18.	Ethers	197
19.	Aldehydes	49
20.	Ketones	121
21.	Aliphatic acids	131
22.	Aromatic acids	86
23.	Esters	125
24.	Alcohols	164
25.	Arylalkanes	141
26.	Alkanes	133
27.	Alkenes	77
28.	Amino acids and proteins	36
29.	Carbohydrates	21
30.	Steroids	13
31.	Pigments, enzymes, vitamins, nucleosides and misc. compounds	21

Substance Classification: Phenols

Table 2 General format of CICLOPS output

Substance	Sample details		Concentration (µg/kg solids) (G/L waters)	Laboratory	Date of sampling	Analysis/ estimation	Ref
Chlorocresol(4-) isomers							
1DAA £	Tap water	The Netherlands	From R.Rhine. Also storage, chlorination, coagulation, filtr.	RID			
1DAB £	Tap water	Ozonisation The Netherlands	Not detected	RID			
1DAC £	Tap water	Ozonisation The Netherlands	(Water from R.Rhine.) Also sand infilt, filtration, chlorin.	RID			
1DAD £	Tap water	Ozonisation The Netherlands	Not detected	RID			
1DAE £	Tap water	Activated carbon The Netherlands	(Water from R.Rhine.) Also sand dune infilt. and coagulation.	RID			
			Not detected				
			From R.Rhine. Also sand infilt, aeration, chlorin, filtration.				
Chloroethylcatechol isomers							
1DAF £	Effluent discharge (Industrial)		Sulphite bleachery sewage	CIIR		GLC MS	391
Chlorohydroxybenzophenone isomers							
1DB0	Tap water		Receiving water from River Ohio (USA)	EPA		MS	47
1DB1	Tap water	USA		EPA		GLC MS	
Chloromethylphenol isomers							
1DB2	Effluent discharge (Industrial)	USA	<10-100E-06	EPA		GLC MS	354
Chlorophenol isomers							
1DB3 £	Surface water	Lowland river UK rivers		WRC	1977	GLC MS	185
1DB4 £	Effluent discharge	Waste water Crude sewage		RIV		GLC MS	354
1DB5 £	Effluent discharge (Industrial)	USA	10->100E-06	EPA		GLC MS	
1DB6 £	Tap water	Chlorination Germany					
Chlorophenol(2-) WLN:QR RG							
1DB7 £	Surface water	River water Tamagewa Japan	3.6E-06(av)	RID	1973-74	TLC	299
			19E-06(max)				
1DB8	Surface water	River water German rivers		EPA		GLC MS	233
1DB9	Effluent discharge	Waste water Chlorinated bio-treated	1.7E-06	EPA		GLC	157
1DBA £	Tap water	The Netherlands	2.2E-06	RID			
1DBB £	Tap water	The Netherlands	<2E-06	RID		GLC MS	184
1DBC £	Tap water	The Netherlands	0.3E-06(av)	RID	1973-74	TLC	299
			2.2E-06(max				
1DBD £	Tap water	The Netherlands	1.0E-06	RID			
Chlorophenol(3- & 4-)							
1DBE £	Surface water	River water German rivers		EPA		GLC MS	233
Chlorophenol(3-) WLN:QR CG							
1DBF	Effluent discharge	Waste water Chlorinated bio-treated	0.51E-06	EPA		GLC	157
Chlorophenol(4-) WLN:QR DG							
1DC0	Effluent discharge	Waste water Chlorinated bio-treated	1.7E-06	EPA		GLC	157
1DC1 £	Subterranean water	Melbourne, Australia	Contaminated by industrial effluent			GLC MS	313

- 233 -

The inventory is not intended as a critical appraisal of the available information on the subject but as a summary and guide towards more detailed data. However, in cases where there is doubt about the identity of organic compounds because of the use of non-systematic nomenclature, the data have not been included.

The data have been compiled from information contributed by the laboratories participating in the COST Project 64b bis "Micropollutants" and from literature dating from 1960 onwards. All data have been cleared for distribution by the contributing laboratory or are freely available in open literature. However, some important data published prior to that date have been included, especially if there is a scarcity of information available after 1960. The list is not fully comprehensive in certain cases where there is a wealth of repetitive information, eg pesticides and PCBs. Here a representative selection of data have been taken or the available data condensed into ranges of concentration.

4. DEVELOPMENT OF THE COMPUTERISED DATA BASE

In 1972 when work on the compilation of the inventory commenced, it was a relatively easy task to collate all of the available information and to continue to keep the inventory up to date. However, the amount of available information has grown exponentially so that the burden of work has grown more onerous. The situation was eased considerably in 1976 when the inventory was converted to a computer format on the Water Research Centre's Rank Xerox 530 computer. Programmes were written the main intention of which was to produce hard copies of the data base in exactly the same format as that produced in previous manual versions. The computer file was called CICLOPS, short for Computer Interrogation of a Comprehensive List of Organic Pollutants, and it allowed selected information to be extracted from the data base as well as producing a full printout. Only information which satisfied a predefined specification was selected by the search program. The restrictions which could be placed upon the search were: classification of compound, nature of sample, location of sample, records which include quantitative data on concentrations found, and records which are recent additions to the data base.

Towards the end of the 1970s certain limitations in the system became apparent. Restricted lists were produced by a core-storage rather than a disc or magnetic tape search and limitations in the size of the computer meant that in 1978 the data base had to be split into two files which were searched separately. The output was produced in a two-volume version. Limitations in storage space also meant that programmes were written in a hybrid language of FORTRAN and machine code. This means that the package as a whole could not be modified to meet changing circumstances. In 1980 WRC acquired a larger and more efficient DEC VAX 11/780 computer. The CICLOPS programmes were re-written entirely in FORTRAN and towards the end of 1980 set up on the new computer. The data base is now again on one file and all the original search options are available.

5. PRESENT ACTIVITIES

A serious limitation of the present computer programmes is that they do not allow ordering or searching of the data base by chemical structure. In order to change this state of affairs a suite of programmes from the CROSSBOW (Computerised Retrieval of Structures Based on Wiswesser) package has been purchased and commissioned[1]. CROSSBOW is a computer software package for the storage, retrieval and analysis of chemical structure information. The programmes installed will allow the creation of a master file of organic pollutants, syntax and other checks on the WLN codings, automatic fragment generation (for simpler and more efficient file searches), and bit and string search (position and context taken into account) of all major fields on the data base. Work on the creation of a master file from the existing CICLOPS data base has just commenced. Initially the two systems CICLOPS and CROSSBOW will be used independently and cross reference made manually. However, an associated CROSSBOW programme has also been purchased which creates and updates text data files. This file could be converted to hold the data base records of identification, although a considerable amount of rewriting of basic programmes would be involved.

In order to make CICLOPS ultimately more efficiently accessible to participating laboratories and other establishments with a legitimate interest, its conversion and loading as a subset of the environmental matrices files of the ECDIN (Environmental Chemicals Data and Information Network) data base is being investigated[2]. The ECDIN file is linked independently by the ADABAS data base management system to a compound file, a bibliographic file and eventually to a geographic location file[3]. Some problems have been encountered in transferring a sample set of data to ECDIN and some further work in order to define more precisely the data fields in CICLOPS will be required. Ways and means of solving the problem most efficiently are at present being considered.

A similar data base[4] commissioned by the Environmental Protection Agency is now accessible via the EPA/NIH computer network. This data file is known as Water DROP (Distribtuion Register of Organic Pollutants) and contains mainly data of US origin. It is hoped that its development will allow the present effort to be concentrated on compilation of data other than that of US origin.

6. GROWTH OF FILE

During a period of abeyance of the COST Project programme (1975-78) a limited amount of effort continued to be devoted to the compilation of the inventory. This was sufficient to continue the literature search for information but not sufficient for its entry into the computer data base. At present data compilation is up to date and except for that of US origin all data up to approximately December 1979 have been entered. The data base, now contains information on approximately 3,750 organic compounds with over 13,000 individual entries.

The growth of the file based on number of individual entries is shown in Fig. 1. The plateau corresponds to the period of abeyance of the project from 1975 until 1978. Growth rate as indicated by the slope of the curve has apparently been fairly constant during the years when data was actually being entered on file. However, this graph reflects the amount of effort which has been put into the project rather than total amount of data available for collation. Although no accurate measurement can be made it is estimated that the amount of available data doubles each year. Projected until the end of 1983, it is estimated that the data list will have more than doubled its present size with probably more than 25,000 individual entries by that time.

7. FUTURE OBJECTIVES

Some proposed objectives and completion dates are shown in Table 3. The Concerted Action Programme is scheduled to end in December 1983 and this is therefore the date by which ideally all objectives should have been achieved. However, the effort at present being expended on this work may not be sufficient to meet all of the objectives connected with entry of data into CICLOPS. Because of this, data have been classified on a priority basis according to geographical location and novelty of compounds identified. This will ensure that available effort is devoted to entry of data of greatest importance to the Concerted Action Group.

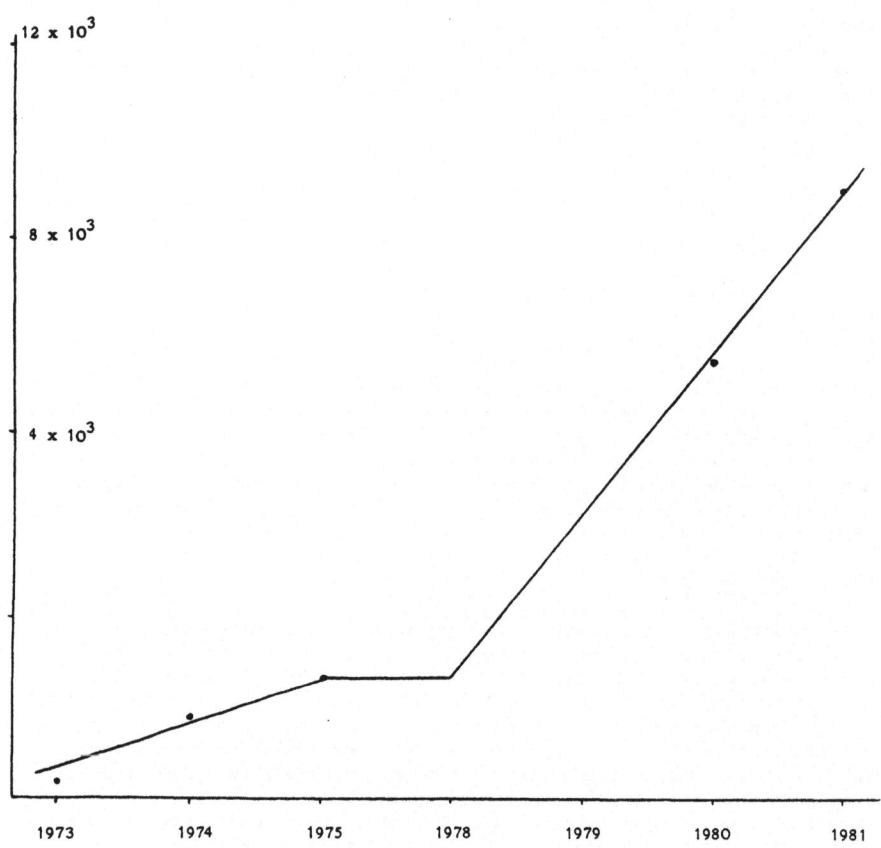

Fig. 1   Rate of growth of ICLOPS

Objectives 1 and 4 in Table 3 have already been achieved.

The RTECS (Register of Toxic Effects of Chemical Substances) referred to in objective 8 (Table 3) is available on magnetic tape. It is hoped that it may be edited in order to produce a third volume of additional data on the organic compounds already in CICLOPS. This might include synonyms, molecular formula, molecular weight, chemical abstracts systems numbers, etc.

Table 3. Proposed objectives and completion dates

	Objective	Completion date
1.	Transfer search and retrieval programs from Rank Xerox 530 to DEC VAX 11/780 computer	June 1981
2.	Data collection	December 1983
3.	Data storage	December 1983
4.	Commission CROSSBOW programs and put into operation	June 1981
5.	Enter all structures onto CROSSBOW	December 1981
6.	Editing work	December 1983
7.	Transfer data base to ECDIN system	December 1983
8.	Editing of RTECS for provision of additional data	December 1983

REFERENCES

1. Eakin, D.R., Hyde, E., and Palmer, G. The use of computers with chemical structural information: the ICI CROSSBOW system. Pestic. Sci., 5, (1974), 319-326.
2. Town, W.G., and Penning, W. Integration of the COST 64b list of organic micropollutants into the ECDIN data base. Proceedings of 4th Technical Symposium of COST project 64b, "micropollutants", Berlin, 11-13 December, 1979.
3. Ott, H., Geiss, F., and Town, W.G. The Environmental Chemicals Data and Information Network (ECDIN) and related activities of the European Communities. In Aquatic Pollutants: Transformations and Biological Effects. Proceedings of the 2nd International Symposium on Aquatic Pollutants, Amsterdam, 26-28 September, 1977, 33-38. Ed. Hutzinger O. Pergamon Press.
4. Garrison, A.W., Keith, L.H., Shackleford, W.M. Occurrence, registry and classification of organic pollutants in water, with development of a master scheme for their analysis. In Aquatic Pollutants: Transformations and Biological Effects. Proceedings of the 2nd International Symposium on Aquatic Pollutants, Amsterdam, 26-28 September, 1977, 39-68. Ed. Hutzinger O. Pergamon Press.

STATUS OF THE COMPUTERIZED COMPILATION OF MASS-SPECTRA
OF ORGANIC POLLUTANTS.

P. GROLL

Institut für Heiße Chemie, Kernforschungszentrum Karlsruhe

Postfach 3640, D-7500 Karlsruhe

Bundesrepublik Deutschland

Summary

The objective and the status of working party 6 of the Concerted
Action 64b-bis of the European Community is summarized. Till now
there are 1383 different mass-spectra of identical or different
organic pollutants collected. Further activities will be to
continue this collection and to characterize the advantages and
disadvantages of the different library search methods.

1. INTRODUCTION

Mass-spectrometry is the most used analytical tool for the
identification of water-pollutants, which have been separated by
gas-chromatography, preceeded by a more or less sophisticated
sampling and preparation step. The evaluation of data and the
interpretation of mass-spectra of up to one hundred or more
substances per gas-chromatogramm is a very slow and time consuming
job.

To facilitate the evaluation of mass-spectra these have been
collected since the fifties or earlier. In the early period this
collections have been books. When computers have been become
available and cheaper, computerized retrievals have been developed.
This software has been developed more and more, starting from very
primitive algorithm. Many of this software-packages are used at

different analytical laboratories with more or less success. Dr. Henneberg will give an overview of the different systems and discuss their advantages in his paper.

2. STATUS OF SPECTRA COLLECTION

To increase the probability of successful use of this retrieval software it is necessary to develop it further. On the other hand the collection of mass spectra has to be continued, for most of these algorithm are using data collections as a base for their work. 10000 to 30000 mass spectra of organics are, compared with the number of substances possibly polluting surface water, only a small percentage.

When we started in 1973 the work in the COST-action 64 b, the work should be terminated within 2 years. Under this condition development of retrieval interpretative software has been impossible, for this work would take much more time. The only work which could be successfully terminated within this time was the collection of mass-spectra of substances, which had been analyzed already as pollutant of water. During the past time about 2000 mass-spectra of organic substances have been collected.

In this collection there has been quite a lot of similar mass-spectra of identical substances. The elimination of these data reduced the collection down to 1383 mass-spectra. The origin of the spectra of the resulting collection is given in the following table.

Laboratory	Number of Spectra
National Food Institute	
Soeborg, Denmark	256
Kernforschungszentrum Karlsruhe	
Karlsruhe, Fed. Rep. of Germany	31
Centre d'Etudes Nucleaires de Grenoble	
Grenoble, France	326

Laboratory	Number of Spectra
Centre International de Recherche sur la Cancer	
Lyon, France	23
Cerchar	
Creil, France	86
Laboratoire de Physico-Chimie Instrumentale	
Paris, France	59
Universite Paris VI - Batiment F	
Paris, France	33
CCR Euratom	
Ispra, Italy	44
Institut "Rudjer Boscovic"	
Zagreb, Jugoslawia	28
Institut za Medicinska - Istrazivanja Jazu	
Zagreb, Jugoslawia	11
Zawod za Hemiju PMF	
Novi Sad, Jugoslawia	1
National Institute of Public Health	
Bilthoven, Netherlands	330
Rijks Instituut voor Drinkwatervoorziening	
Den Haag, Netherlands	7
Central Institute for Industrial Research	
Oslo, Norway	30
Centro des Estudios Hidrograficos	
Madrid, Spain	27
Stockholms Universitet - Lilla Frescatti	
Stockholm, Sweden	31
EAWAG	
Dübendorf, Swiss	25
Water Research Center	
Medmenham, United Kingdom	34
Water Research Center	
Stevenage, United Kingdom	1

2. FURTHER ACTIVITIES

To facilitate the comparison of different search algorithm a test is performed. The following persons or institutions are taking part:

D. Henneberg, Max-Planck-Institut für Kohleforschung
D-6433 Mühlheim/Ruhr, Federal Republik of Germany

F.W. Mc.Lafferty, Cornell University, Dep. of Chem.,
Ithaca, New York14853, U.S.A.

St.R. Heller, U.S. Enviromental Protection Agency
MIDSD, PM-218, Washington, D.C.20460, U.S.A.

FINNIGAN GmbH, Benzstraße 28,
D-8039 Puchheim bei München, Federal Republic of Germany

VG Anal. Ltd., Tudor Road 28, Altrincham, Cheshire WA145RZ
Great Britain

The result will be discussed with the participating laboratories in spring 1982.

CURRENT STATUS OF THE COMPILATION OF REFERENCE DATA

A. CORNU
Commissariat à l'Energie Atomique
Centre d'Etudes Nucléaires de Grenoble

Summary

1. Compilation of organic micropollutants mass spectra.

   Our file contains 1884 mass spectra; a detailed list is given.
   A substantial increase of contributions is strongly desirable.

2. Index of the same mass spectra.

   It will be shortly completed, for a convenient access to the
   compilated data, through three entries.

3. For the French data collection, we did include it in the Pluridata
   Bank.

4. Compilation of collision induced mass spectra.

   It would be interesting to decide if, or not, we undertake the
   compilation of such spectra.

5. International exchange.

   Prof. Mac LAFFERTY asks for the communication of our spectra for
   including them into the CORNELL 55000 Spectra Collection.

1. COMPILATION OF THE MASS SPECTRA OF ORGANIC MICROPOLLUTANTS IN THE FORM OF STANDARDIZED TABLES

Compilation work is continuing. Our collection now comprises 1884 mass
spectra produced by the laboratories listed in Table No 1.

It should be pointed out that the contribution of the various laboratories
is losing momentum and it would be desirable for the laboratories of all
the participating countries working in this field systematically to
send us the mass spectra of pure compounds so as to increase this collection,
which is proving extremely useful in a large number of laboratories.

Since the beginning of 1981 these spectra have been disseminated by
the Committee Secretariat. The laboratories greatly appreciate this wider
distribution and would appreciate this reference document, which is bound
to be handled very frequently in the working area next to the mass
spectrometers, being reproduced in a more easily readable and more
durable form.

2. PREPARATION OF AN INDEX OF MASS SPECTRA

In view of its size, it is obvious that our collection can be used only if there is an effective means of gaining access to the data it comprises — a sort of sophisticated table of contents.

In line with the wish recently expressed by the Committee, we have resumed work on preparing an exhaustive index of our collection, comprising, for each of the 1800 spectra :

- the name of the compound;

- its COST reference number;

- its molecular mass;

- its empirical formula;

- a list of the 10 highest peaks classified in decreasing order of height and the corresponding heights.

This index consists of three parts, each with a different entry :

- the molecular mass of the compound;

- the empirical formula of the compound;

- the mass of the highest peaks in the spectrum.

This document, which is easy to use and always immediately available on the work bench, is greatly appreciated by users as an initial and rapid approach to the identification of pollutants.

The information thus obtained can then be usefully supplemented by referring to the full data, which have been put on computer by Mr. Groll.

As in the case of the spectra collection, this index will be distributed by the Committee Secretariat.

I would like to take the opportunity of pointing out that this index makes it possible to distinguish easily between the spectra of an identical substance produced by different laboratories. The question of whether it was useful to have such publications was asked on a number of occasions; experience shows that a data bank is in fact more reliable where it can take into consideration the relative varia- tions in the peak intensities caused by the experimental conditions and in particular by the different types of magnetic or quadrupolar spectro- meters, since it makes automatic library search methods more effective. This is, furthermore, the opinion of Prof. MacLafferty, who has devoted a great deal of study to this question in the USA.

3. AS REGARDS THE FRENCH LABORATORIES, we took the decision to incorporate

our entire collection of spectra into the DARC-Pluridata data bank managed by the Université de Paris VII under the leadership of Prof. Dubois.

Table 1: List of laboratories submitting mass-spectra on July 6, 1981

			Number of spectra
CH 1	E.A.W.A.G.		nil
D 1		Bundesgesundheitsamt	nil
D 2	WABOLU		nil
D 3		Gesellschaft für Kernforschung	31
DK 1		Institut for Farmacologi og Toksicologi den Kgl. Veterinaer og Landbohoejskole	nil
DK 2	N F I	National Food Institut	234
E		Centro de Estudios Hidrográficos	41
EIR 1		An Foras Forbatha	nil
EUR 1	C.C.R.	Centre Commun de Recherches (Ispra)	44
F 1	C.E.N.G.	Centre d'Etudes Nucléaires de Grenoble	434
F 2	CERCHAR	Centre Etudes et Recherche des Charbonnages de France.	246
F 3		Laboratoire de Recherches de Spectrochimie Moléculaire. Faculté des Sciences - Paris	6
F 4		Laboratoire de Physico-Chimie Instrumentale - Paris VII	60
F 5		Laboratoire de Chimie Organique Structurale - Paris VI	31
F 6		Service des Cancérogènes de l'Environnement Centre International de Recherche sur le Cancer - Lyon	27
F 7		Laboratoire de Chimie Structurale - Université de Nice	55
F 8	C.N.R.S.	Institut de Chimie des Substances Naturelles Gif-sur-Yvette	7
F 9		Laboratoire de Chimie Organique Hétérocyclique et Laboratoire de Synthèse et d'Etudes Physico-Chimiques d'Hétérocycles azotés - Université de Montpellier	61
F 10		Laboratoire de Chimie Organique de Synthèse - Université Paris-Sud-Orsay et Laboratoire de Synthèse Organique - Ecole Polytechnique - Palaiseau	19
F 11		Laboratoire de Chimie Appliquée - Université de Franche Comté - Faculté des Sciences et Techniques - Besançon et Département de Chimie Organique - Université Claude Bernard - Villeurbanne	15
GB 1	W.R.A.		35
GB 2	W.R.C.	Stevenage Laboratory - Stevenage, Herts	140
I 1	C.N.R.	Instituto di Ricerca sulle Acque	nil
I 2		Institute of Analytical Chemistry	nil
N 1		Sentralinstitutt for Industriell Forskning	30
NL 1		Rijks Instituut voor de Volksgezondheid	319
NL 2		Rijks Instituut voor Drinkwatervoorziening	6
P 1		Instituto Hidrografico	nil
S 1		Wallenberglaboratoriet - Stockholms Universiteit	31
Y		Institute for Medical Research and Occupational Health	12

1884 spectra

This integration of our mass spectra enables us to exploit the other
data stored in this bank (NMR spectra and crystallographic data)
and to conduct structural research using the DARC method. This could
prove to be of value for the study of degradation products and metabolites,
in the search for the initial substances of a pollutant that has been
found or, at the other end of the scale, the prediction of the possible
toxic by-products. Prof. Dubois is, moreover, to present a paper on this
subject to this Symposium.

4. COMPILATION OF COLLISION SPECTRA

The various methods for identifying the constituents of a complex
mixture without prior chromatographic separation are being developed
throughout the world under the general designation of "MS/MS techniques".

The final identification is based on the observation of a mass spectrum
produced by collision-induced disintegrations (CID). These collision
spectra are similar but not identical to the spectra obtained by electron
impact. It will therefore have to be decided whether we are to undertake
also to compile collision spectra for the molecules of the organic
pollutants concerned in our project.

5. INTERNATIONAL EXCHANGES OF DATA BY THE COST COMMITTEE

We have received a request from Prof. MacLafferty to send a copy of our
collection to Cornell University. The action to be taken is to be discussed at the next meeting of the Committee.

6. GAS CHROMATOGRAPHY REFERENCE DATA

The study group on gas chromatography has not yet determined the way
in which chromatographic data are to be standardized and presented; this
is why progress has not yet been made in this sector.

7. CONCLUSION

The work on the compilation of the mass spectra of organic pollutants and
of all related substances (precursors, degradation products and metabolites)
is progressing satisfactorily. Thanks to the wider distribution of these
documents by the Committee Secretariat, this information is being used
by an increasing number of laboratories and providing a concrete example
of the international cooperation that we have established.

SESSION VI - SPECIFIC ANALYTICAL PROBLEMS

A. Organic halogens

Chairman: F. ZUERCHER, EAWAG, Switzerland

Review paper :

- Determination of organic halogens; a critical review of sum parameters

Short paper :

- Volatile halogenated hydrocarbons in river water, ground water, drinking water and swimming-pool water in the Federal Republic of of Germany

Poster papers :

- Simultaneous determination of total purgable organo-chlorine, -bromine and -fluorine compounds in water by ion-chromatography

- Etude des trihalomethanes dans l'eau potable de Barcelone - Evolution de l'efficacité des filtres à charbon actif

- Analysis of trihalomethanes formed during drinking water chlorination

B. Phenolic and other compounds

Chairman: G. SUNDSTROM, National Swedish Environment Protection Board, Sweden

Review paper :

- Determination of phenolics in the aquatic environment

Short papers :

- Analysis of alkylphenols in an aqueous matrix containing aromatic hydrocarbons

- Analysis of nitrogenous organic substances in water

- The electroanalysis of organic pollutants in aquatic matrices

Poster papers :

- Determination of nonylphenols and nonylphenolethoxylates in secondary sewage effluents

- Determination of phenols in water by HPLC

- Surfactants survey in the surface waters in the Paris-area

DETERMINATION OF ORGANIC HALOGENS; A CRITICAL REVIEW OF SUM PARAMETERS

R.C.C.Wegman
Unit for Residue Analysis, National Institute of Public Health,
3720 BA Bilthoven, the Netherlands

Summary

A review is given of analytical methods which are used for the determination of the sum of halogenated organic compounds in water samples. Almost all of the analytical procedures mentioned in the literature e.g. total organic chlorine (TOCl), extractable organic chlorine (EOCl) and purgeable organic halogen (POX) include three steps, namely, enrichment, conversion of organic-bound halogen into halide-ion and detection of halide-ion. For the enrichment step several methods are used such as liquid-liquid extraction, adsorption and the purge technique. Conversion is carried out by combustion in an oxygen atmosphere in a furnace at a high temperature, by combustion in a hydrogen/oxygen flame and by saponification with sodium. Detection is carried out by means of the microcoulometer, ion selective electrodes, titration, conductive measurement after ion chromatography and neutron activation analysis. This paper reports the present state of knowledge concerning these sum parameters and it gives a critique of the methods used.

1. INTRODUCTION

Organic compounds containing halogen can be expected in "chemically polluted waters"; they rarely occur naturally. The majority of halogenated organic compounds in surface water are either chemicals discharged by various industries like chemical plants, paper mills, etc. or are agriculturally related compounds, such as pesticides. Volatile halogenated organic compounds are used as solvents, and cleaning agents and they are intermediate in the production of various end-products.

Since the discovery of haloforms in drinking water (1) there has been intense interest in their source. The halomethanes have been shown to be formed during the chlorination process, used in water treatment plants.

The presence of halogenated organic compounds must be regarded as undesirable because many of these compounds are considered to be toxic, mutagenic and carcinogenic and hence of concern for public health. Their importance is reflected in the fact that more than half of all priority pollutants, as designated by the U.S. EPA, are halogenated organics (29). Sophisticated and expensive analytical techniques have been developed to detect and identify these contaminants. A lot of halogenated organic compounds have been identified in surface water, discharge water, ground water and drinking water. Despite this a large portion of the halogenated organic compounds are still unidentified. Thus, while their complete identification and the determination of their amounts is desirable it appears that such comprehensive information is neither technically nor economically feasible as a routine evaluation and control tool.

Group parameters are especially useful in determining the presence amount of halogen compounds since they can be determined within a short time with relatively simple equipment. A disadvantage of this approach is the lack of information it gives concerning the multitude of specific halogenated organic compounds present.

Almost all the analytical procedures mentioned in the literature include three steps, namely: 1) enrichment 2) conversion of organic-bound

Table 1 Summary of abbreviations and descriptions of methods for the determination of the sum of organohalogen compounds.

Reference Author	Year	Abbreviation	Description	Examples of compounds covered (recovery >80%)
Greve (3)	1972		organic bound halogen	not given
Kühn (4,5)	1973/77	TOCl	total organic chlorine	CHCl₃, dichlorophenol
Lunde (6)	1975		chlorinated and brominated non-polar hydrocarbons	not given
Wegman (7)	1977	EOCl	extractable organic chlorine	HCB
Glaze (8)	1977	TOX	total organic halogen	CHBr₂Cl, dichlorobenzene, 4-chlorophenol, 2,2'-dichloroisopropylether
Dressman (9)	1977	CAOX	carbon adsorbable organohalides	PCP
Selenka (10)	1978		organic bound chlorine	CHCl₃, C₂Cl₄, C₂HCl₃, CCl₄'
Fritschi (11)	1978		a) non-volatile organochlorine compounds	HCB
			b) volatile organochlorine compounds	CHCl₃, CCl₄, tri- and tetrachlorobenzene
Kussmaul (12)	1979		1) extractable non-volatile fluoro-, chloro-, and bromocarbon compounds	not given
			2) volatile fluoro-, chloro-, and bromocarbon compounds	not given
Wegman (13)	1979	VOCl	volatile organic chlorine	CHCl₃, CHBrCl₂, CH₂Cl₂
Takahashi (14)	1979	TOX	total organic halides	not given
Dressman (15)	1979	OX	total organohalides	CHCl₃, CHBr₂Cl, CHBr₂Cl, PCP, bis(2-chloroethyl)ether
		POX/NPOX	purgeable and non-purgeable fraction of OX	not given
Jekel (16)	1980	POX	purgeable organic halogen	CHCl₃, CCl₄
		NPOX	non-purgeable organic halogen	bromophenol, dichlorophenol
		TOX	POX + NPOX	
Zürcher (17)	1980	HEOCl	hexane extractable organic chlorine	HCB, PCB's

Continuation Table 1.

Summary of abbreviations and descriptions of methods for the determination of the sum of organohalogen compounds.

Reference	Year	Abbreviation	Description	Examples of compounds covered (recovered >80%)
Author				
Zürcher (18,19)	1980	VOCl/VOBr/VOF	volatile organochlorine-, bromine- and fluorine compounds	$CHCl_3$, CH_2Cl_2, C_2Cl_4, chlorobenzene, dichlorobenzene
EPA (20)	1980	TOX	total organic halide	$CHCl_3$, $CHBrCl_2$, $CHBr_3$, $CHBr_2Cl$, PCP
Van Steenderen (21)	1980	TOH	total organohalogen	$CHCl_3$, $CHBr_2Cl_2$, $CHBrCl_2$, $CHBr_3$, CCl_4
Jaeger (22)	1980	EOCl	total extractable organochlorine compounds	$CHCl_3$, C_2Cl_4, $C_2H_2Cl_2$, CCl_4
Veenendaal (23)	1981	VOCl	volatile organic chlorine	$CHCl_3$, C_2Cl_4, CH_2Cl_2, chlorobenzene
Smeenk (24)	1981	AOCl	adsorbable organic chlorine	2,7-dichlorofluorescene

Besides the methods in the literature which were described in detail, some other methods and their abbreviations were mentioned by Kühn et al. (26): DOCl (dissolved organic chlorine = TOCl in earlier work) and DOClN (dissolved chlorine-non-polar = EOCl) and by Dressman et al. (15): PCAOX (purgeable carbon adsorbable organohalides, method under investigation), NPCAOX (non-purgeable carbon adsorbable organohalides and ATOX (appearent organohalides = POX + NPCAOX).

Table 2 Methods with an extraction step.

Reference Author	Year	Sample volume	pH	Extraction	Concentration	Combustion	Detection	Detection limit ($\mu g\ l^{-1}$)		Remarks
Greve (3)	1972	1000		3 x 100 ml of PE (40-60) or 3 x 100 ml CH_2Cl_2 by shake	Kuderna-Danish evaporation Kuderna-Danish evaporation	O_2/Ar at 850°C O_2/Ar at 850°C	microcoulometer microcoulometer	0.3 0.3	Cl Cl	
Wegman/Greve (7)	1977	1000		1 x 200 + 2 x 100 ml of PE by shake	Kuderna-Danish	O_2/Ar at 850°C	microcoulometer	1	Cl	
Fritschi (11)	1978	2300		160 ml of diiso-propyl ether by rotation-perforator	waterbath at 40°C and N_2	O_2/N_2 at 950°C	microcoulometer	0.5	Cl	Cl_2 in sample removed by sodium-sulfite
Kussmaul (12)	1979	1000-5000		160 ml of diiso-propyl ether by rotation perforator	waterbath 40°C and N_2	O_2, Pt quartz-wool at 1100°C	ion-selective electrode	2 0.2 0.2	Cl Br F	
Van Steenderen (21)	1980	125		2 ml of PE by tumbling machine	none	Ar/O_2 at 850°C	microcoulometer	10	Cl	
Jaeger (22)	1980	1000		2 x 30 ml of PE (30-50) by magnetic stirrer dried; addition of 10 ml diiso-propanol	extract saponified with 0.5 g Na and concentrated at rotary evaporator	none	a) titration b) microcoulometer	100 10	Cl Cl	sulfite-ion removed by H_2O_2

Continuation Table 2.

Methods with an extraction step.

Reference Author	Year	Sample volume	pH	Extraction	Concentration	Combustion	Detection	Detection limit ($\mu g \ l^{-1}$)	Remarks
Zürcher (17)	1980	1000–5000	5	flocculation by aluminum sulfate next pH 7; filtration (0.45 µm); dried at 105°C soxhlet extraction with 25 ml of hexane	at rotary evaporator	Wickbold apparatus (O_2/H_2) flame	titration	1	Cl before extraction purging with N_2 at 30°C
Crane (25)	1981	2000		30–50 ml of PE by shake	evaporation	O_2/Ar at 850°C	microcoulometer	0.5	
				250 ml of ethyl ether by shake		O_2/Ar at 850°C	microcoulometer	0.5	
Lunde (6)	1975	200000		200 ml of cyclohexane by continuous extraction	Kuderna-Danish	none	neutron activation analysis	0.05	CL Cl^- in extract was
								0.01	Br removed by H_2O washing

PE = petroleum ether (br. 40–60°C)

halogen into halide ions 3) detection of halide ions.

For the enrichment step several methods are used, such as purge techniques, liquid-liquid extraction and adsorption.

The conversion of organic-bound halogen into halide ions is carried out by a) pyro (hydro) lysis at high temperatures in a furnace b) combustion in a H_2/O_2 flame c) saponification with sodium.

The detection of halide ions is carried out by a) microcoulometer b) ion selective electrodes c) titration d) ion chromatography with conductance measurement. Neutron activation analysis can be used without the conversion of organic-bound halogen into halide ions.

This paper will report the present state of knowledge concerning the sum parameter for the determination of organic halogens in water samples and it will give a critique of the methods used.

2. PRESENT OF STATE OF THE TECHNIQUES

In the literature a great number of methods for the determination of the sum of halogenated organic compounds are described. This growth of methodology has also led to a great number of abbreviations, like TOCl, TOX, EOCl, etc. In Table 1 a summary of abbreviations, a short description of each of the methods and some examples of compounds which are covered with a recovery of over 80% are given. In order to compare the different methods, the analytical procedure and experimental conditions are summarized in three tables. The methods are arranged according to the enrichment technique.

2.1. METHODS WITH AN EXTRACTION STEP (see Table 2)

The first technique for the determination of the sum of halogenated organic compounds in water samples was that described by Greve and Haring (3) in 1972. The first step involved extraction by petroleum ether or dichloromethane. After concentration by means of a Kuderna-Danish apparatus a small amount of the concentrated extract was pyrolysed to convert the organohalogens into halide ions. These ions were measured with a microcoulometer. Wegman and Greve (7) tested an improved the method and found that the recoveries for bromine-, jodine- and fluorine-compounds are poor. Fritschi et al. (11) used another extraction solvent namely diisopropylether. Van Steenderen (21) diminished the volume of the extraction solvent/water sample ratio and avoided the concentration step. With his method it was also possible to measure volatile halogenated organic compounds. However the extraction could take a long time because of the slow kinetics to reach equilibrium. Kussmaul and Hegazi (12) selected another detection method namely one with ion-selective electrodes, in order to be able to discriminate between the halogens.

Jaeger and Hagenmaier (22) replaced the pyrolysis of the extract by a saponification with sodium. The detection was carried out by titration and microcoulometer.

Zürcher (17) was only interested in the non-volatile, lipophilic compounds. The water sample was purged with nitrogen and a flocculation was then carried out at pH 5. After filtration the solids werd dried at 105°C and extracted with hexane. Conversion of the organic-bound halogens was carried out in a H_2/O_2 flame (Wickbold apparatus).

A quite different detection technique was used by Lund and Gether (6). After extraction with cyclohexane and concentration, Cl and Br were measured by neutron activation analysis.

2.2. METHODS WITH AN ADSORPTION STEP (see Table 3)

The first attempt to measure total organic chlorine (TOCl) as an indicator of water quality was described by Kuhn and Sontheimer (4). Their initial work was directed at finding a method for determing the TOCl concentration adsorbed onto samples of granular activated carbon taken from the full-scale activated carbon of sampler units. Later the method was modified for the analysis of water samples (5). The first step involved the adsorption of the organic compounds from the water onto ground granular activated carbon. This was accomplished by adding 2 g of activated carbon to a 20 l water sample and stirring for 1 h. The carbon was recovered by a process of flocculation, sedimentation and membrane filtration and it was then pyrohydrolyzed to convert organic chlorine into chloride-ion. Cl^- was measured by means of an ion-selective electrode. To account for the interference from inorganic Cl^- adsorbed onto the activated carbon and present in the pore water a duplicate sample was extracted in the same way and the recovered activated carbon was washed in a solution of sodium nitrate. In this wash process NO_3^- displaces inorganic Cl^- from the carbon. The Cl^- in the wash water is measured. The amount of Cl^- from organic bound chlorine was then calculated by substrating the amount of Cl^- measured in the wash water from the amount of Cl^- measured in the pyrohydrolyzate ("Difference method").

Dressman et al. (9) repeated the Kuhn-method and reported some improvements namely:
- adsorbed Cl^- as well as the Cl^- in the pore water of the carbon were removed by washing with nitrate solution prior to pyrohydrolysis ("direct method")
- the halide ion measurement was carried out with a microcoulometer because they found that the chloride selective electrode gave erroneously high Cl^- readings and was affected by addition of $CuSO_4$ and H_2O_2
- the carbon adsorbable organic halides (CAOX) method was applied to the non-purgeable fraction only because the recoveries for halomethanes were low.

Takahashi (14) shortened the length of the carbon adsorption step drastically and increased the response for organobromide compounds. The improvements were:
- adsorption of organic compounds from water onto granular activated carbon packed in a micro column
- simple nitrate working of the column to remove inorganic Cl^-
- combustion of sorbed organic compounds along with the carbon at changed combustion conditions namely the carbon is heated at 200°C with a CO_2 carrier gas, then the carbon is heated at 800°C with CO_2 and finally the carbon is completely oxidized with O_2 at 800°C.

Under these combustion conditions brominated organic compounds also had a quantitative recovery. With this method the extraction is scaled down to a 100 ml sample.

Dressman et al. (15) compared the method improved by Takahashi with the Kuhn/Dressman method and concluded:
- adsorption efficiency for 2 microcolumns is similar to two batch extractions
- average recovery percentages are very similar
- the simple nitrate wash of the micro column effectively removes inorganic Cl^-
- the adsorption of halogenated organic compounds is not complete, especially at high concentrations of organically bound halogen (greater than 200 μg l^{-1}) and the adsorption efficiency of volatile halogenated organic compounds decrease.

These conclusions indicated a need for a special methodology to account for volatile halogenated organic compounds when the total organohalide concentration is greater than 150 μg l^{-1}.

Smeenk (24) purged the water samples before the adsorption step to remove the volatile organohalogen compounds which have a low recovery.

Glaze et al. (8) used a macro reticular resin (XAD) to adsorp organic compounds from water and then eluted these organic compounds with a small amount of ether. In this method XAD is also packed in a microcolumn, a small water sample is forced through the column and finally, the XAD column is washed with ether. The recovery of $CHCl_3$ and CCl_4 were below 80% and those for dodecylchloride and hexadecylchloride were much lower. With the exception of the Kuhn-method all the methods with an adsorption step included the determination of the halide ions by microcoulometer.

Table 3 Methods with an adsorption step.

Reference Author	Year	Sample volume (ml)	pH	Adsorbent	Adsorption Time (min)	Adsorption Procedure	Elution (el.)/combustion (comb.)	Detection	Detection limit ($\mu g\ l^{-1}$ Cl)	Remarks
Kühn (5)	1977	20000	5	2 x 2 g of powdered carbon (<60 μm)	60	adsorption; flocculation by aluminum sulfate and polyacrylamide at pH 6.5	comb. O_2/steam at 700/1000°C	ion selective electrode	20	nitrate wash*
Glaze (8)	1977	120		0.2 g of 100-120 mesh XAD-2 or XAD-4	5	column	el. with 1 ml of ethyl ether	microcoulometer	2	addition of sulfite**
Dressman (9)	1977	10000	6.5-7.5	2 x 1 g of ground granular act. carbon	60	adsorption; flocculation by aluminum sulfate and polyacrylamide flocc. aid	comb. O_2/steam at 700/1000°C	microcoulometer	10	nitrate wash* interfering H_2S is removed
Takahashi (14)	1979	25-100		2 x 40 mg of 100-200 mesh act. carbon	15-30	column	comb. 1st CO_2/steam at 200°C; 2nd CO_2/steam at 800°C; 3rd O_2/steam at 800°C	microcoulometer	2	nitrate wash*

Continuation Table 3.

Methods with an adsorption step.

Reference Author	Year	Sample volume (ml)	pH	Adsorbent	Adsorption Time (min)	Adsorption Procedure	Elution (el.)/combustion (comb.)	Detection	Detection limit (μg l^{-1} Cl)	Remarks
Dressman (15)	1979	120	2	40 mg of 100-200 mesh activated carbon	40-50	column	comb. 1st CO_2/ at 200°C;2nd CO_2/O_2 at 800°C	microcoulometer	5	nitrate wash* addition of sulfite**
Jekel (16)	1980	40	3	2 x 20 mg of 100-200 mesh act. carbon	45	filtration of carbon 0.4 μm membrane	comb. O_2/Ar (1:1) at 850°C	microcoulometer	35	combined with purge-technique nitrate wash*
EPA (20)	1980	100	2	2 x 40 mg of 100-200 mesh act. carbon	35	column	comb. 1st CO_2/ steam at 200°C; 2nd O_2 at 1000°C	microcoulometer	5	nitrate wash* addition of sulfite**
Smeenk (24)	1981	1500	2	2 x 150 mg powder carbon		before adsorption N_2 purging and addition of nitrate sol.	comb. O_2/steam at 900°C	microcoulometer	10	nitrate wash* SO_2 is removed by H_2O_2

*. inorganic Cl$^-$ is removed by nitrate wash
** sulfite is used for removal of Cl_2 from water sample

2.3. METHODS WITH A PURGE TECHNIQUE STEP (see Table 4)

In 1974 Bellar and Lichtenberg (2) reported an attractive technique for detecting volatile compounds which involves gas stripping of the water sample and adsorption of the organic compounds on a sorbent medium such as Tenax-GC, followed by thermal desorption and analysis by gas chromatography.

Wegman and Hofstee (13) and Veenendaal (23) used this purging and trap technique in combination with a combustion furnace and a microcoulometer. Wegman purged the water samples at 60°C, Veenendaal at 90°C. To avoid excessive moistening of the Tenax-GC at 90°C, a cooler was placed before the sorbent. Selenka and Bauer (10) reported the use of XAD-2 as sorbent for the purged halogenated organic compounds. Desorption was carried out by elution with pentane.

Fritschi et al. (11) and Jekel and Roberts (16) reported a method which involves a purging step directly followed by combustion and measurement with a microcoulometer. Kussmaul carried out the same method but replaced the microcoulometer by ion-selective electrodes in order to discriminate between the halogens Cl, Br and F.

Zürcher (19) replaced measurement with the microcoulometer by a potentiometric measurement.

Zürcher (18) also reported a method which involved purging of the water sample and the combustion and measurement of Cl, Br and F after ion chromatographic separation.

Steadily with these various methods more volatile halogenated organic compounds were determined.

3. DISCUSSION

3.1. LIQUID-LIQUID EXTRACTION

For enrichment the liquid-liquid extraction is a simple and conventional technique. However the types of compounds which can be extracted from water samples depends on a great number of factors, such as, type of water sample (presence of particles, ionic strength of the water), pH, type of extraction solvent (polarity of solvent is a main factor), solvent-water ratio and extraction procedure. After the extraction a great loss of more volatile compounds is possible if the method includes a concentration step.

All authors mentioned in Table 2 with the exception of Zürcher (17) carried out an extraction with the water samples as such without adjusting the pH. Extraction at approximately pH 7 can be inefficient for some classes of compounds e.g. phenols, which means that consecutive extraction e.g. at pH $\leqslant 2$ and pH $\geqslant 11$ may be advisable.

Petroleum ether as extraction solvent had the advantage that formation of emulsion was rarely observed (7, 21 and 22). However only the more lipophilic compounds can be extracted.

A low solvent/water ratio can lead to the formation of emulsion and extraction may take a long time because of the slow kinetics to reach equilibrium, especially for high molecular weight organic compounds. If a high solvent/water ratio is applied a concentration step is usually necessary in order to make the method sensitive. Concentration by evaporation led to major losses of the more volatile compounds and could lead to the concentration of impurities in the extraction solvent. Especially evaporation to dryness gave poor results.

Greve and Haring (3), Lunde and Gether (6), Wegman and Hofstee (7), Fritschi et al. (11), Kussmaul and Hegazi (12) and Crane (25) reported concentration by evaporation.

Comparison of the methods summarized in Table 2 led to the conclusion that the extraction and concentration procedure described by Jaeger and Hagenmaier (22) which includes a liquid-liquid extraction and then a concentration step after the conversion of volatile organohalogen compounds into halide ion; probably gave the best results. The method described by

Table 4 Methods with a purge technique step.

Reference Author	Year	Sample volume (ml)	Purge T (°C)	Purge time min	Purging gas, volume (ml min⁻¹)	Sorbent	Desorption/ elution	Combustion	Detection	Detection limit (µg l⁻¹)	Remarks
Fritschi (11)	1978	5	20	5	N_2, 150	none	none	O_2/N_2 at 950°C	microcoulometer	0.5 Cl	
Selenka (10)	1978	2000	60	60	N_2, 180	1.5 g of XAD-2 at 0°C	10 ml of pentane	O_2/N_2 at 950°C	microcoulometer	1.0 Cl	
Wegman (13)	1979	60	60	2 x 20	He, 40	0.3 g of Tenax	thermal at 190°C	O_2/Ar at 850°C	microcoulometer	0.05 Cl	
Kussmaul (12)	1979	1000	70	60	N_2, 150	none	none	O_2 at 1100°C	ion selective electrode	0.2 F 2 Cl 0.2 Br	
Zürcher (19)	1980	1000	60	30	N_2/air, 1000	none	none	N_2/air at 1000°C	potentrometric	0.6 Cl	Cl_2 removed by sodium thiosulfate
Zürcher (18)	1980	100–1000	60	10–30	O_2, 1000	none	none	O_2 at 950°C	conductance measurements after ion-chromatographic separation	0.02 F 0.02 Br 0.05 Cl	
Jekel (16)	1980	50	±20	10	O_2/Ar, 200	none	none	O_2/Ar at 850°C	microcoulometer	1.0	combined with AOCl
Veenendaal (23)	1981	100	90	15	N_2, 40	0.3 g of Tenax	thermal at 300°C	O_2/Ar at 850°C	microcoulometer	0.10	cooling between purge vessel and sorbent

Lunde and Gether (6) does not need a concentration step but neutron activation analysis is not a simple method for routine analysis.

3.2. ABSORPTION

From the methods summarized in Table 3 it follows that only carbon (various quality) and XAD-2 and 4 were used as sorbent. Desorption of the organic compounds from carbon was carried out by combustion, the desorption from XAD by elution with ethylether. Commercially 5 types of XAD are available namely XAD-1, 2, 7 and 8. However only XAD-2 and 4 were used for the sum determination of halogens.

Melton et al. (27) reported a comparison of experimental results from the analysis of drinking water before and after water treatment with granular activated carbon at the Cincinnati Water Works. The following methods of organic analysis were used a) Grob closed-loop stripping analysis (CLSA) b) Bellar purge and trap (P&T) c) batch liquid-liquid extraction with dichloromethane (BLLE) d) XAD-2 adsorption-ethylether solution (XAD-EEE). The individual compounds were measured bu the capillary GC/MS. Of the 183 different organic compounds which were measured by the four methods, 6 organic compounds were detected by the P&T method, 107 by the CLSA method, 90 by BLLE and 58 by XAD-EEE analysis. The CLSA method with the carbon trap (with CS_2 elution) gave the best results except that the recovery for highly volatile compounds, such as chloromethane, vinylchloride, dichloromethane and chloroform was poor. Noordsij (28) who reviewed all recovery data for compounds isolated by XAD found that the recoveries were poor for many organic compounds.

From all these studies and from the methods summarized in Table 3 it can be concluded that the best method is the procedure described by Takahashi (14) and tested by Dressman (15) but extended with a purge technique step for determination of the more volatile compounds. However there are still some faulty aspects of this technique:
- the repeatability of the blank value
- the fact that specific requirements are necessary to prevent the introduction of contaminations.

3.3. PURGE TECHNIQUE

For enrichment the purge technique is a very fast method to use for compounds which have low water solubility (<1%, with exception of CH_2Cl_2), a low boiling point (<180°C), a high density (d_{20} mostly >1.0) and a Henry coefficient (=concentration in air/concentration in water at 25°C) H >0.05 (19).

Important experimental conditions are: purge temperature, purged gas volume/water sample volume ratio and, when using a sorbent, the type and quantity of this sorbent and the desorption or elution of the adsorbed compounds.

From the methods summarized in Table 4 it can be concluded that:
- a purge temperature of 60°C is usually applied
- a sorbent is not strictly necessary
- it must be possible to come to one normalised method for a purge technique because there is only a small variation in purging condition for most of the methods described.

3.4. DETECTION

The methods summarized in Tables 2-4 include the following detection techniques:
- microcoulometer
- ion selective electrodes
- titration
- conductance measurement after ion chromatographic separation
- neutron activation analysis.

With the exception of the non-destructive neutron activation analysis all methods used a conversion step of organohalogens into halide ions. The methods for conversion are: pyrolysis, saponification with sodium and Wickbold combustion. The advantage and disadvantages for these detection techniques are summarized in Table 5.

Table 5 The advantages and disadvantages for detection techniques.

Detection	Advantage	Disadvantage
microcoulometer	high sensitivity ng-range fast	low recovery for Br, J and F, no discrimination between halogens interference of high sulfur concentrations
ion-selective electrodes	- discrimination between Cl, Br and F (however restricted!)	- interference of sulfur compounds - non reliable results (9)
titration	instrumentation is · cheap	no discrimination between halogens labour intensive
conductance measurement after ion-chromatography	simultaneous analysis of Cl, Br, F, nitrogen and sulfur	interference of N_2O expensive
neutron activation analysis	- non-destructive - high sensitivity because large aliquot of sample can be used for measurement - discrimination between Cl and Br	expensive, not suitable for routine analysis

None of the authors used photometric detection. Of the detection techniques mentioned above conductance measurement after ion chromatography seems to be the most attractive.

4. CONCLUSIONS

- It is not possible to determine the sum of halogenated organic compounds in water samples with any of the methods described in the literature.
- A reliable method for the determination of the sum of halogenated organic compounds in water samples seemed to be a method with a) a purge step b) an adsorption step with carbon as sorbent c) a combustion step d) conductance measurement after ion-chromatographic separation of each of the halogens.
- In future work attention must be given to particles in the water samples. Many compounds are adsorbed to these particles.
- The number of abbreviations must be kept to a minimum.

5. REFERENCES

1) Rook, J.J. (1974) Formation of haloforms during chlorination of natural waters. J.Water Treatm.Examin., 23, 234-238
2) Bellar, T.A. and Lichtenberg J.J. (1974) Determining volatile organics at microgram-per-litre levels by gas chromatography, J.A.W.W.A. 66, 739-744
3) Greve, P.A. und Haring, B.J.A. (1972) Die mikrocoulometrische Bestimmung von organisch gebundenem Halogen in Oberflächen- und anderen Gewässern. Schr.R.Ver.Wass.-, Boden- und Lufthyg. 37, 59-64
4) Kühn, W. und Sontheimer, H. (1973) Einige Undersuchungen zur Bestimmung von organischen Chlorverbindungen auf Aktivkohlen. Vom Wasser 41, 65-79
5) Kühn, W., Fuchs, F. und Sontheimer, H. (1977) Untersuchungen zur Bestim-

mung des organisch gebundenen Chlors mit Hilfe eines neuartigen Anrei-
cherungsverfahrens. Z.f.Wasser- und Abwasser-Forschung 10, 192-194

6) Lunde, G. and Gether, J. (1975) The sum of chlorinated and of brominated
non-polar hydrocarbons   water. Bull.Environm.Contam.Toxicol. 13, 656-661

7) Wegman, R.C.C. and Greve, P.A. (1977) The microcoulometric determination
of extractable organic halogen in surface water: application to surface
waters of the Netherlands. Sci.Total Environm. 7, 235-245

8) Glaze, W.H., Peyton, G.R. and Rawley, R. (1977) Total organic chlorine as
water quality parameter. Adsorption/Microcoulometric method. Environm.
Sci.Technol. 11, 685-690

9) Dressman, R.C., McFarren, E.F. and Simons, J.F. (1977) An evaluation of
the determination of total organic chlorine (TOCl) in water by adsorption
onto ground granular activated carbon, pyrolysis, and chloride-ion measurement. Paper presented at the Water Quality Technology Conference at
Kansas City, Missoury, December 5 and 6, 1977

10) Selenka, F. und Bauer, U (1978) Erfassung leicht flüchtiger organischer
Chlorverbindungen in Wasser. In Organische Verunreinigungen in der Umwelt.
Erkennen, Bewerten, Vermindern. K.Aurand et al., Ericht Schmidt Verlag,
Berlin, 242-255

11) Fritschi, U. Fritschi, G. und Kussmaul, H. (1978) Mikrocoulometrische
Summenbestimmung von schwer- und leicht flüchtigen Organochlorverbindungen im Wasser. Z.f.Wasser- und Abwasser-Forschung 11, 165-170

12) Kussmaul, H. und Hegazi, M. (1979) Summenbestimmung leicht- und schwerflüchtiger fluorierter, chlorierter, und bromierter Kohlenwasserstoffe.
Paper presented at the First European Symposium "Analysis of organic micropollutants in water" Berlin

13) Wegman, R.C.C. and Hofstee, A.W.M. (1979) The microcoulometric determina-
tion of volatile organic halogen in water samples. Paper presented at the
First European Symposium "Analysis of organic micropollutants in water"
Berlin

14) Takahashi, Y. (1979) A review of analysis techniques for organic carbon
and organic halide in drinking water. Conference on practical application
of adsorption techniques in drinking water. Reston, V.A., USA,

15) Dressman, R.C., Najar, B.A. and Redzikowski, R. (1979) The analysis of
organohalides (OX) in water as a group parameter. Paper presented at the
Water Quality Technology Conference, American Water Works Association,
Philadelphia, P.A.

16) Jekel, M.R. and Roberts, P.V. (1980) Total organic halogen as a parameter
for the characterization of reclaimed waters: measurement, occurrence,
formation and removal. Environm.Sci. & Technol. 14, 970-875

17) Zürcher, F. (1980) Swiss Standard method 50. Lipophile schwerflüchtige
organische Chlorverbindungen (Entwurf). Schweiser-Standardmethode für
Oberflächenwasser 3A

18) Zürcher, F. and Lutz, F. (1980) Simultaneous determination of total volatile organochlorine, -bromide and -fluorine compounds in water by ion-
chromatography. Workshop on Ion-chromatography June 2-3, Petten, the
Netherlands

19) Zürcher, F. (1980) (Swiss Standard Method 49) Chlorierte Lösungsmittel
(Flüchtige Organochlorverbindungen), Schweizer-Standardmethode für Oberflächenwasser 2A

20) U.S. Environmental Protection Agency (1980) Total organic halide. Method
450.1 interim

21) Steenderen, R.A.van (1980) The construction of a total organohalogen analyser system. Lab.Pract. 29, 380-385

22) Jaeger, W.von and Hagenmaier, H. (1980) Einfache Bestimmung von Organochlorverbindungen als Summenparameter (EOCl) in Abwasser.Z.f.Wasser-, und
Abwasser Forsch. 13, 66-69

23) Veenendaal, G. (1981) Personal communication

24) Smeenk, H. (1981) Personal communication

25) Crane, R.I. (1981) Organohalide measurements at the Water Research Centre.
Paper presented at the Cost 64b Working Party 8; Workshop on petroleum

hydrocarbons and organic halogens. Brussels.

26) Kühn, W., Sontheimer, H., Stieglitz, L., Maier, D. and Kurz, R. (1978) Use of ozone and chlorine in water utilities in the Federal Republic of Germany, J.Am.Water Works.Ass. 326-331

27) Melton,R.G., Coleman, W.E., Slater, R.W., Kopfler, F.C., Allen, W.K., Aurand, T.A., Mitchell, D.E., Voto, S.J., Lucas, S.V. and Watson, S.C. (1981) Comparison of Grob Closed-Loop-Stripping Analysis (CLSA) to other trace organic methods. Advances in the identification and analysis of organic pollutants in water, II, (L.H.Keith ed.), Ann.Arbor.Sci.Pub.Inc., Ann Arbor, MI

28) Noordsij, A. (1981) Rendementen van XAD-isolatie. Literatuurgegevens en KIWA-onderzoek. KIWA-report no 258

29) U.S. Environmental Protection Agency, "Recommended List of Priority Pollutants", Washington, D.C., Effluents Guidelines Division, 1977

VOLATILE HALOGENATED HYDROCARBONS IN RIVER WATER,

GROUND WATER, DRINKING WATER AND SWIMMING-POOL WATER

IN THE FEDERAL REPUBLIC OF GERMANY

M. SONNEBORN, S. GERDES, R. SCHWABE
Institute of Water, Soil and Air Hygiene
Federal Health Office
Berlin, Federal Republic of Germany

Summary

With increasing shortage of ground water it becomes
more and more necessary to use surface water as a re-
source for drinking water and swimming-pool water preparation. In the judgement of water quality regarding
the methods to be applied in its preparation and of the
quality the content of halogenated hydrocarbons is one
of the main criteria.
Water sample of the river Weser and Lippe were analysed
for the content of volatile halogenated hydrocarbons
with the aim to get a profile of these compounds along
the river. For checking the quality of drinking water
85 samples from the Federal Republic of Germany were
investigated; additional chlorinated swimming-pool
water samples were taken from public indoor swimming-
pools.
Examinations of ground water showed that even in this
water the determination of halogenated hydrocarbons is
possible, which can in particular circumstances affect
the preparation of drinking water.
For the determination of the halogenated hydrocarbons,
representing a widespread range of concentrations, GC
and GC/MS were applied.

1. INTRODUCTION

The occurence of volatile halogenated hydrocarbons in water
has been allocated a peculiar importence since some epidemio-
logic studies in the USA showed that some of these compounds
may be the reason of cancerogenic effects of drinking water.
Until now a clear statement concerning the cancerogenic effect of these compound has not been given; a survey
and a critical valuation of the literature is provided by
SCHÖN(1).
In the judgement of water quality regarding the methods
to be applied in its preparation and of the quality of the
drinking water after preparation the content of halogenated
hydrocarbons is one of the main criteria.
Only a small number of volatile halogenated hydrocarbons reach the water as anthropogenic pollutants during the

perparation of surface water. Today only two volatile halogenated compounds (trichloroethylene,tetrachloroethylene)
can be regarded as compounds which appear in surface water
and in ground water and which can reach drinking water in
higher concentrations (2,3). In particular, trihalogenmethanes (haloforms) from naturally occuring humin compounds
are not formed   until the stage of oxidative water preparation and disinfection (chlorination, ozonation)(4,5).
     Since chlorination can produce in higher doses tri-
halogenmethanes, whereby chloroform and, in the presence of
bromid , bromoform can appear as the main reaction products,
the question for us was how far swimming-pools, in which
the water undergoes additional chlorination treatment, are
charged by these compounds. Although incontrast to other
countries chlorination is applied reservedly or in minor
doses in the preparation of drinking water in the Federal
Republic of Germany, higher concentrations of chloroform
and similar compounds can be expected in case with the chlorination conditions in swimming-pools.
     Furthermore to get a general view of the content of vo-
latile halogenated hydrocarbons in surface water, samples
were taken along the rivers Weser and Lippe and waste water
was analysed. For the judgement of the water quality of
drinking water the contents of volatile halogenated hydrocarbons in 85 drinking water samples from the Federal Republic of Germany were determined.
     Examinations of ground water showed that even here the
evidence of halogenated hydrocarbons is possible, which can
in particular circumstances affect the preparation of drinking water.
     For the determination of the halogenated hydrocarbons,
representing a widespread range of concentrations, GC and
GC/MS were applied.

2. METHODS

For the investigation samples were scooped from some public
swimming-pools in Berlin, from the rivers, the waste water
and taken from the taps for drinking water. The 250 ml bottles were filled bubble-free and were plugged with a glass
stopper. The samples were stored at a temperature of 4^0 C
until analysed.
     For the gaschromatography analysis of the samples the
purge-trap system (6) by Hewlett-Packard (Mod. 7675 A) with
a direct connection to the GC was applied. 3 ml of the
water sample was degassed with nitrogen as inert gas over a
period of 5 min. at room temperature.The degassed volatile
compounds were absorbed in a Tenax column (100 mm, 6.4 mm $\emptyset$,
1 g Tenax). The following desorption was carried out by a
temperature program reaching 150 ^{0}C over a period of 5 min.
     To exclude a break-through of the compounds to be ana-
lysed, only packed columns can be used at the connected GC.
(Chromosorb 101 80-100 mesh, Column 2.70 m, 2 mm $\emptyset$). Nitrogen was used as carrier gas. The detection was carried
out by a Flame Ionisation Detector (FID) for all organic
compounds and at the same time by an Electron Capture Detector for the specific determination of halogenated compounds.

The signals of the GC were processed by a laboratory data system (Mod. HAL 3, Hewlett Packard) for the quantitative determination. To be on the safe side in respect of the qualitative determination, control samples were investigated by MS.

3. RESULTS AND DISCUSSION

Samples from five different swimming-pools (A,B,C,E and F) in Berlin were investigated, the results are shown in Table I and II. Detectable were the compounds chloroform, bromdichloromethane, dibromchloromethane, dichloromethane,trichloroethane,trichloroethylene and tetrachloroethylene.Other volatile halogenated hydrocarbons were absent or below the limit of determination of 0.1 μg/l.

If the limit of 25 μg/l recommended by the Federal Health Office is taken, an excess can be seen only at swimming-pools C and E (Tab. I)with a maximum of 65 μg/l. All concentrations are clearly higher than the concentrations found in the mains drinking water of the corresponding city districts (Tab. III). As expected, the additional chlorination has caused an increase in the concentration of trihalogenmethanes. Nevertheless the concentrations determined are relatively low, and it must also be considered that,for health reasons, the use of chlorination cannot be dispensed with and that this water will not be used as drinking water.

In three public swimming-pools samples were taken at different times during the day with the aim of registering the fluctuations in the concentrations of organohalogenated compounds. The concentrations so determined were not significantly different from each other, the fluctuation being within the analytical accuracy. Significant differences in the case of the contents in samples taken in different places (Fig. I) of the water circuit could not be found: for all detected halogenated compounds the concentrations were approximately equal in the pool, in the drain, in the supply, before and after filtration and after chlorination.

Tab. IV shows a summary of measured concentrations of volatile halogenated hydrocarbons in surface water, waste water, and drinking water ((a) from(8); (b) from (2)). Compared with the rivers Lippe and Weser, the concentrations in the river Main (8) are clearly higher. In the investigated river waters and also in waste water and drinking water the concentrations of chloroform and bromdichloromethane were distinctly lower when compared with the measured concentrations in swimming-pool water, which can be attributed to the additional chlorination of this water.

In drinking water from the areas around Frankfurt (2) an increase in the concentrations of trichloroethylene and tetrachloroethylene could be found when a pump station was working. Investigations of the ground water around the pump station led to the conclusion that large amounts of trichloroethylene and tetrachloroethylene , used at the nearby airport, had got into the waste water of the airport and in this way contaminated the ground water and the drinking water.

4. LITERATURE

(1) Schön, D.: Trihalomethane im Trinkwasser und die Häufigkeit von Krebs. SozEp-Berichte 6/1981, Dietrich Reimer Verlag, Berlin, 1981

(2) Fritschi, G., Neumayer, V. and Schinz, V.: Tetrachlorethylen und Trichlorethylen im Trink- und Grundwasser - WaBoLu-Berichte 1/1979, Dietrich Reimer Verlag, Berlin, 1979

(3) Aurand, K. and Fischer, M. (Edt.): Gefährdung von Grund- und Trinkwasser durch leichtflüchtige Chlorkohlenwasserstoffe - WaBoLu-Berichte 3/1981, Dietrich Reimer Verlag, Berlin, 1981

(4) Sonneborn, M. and Bohn, B.: Formation and Occurence of Haloforms in Drinking Water in the Federal Republic of Germany - Water Chlorination-Environmental Impact and Health Effects 2 (1978) 537-542

(5) Cotruvo, J.A.: Trihalomethanes in Drinking Water - Env. Sci. & Techn 15 (1981) 268-274

(6) Sydor, R. and Pietrzyk, D.J.: Comparison of Porous Copolymers and Related Absorbents for the Stripping of Low Molecular Weight Compounds from a Flowing Air Stream - Anal. Chem. 50 (1978) 1842-1847

(7) Bundesgesundheitsamt: Empfehlungen des BGA zum Problem "Trihalogenmethane im Trinkwasser - Bundesgesundheitsbl. 22 (1979) 102

(8) Aurand, K. (Edt.): Arbeiten zur Umwelthygiene im Rhein-Main-Gebiet WaBoLu-Berichte 7/1980, Dietrich Reimer Verlag, Berlin, 1980

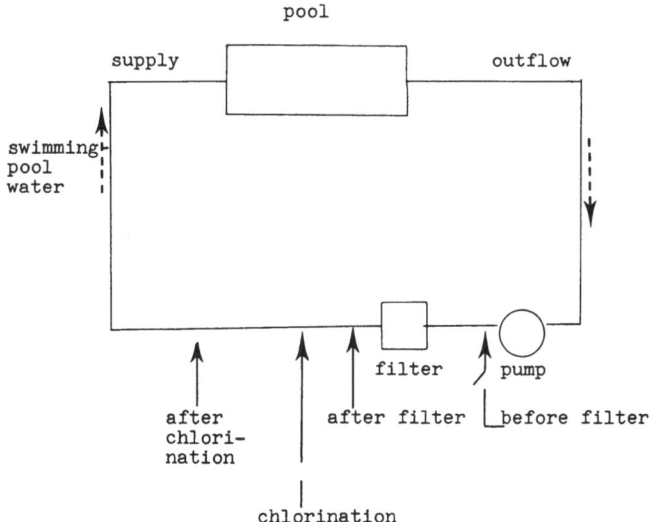

Fig. 1 : places of sampling in the water circuit of the swimming-pools

Tab . I : Concentrations of different trihalogenmethanes in swimming-pool waters

(conc. in µg/l; $\bar{x}$ = mean value; s = standard deviation)

swimming-pool	place of sampling	number of events	Chloroform $\bar{x}$	Chloroform s	Bromdichloromethane $\bar{x}$	Bromdichloromethane s	Dibromchloromethane $\bar{x}$	Dibromchloromethane s	Sum of Trihalogenmethanes
A	a	1	14.7		3.3		0.1		18.0
	b	1	11.3		3.0		0.1		14.3
	c	1	10.3		2.6		0.1		12.9
B	a	1	17.2		9.9		2.6		27.1
	b	1	20.9		12.5		0.1		33.4
	c	1	21.2		11.5		0.1		32.7
C	b	15	28.7	±0.9	22.6	±3.3	3.94	±1.63	55.2
	c	15	28.3	±0.6	20.5	±3.7	2.97	±0.73	51.8
	d	3	31.3	±0.7	25.2	±2.7	4.35	±1.16	61.3
	e	12	30.3	±0.6	23.6	±3.4	4.11	±1.90	58.0
E	d	13	57.6	±5.6	0.1		0.1		
	e	13	65.2	±10.8	0.1		0.1		
	f	13	63.1	±7.4	0.1		0.1		
F	d	11	16.2	±2.7	0.1		0.1		
	e	11	10.4	±1.5	0.1		0.1		
	f	11	16.6	±9.5	0.1		0.1		

*) a) pool   b) supply   c) outflow   d) before filter
   e) after filter, before chlorination   f) after filter, after chlorination

Tab. II : Concentrations of further halogenated hydrocarbons in swimming-pool water

(conc. in μg/l; $\bar{x}$ = mean value; s = standard deviation)

swimming-pool	place of sampling (Fig. 1)	number of events	Dichloro-methane $\bar{x}$	s	Trichloro-methane $\bar{x}$	s	Trichloro-ethylene $\bar{x}$	s	Tetrachloro-ethylene $\bar{x}$	s
C	b	15	0.59	± 0.22	0.34	± 0.08	0.30	± 0.25	0.12	± 0.11
	c	15	0.58	± 0.15	0.36	± 0.11	0.22	± 0.11	0.10	± 0.05
	d	3	0.68	± 0.15	1.09	± 1.15	0.30	± 0.08	0.14	± 0.05
	e	12	0.56	± 0.12	0.37	± 0.21	0.25	± 0.21	0.13	± 0.11
E	d	13	0.46	± 0.32	0.34	± 0.08	0.1		0.1	
	e	13	0.74	± 0.59	0.37	± 0.07	0.1		0.1	
	f	13	0.89	± 0.83	0.47	± 0.40	0.1		0.1	
F	d	11	0.1		0.1		0.1		1.87	± 0.65
	e	11	0.1		0.1		0.1		0.90	± 0.39
	f	5			0.1		0.1		1.07	± 0.20

*) a) surface of the pool   b) supply   c) outflow   d) before filter
   e) after filter, before chlorination   f) after filter, after chlorination

Note: For swimming-pool A and B all concentrations were below 0.1 ug/l

Tab. III : Volatile halogenated hydrocarbons in the mains of the investigated swimming-pools

(conc. in μg/l)

Swimming-pool	Trihalogen-methanes	Trichloro-ethylene	Other halogenated hydrocarbons
A	0.1	0.1	0.1
B	1.9	0.1	0.1
C	0.5	6.6	Dichloroethan: 0.4
E	0.5	6.6	" 0.4
F	0.4	0.1	0.2

Tab. IV : Concentrations of some volatile halogenated hydrocarbons in river water, drinking water, waste water and ground water in comparison to swimming-pool water; conc. in μg/l. $\bar{x}$ = mean value ; s = standard deviation; (n) = number of events;

Water	Chloroform			Bromdichloro-methane			Dibromchloro-methane			Dichloromethane		
	$\bar{x}$	s	(n)	$\bar{x}$	s	(n)	$\bar{x}$	s	(n)	$\bar{x}$	s	(n)
River water (Lippe, 1980)	0.72	± 0.49	(13)	0.04	± 0.01	(6)		–		0.77	± 0.28	(16)
River water (Weser, 1980)	1.00	± 1.49	(16)		–			–			–	
Waste water (1979)	1.55	± 1.50	(21)	0.07	± 0.05	(10)		–		0.37	± 0.29	(23)
Drinking water (1979)	1.52	± 2.47	(85)	0.62	± 0.75	(22)		–		0.17	± 0.13	(31)
Main (1980) (a)	53.76	± 174.3	(36)		–			–			–	
Drinking water (Frankfurt) no pumping (b)		–			–			–			–	
Drinking water (Frankfurt) pumping (b)		–			–			–			–	
Ground water (pumping-work)(b)		–			–			–			–	
Ground water (airport) (b)		–			–			–			–	
Waste water (airport) (b)		–			–			–			–	
Swimming-pool water (1980)	34.96	± 20.27	(123)	20.56	± 6.14	(51)	3.67	± 1.47	(46)	0.64	± 0.46	(84)

Tab. IV : Concentrations of some volatile halogenated hydrocarbons in river water, drinking water, waste water and ground water in comparison to swimming-pool water; (n) = number of events; conc. in µg/l. x̄ = mean value ; s = standard deviation

Water	Trichloroethane			Trichloroethylene			Dichloroethane			Tetrachloroethylene		
	x̄	s	(n)	x̄	s	(n)	x̄	s	(n)	x̄	s	(n)
River water (Lippe,1980)	0.57 ±	0.53	(11)	0.06 ±	0.04	(9)	0.06 ±	0.03	(7)	4.90 ±	4.50	(11)
River water (Weser,1980)	0.29 ±	0.25	(12)	0.34 ±	0.21	(11)	-			-		
Waste water (1979)	0.17 ±	0.28	(17)	0.16 ±	0.21	(20)	-			-		
Drinking water (1979)	0.60 ±	0.60	(5)	0.17 ±	0.59	(45)	7.42 ±	0.72	(4)	-		
Main (1980) (a)	7.36 ±	18.10	(28)	2.98 ±	2.92	(36)	-			8.20 ±	19.25	(36)
Drinking water (Frankfurt) no pumping (b)	-			1.18 ±	0.55	(33)	-			1.09 ±	0.74	(33)
Drinking water (Frankfurt) pumping (b)	-			10.69 ±	21.50	(9)	-			109.0 ±	248.3	(9)
Ground water (pumping-work) (b)	-			19.43 ±	36.02	(39)	-			110.0 ±	258.5	(39)
Ground water (airport) (b)	-			6.19 ±	6.57	(17)	-			7.18 ±	9.66	(16)
Waste water (airport) (b)	-			128.0 ±	438.6	(21)	-			34989 ±	163103	(23)
Swimming-pool water (1980)	0.39 ±	0.29	(84)	0.29 ±	0.21	(44)	-			3.67 ±	1.47	(46)

SIMULTANEOUS DETERMINATION OF TOTAL PURGABLE ORGANO-CHLORINE, -BROMINE AND -FLUORINE COMPOUNDS IN WATER BY ION-CHROMATOGRAPHY

FRITZ ZUERCHER

Federal Institute for Water Resources and Water Pollution Control,
CH-8600 Dübendorf, Switzerland

SUMMARY

A method for the determination of total purgable organohalogen compounds in water is described. The halogenated organic compounds are purged from the water samples (typically 1 liter) with oxygen for 30 minutes at 60°C. The purged compounds are continuously combusted at 950°C and the produced halides (Cl^-, Br^-, F^-) are trapped and quantified by ion-chromatography.
The method gives the total amount of each group of purgable organohalides and can be applied to a wide range of water samples. It offers the simultaneous determination of low molecular weight chlorinated and brominated secondary products formed during drinking water chlorination. The detection limit of the described method is found to be below 0.1 μg of halide equivalent per liter.
The potential of the method is illustrated by studies on water pollution. It proves to be a valuable supplement to the analyses of individual pollutants determined by gas chromatography and is useful in surveying for heavy pollutant loads.

1. Introduction

A large number of low molecular weight halogenated hydrocarbons are found in natural waters and drinking waters [1]. These compounds (chlorinated solvents, haloforms, chlorobenzenes, etc.) originate from sources such as industrial and municipal discharges landfill leachates and drinking water treatment.

Even at concentrations in the low μg/L range, aggregate exposure to such chemicals is found to be a potential risk to the ecosystem [2].

Because of the great ecotoxicological significance of low molecular weight halogenated hydrocarbons there is a clear need for suitable determination procedures.

Analytical methods using liquid extraction or closed loop stripping combined with high resolution gas chromatography are widely known. However this type of single component analysis measures only a part of the low molecular halogenated hydrocarbons found in water samples. The most comprehensive analysis would be determination of the total low molecular weight halogenated hydrocarbons operationaly defined as purgable organohalogens (POX).

The method which is described here measures individual group of purgable organo-chlorine, -bromine and -fluorine compounds simultaneously [3]. For the development of this procedure we have benefitet from the work of Hadorn [4] and Kussmaul [5].

2. Experimental

Samples are collected in glass bottles (100-1000 ml). The bottles are completely filled so as to eliminate headspace.

In the laboratory samples are purged at 60°C using purified oxygen with a flow rate of 1 liter per minute.(Oxygen, even of high quality shows high blank values. Transfer of purging gas trough cerdioxide at 800°C eliminates organohalogen impurities). The standard purging time is 30 minutes for 1000 ml samples and 5 minutes for 100 ml samples (>95 % recovery for dichloromethane).

To eliminate interferences from free chlorine during, analyses of chlorinated waters, 1 ml of sodiumthiosulfate-solution (1 %) in added to 1 liter of sample before purging.

As shown in Fig. 1, purging is followed directly by combustion at 950°C and subsequent trapping of the produced halides in 2-3 ml of a buffer solution (carbonate/bicarbonate buffer used for the ion chromatographic separation).

For complete recovery of bromine-equivalents hydrazine hydrate is added to the buffer solution (0.01 % hydrazine) to reduce bromate to bromide.

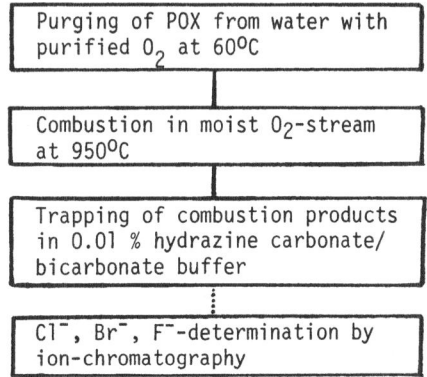

Figure 1   Procedure for the simultaneous determination of total purgable organo-halogens (POX) in water.

The conditions for ion chromatographic separation and detection are the following:

- Analytical column: 3 mm x 500 mm packed with Dionex
  low-capacity anion agglomerated resin
- suppressor column: 3 mm x 50 mm packed with Dionex
  high-capacity cation exchange resin (dual system);
- eluent: 0.003 M NaHCO3/0.0012 M Na2CO3;
- flow rate: 128 ml/hr;
- conductimetric detector: 3 μS/cm, full-scale;
- injection volume: 100-500 μl;
- installation: Dionex 12

The performance characteristics of the method are the following:

Standard deviation:	POCL-conc. (µg/L)	Standard deviation (µg/L)
	0.0	0.02
	1	0.02
	100	1.5

Limits of detection:	POF	0.02	(µg/L)
	POCL	0.06	
	POBR	0.02	

Sensitivity: 1 µg/L POCL → 2 µS/cm

Interferences: Contamination from laboratory-air (HCL)

3. Results

   Practical experience with the method is summarized in Table I. In river water the organochlorinated compounds (POCL) are dominant within the purgable fraction. The organochlorine content of the given random sample from river Rhine (Basel) is above the Swiss water quality goal of 5 µg/L for natural waters. About 34 % of measured POCL was identified as specific substances using closed loop stripping (CLS) combined with gas chromatography. From direct headspace analysis it was concluded that dichloromethane, chloroethane and chloroethylene contributed substantially to the total POCL of the above sample (Table II). River Rhine samples also contained organobromine and organofluorine constituents (Table I).

Table I   Determination of purgable organohalogen-compounds (POX) in water; practical experience

Origin of samples	POX found (µg/L) (determined by IC)		
	Cl	Br	F
River water (Glatt) (Rhein)	0.43 10.2	<0.01 0.3	0.04 0.2
Domestic waste water, raw after secondary treatment	40-65 15-60	n.d.* n.d.	n.d. n.d.
Waste water effluent from chemical plant	2800	<1	3
Groundwater (Glattfelden)	0.45-0.7	<0.01	0.04
Drainage water from sanitary landfill	272	<0.1	16.7
Drinking water	0.2-20	<0.01-3	0.05-0.5

*) n.d.: not determined

   POX compounds of surface waters originate mainly from industrial and municipal waste water discharges. As a consequence of water / gas exchange during secondary waste water treatment (aeration step), the removal of these compounds varies over a wide range (Table I).

Table II   Specific substances identified in river Rhine water (Basel) as a
fraction of total purgable organo-chlorine compounds (POCL).

Purgable organochlorine compounds (POCL)	
34 % identified using CLS/GC method	66 % estimated based on direct headspace/GC analysis
Carbontetrachloride Trichlormethane Trichloroethylene Tetrachloroethylene Chlorobenzene Dichlorobenzene	Dichloromethane Chloroethane Chloroethylene

   The results of the POCL measurements in groundwater indicate that a
significant part of the low molecular weight halogenated hydrocarbons are
removed during the first few meters of bank-infiltration [6]. The persistent
fraction which is not removed during infiltration is readely dispersed
within the groundwater, thus proving its mobility.
   A great potential for groundwater pollution arises from leachates.
Figure 2 represents measured concentrations of purgable organohalogen
compounds (POCL and POF) in leachate from a landfill of solid waste and
sewage sludge. The concentration of purgable organofluoro compounds
decreased significantly during the one year observation period at the field
sampling site.
   The described POX method is also suited for operational control of
processes to minimize the formation and distribution of by-products formed
during drinking water chlorination. Figure 3 shows the removal of
purgable organo-chlorine and organo-bromine compounds (POCL, POBR) during
slow sand filtration. For both organohalogen groups the relative concentration changes during treatment was the same.

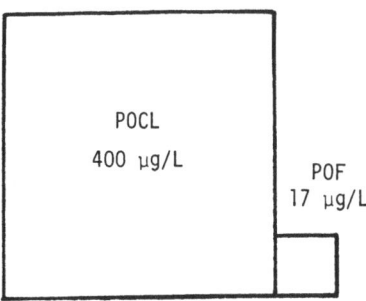

Figure 2   Measured concentration of purgable organohalogen compounds (POX)
in leachate from landfill of solid waste and sewage sludge.

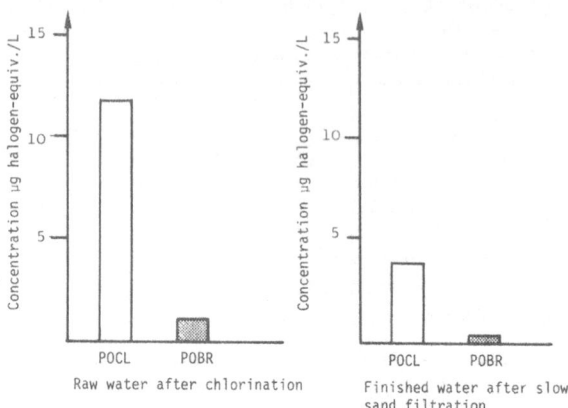

<u>Figure 3</u>   Measured concentration of purgable organo-chlorine and bromine
compounds during water treatment.

Sum determinations using purging combined with ion chromatography offer
a sensitive and rapid method for the determination of an ecotoxicologically
relevant group of organo-halogens. The method is specially valuable for the
surveillance of water quality in different types of water. Furthermore the
sum values are of great importance as quantitative supplements to the
analyses of individual pollutants determined by gas chromatography.

REFERENCES

[1] Commission of the European communities (COST 64b bis). A comprehensive
    list of polluting substances which have been identified in various
    fresh waters, effluent discharges, aquatic animals and plants, and
    bottom sediments. Brussels XII/992,2/79EN, Third Edition 1979.
[2] Environment Committee, Water Management Group; Control of organo-
    chlorinated compounds in drinking water. Organisation for economic
    cooperation and development. Paris 25th March, 1980.
[3] Zürcher, F. and Lutz, F., Simultaneous determination of total volatile
    organochlorine, -bromide and -fluorine compounds in water by ion-
    chromatography. Workshop on Ion-chromatography June 2-3, Petten, the
    Netherlands, 1980.
[4] Hadorn, H., Chlorhaltige Extraktionsmittel-Rückstände in coffeinfreiem
    Kaffee. Mitt. Lebensmittelunters. u. Hygiene 56, 1-16, 1965.
[5] Kussmaul, H. und Hegazi, M., Die Bestimmung von Organofluorverbindungen
    im Wasser. Vom Wasser 48, 143-154, 1977.
[6] Schwarzenbach, R., and Giger, W., Behaviour and fate of halogenated
    hydrocarbons in groundwater. First International Conference on Ground
    Water Quality Research, October 7-10, Houston, Texas, 1981.

ETUDE DES TRIHALOMETHANES DANS L'EAU POTABLE DE BARCELONE. EVOLUTION DE
L'EFFICACITE DES FILTRES A CHARBON ACTIF.

J.Rivera et F.Ventura

Institut de Química Bio-Orgànica

Consejo Superior de Investigaciones Científicas
C/. Jorge Girona Salgado, s/n. BARCELONA-34. (España).

Summary

After three years of monthly monitoring of THM's in different points of
a treatment plant, a study of the efficiency of a granular activated
carbon filter has been performed in connection with different parameters
of raw water. We present here the evolution of the efficiency both for
total THM's and for each compound individually considered. The peculiar
point in this raw water is a very high content of chlorides and bromides.

1. INTRODUCTION

La découverte des trihalomethanes (THM) dans les eaux de boisson (1) et
leurs possibles effets sur la santé humaine (2) ont porté quelques pays a la
limitation des concentrations a 100 ug/l (3) ou encore inférieures.

Dans le cas de la ville de Barcelone, l'eau potable est fournie en partie par le fleuve Llobregat, qui possède un faible débit et se trouve fortement pollué par des eaux résiduelles urbaines, industrielles et surtout par
des décharges importantes de chlorures et bromures (400 tonnes/jour en moyen-
ne) en provenance des mines de potasse existant en amont du fleuve. Cela
mène à des niveaux de THM's bromés très élevés par rapport aux normalement
trouvés dans des eaux avec des valeurs totales de THM's semblables.

Nous disposons actuellement des valeurs mensuelles pour les trois der-
nières années (4) et l'on a trouvé nécessaire d'étudier la dynamique de viei-
llissement des filtres à charbon de l'usine d'épuration en fonction de l'eau
brute dont on dispose.

Le schéma de fonctionnement de l'usine prévoit une préchloration à l'entrée, floculation et décantation après les reactifs, filtrage, à travers 20
filtres à charbon actif d'une surface de 100 m^2 chacun et 1 m de profondeur
de lit. Le temps de contact est de 12 a 15 minutes, une deuxième chloration,
et après distribution aux reservoirs une troisième et dernière chloration
pour garantir un taux de chlore minimum.

2. EXPERIMENTAL

Cette étude a été realisée sur l'un des 20 filtres de l'usine qui ve-
nait d'être régéneré, utilisant du charbon Pittsbourgh Filtrasorb 400 avec
un débit de 250 l/s. Les échantillons ont été pris avant et après passage
par le filtre pendant plus de deux mois, avec au début un échantillon toutes
les 4 heures, pour passer par la suite à 8 heures et à 12 heures. Plus de
300 échantillons ont été analysés, avec les parametres tels que pH, temperature, chlorures, bromures, TOC et dose de chlore utilisés.

L'extraction des échantillons a été réalisée avec n-pentane (25:1) au moment de la prise, et analysés dans les 24 heures. Ces extraits, sans concentration postérieure sont injectés dans une colonne SE-30 au 5% sur Chromosorb W 60/80 mesh (2m x 2mm) a 50°C, les temperatures d'injecteur et détecteur sont de 150°C et 225°C respectivement. Nous utilisons un détecteur à capture d'electrons de ^{63}Ni. Les déterminations quantitatives ont été realisées avec des mélanges des quatre THM's de concentration le plus proche possible au problème et usilisé comme standard externe. Nous avons utilisé les corrections de calcul dues au coefficient de partition à l'extraction décrits par Hu (5).

3. RESULTATS

a) La chloration de l'eau du fleuve Llobregat donne principalment de THM's bromés dans l'ordre $CHCl_2Br$ $CHClBr_2$ $CHBr_3$ $CHCl_3$

b) Les résultats obtenus, pour le filtre décrit, indiquent que le $CHCl_3$ est le THM le moins réténu, puisque le filtre à charbon est epuisé au bout de 3-4 semaines, pour le $CHCl_2Br$ le filtre est efficace pendant 7-8 semaines, plus de 10 semaines pour le $CHClBr_2$ et encore plus pour le $CHBr_3$.

c) L'efficacité du filtre est descendue, au bout de deux mois a 25% de la capacité pour les THM considerés dans sa totalité.

d) Sur la figure 1 on observe entre le 10 et le 15 Mai, les effets d'une crue qui obliga à utiliser des doses très élevées de chlore.Les filtres perdent par la suite une capacité de leur pouvoir d'adsorption.

e) Les valeurs obtenus pour les THM montrent que les niveaux trouvés sont plus hauts pendant la nuit que pendant le jour, en opposition a ce décrit par d'autres auteurs (6).

4. CONCLUSIONS

Des résultats obtenus on peut conclure qu'il est nécéssaire de procéder à des études ultérieures en vue de détérminer l'influence de la teneur en bromure des eaux brutes, ainsi que de la quantité de chlore ajouté.

On procédéra aussi à des essais de variation dans les points de chloration.

Tous ces essais en vue de trouver les conditions de fonctionement de l'usine d'épuration pour minimiser la formation de THM's

Il est évident qu'une solution radicale sérait celle de suprimer la décharge des sels dans le fleuve.

Ce travail a été réalisé en collaboration avec la Sociedad General de Aguas de Barcelona.

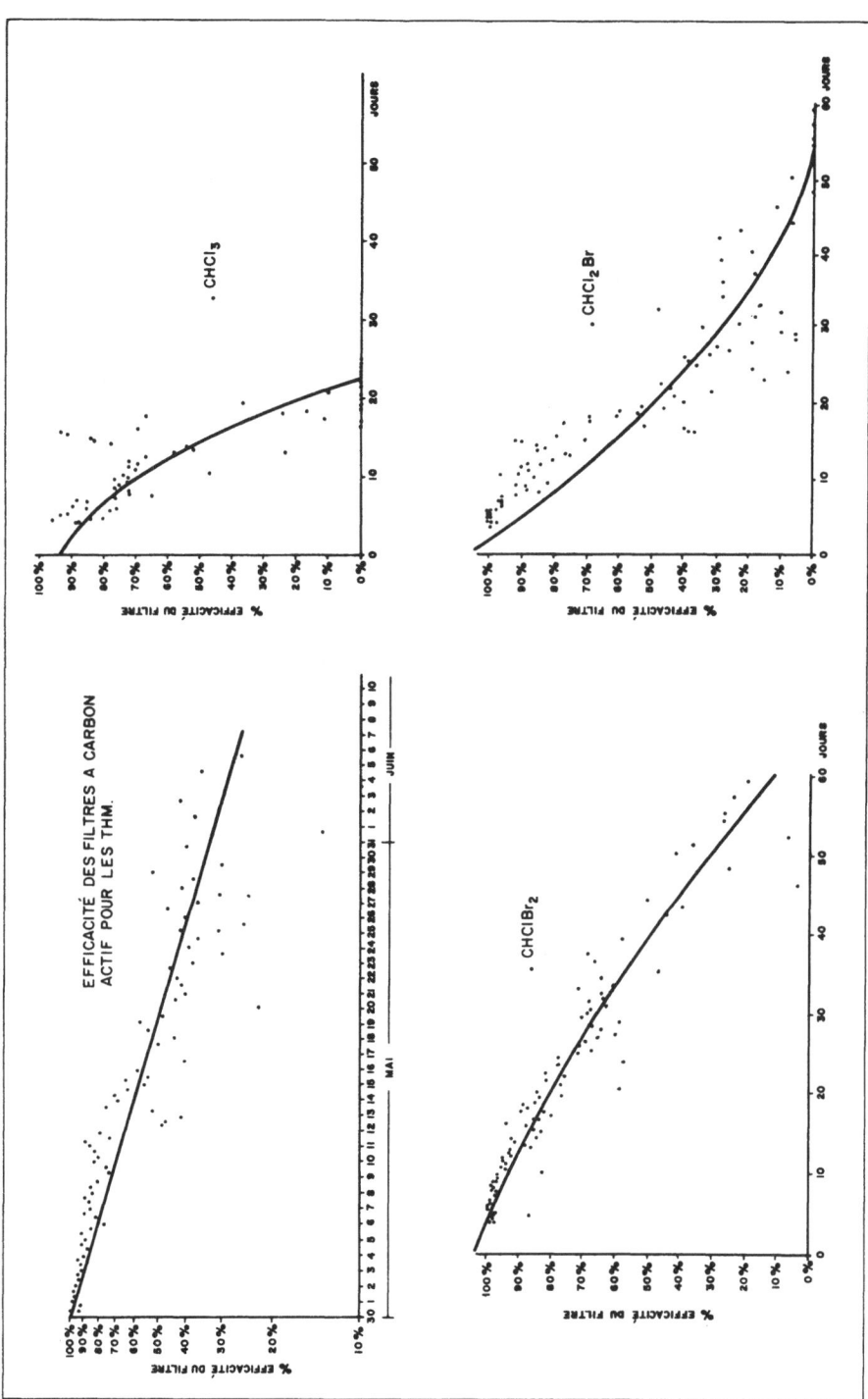

EFFICACITÉ DES FILTRES A CARBON
ACTIF POUR LES THM.

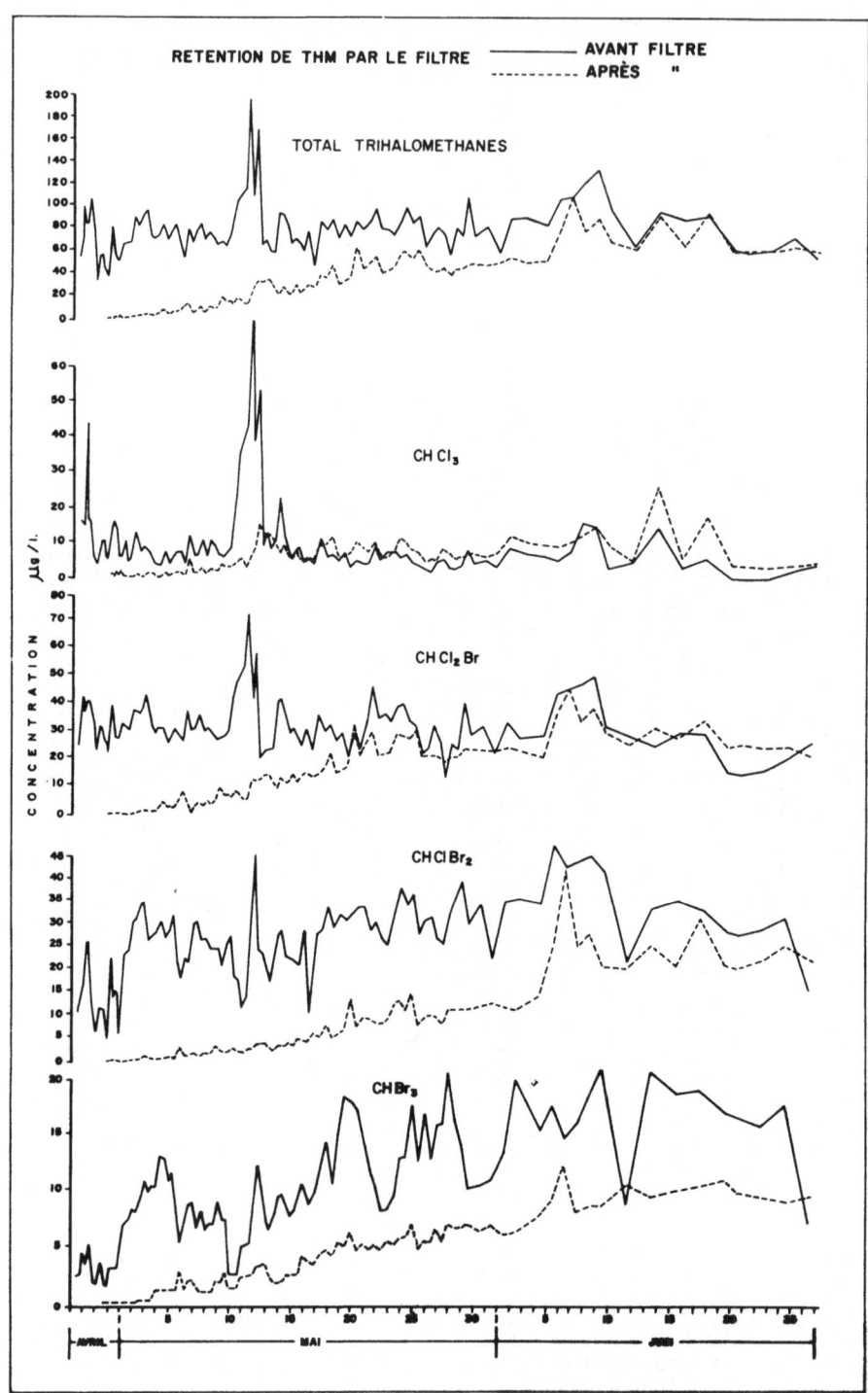

RETENTION DE THM PAR LE FILTRE ——— AVANT FILTRE
------ APRÈS "

TOTAL TRIHALOMETHANES

CH Cl₃

CH Cl₂ Br

CHClBr₂

CHBr₃

CONCENTRATION μg/l.

AVRIL   MAI   JUIN

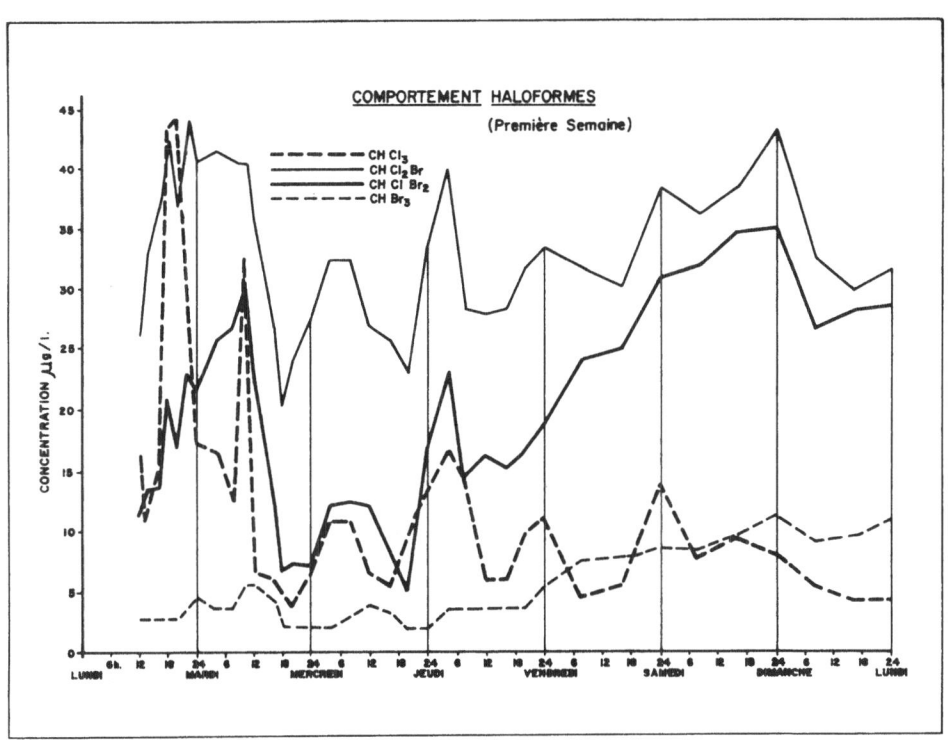

COMPORTEMENT HALOFORMES

(Première Semaine)

5.BIBLIOGRAPHIE

1-Rook J.J. J.Water Treat Exam <u>23</u>, 234-43 (1974)

2-Tardiff R.G. J.Am Water Works Assoc <u>69</u>, 658-61 (1977)

3- U.S.A. Regulations Fed Regist <u>43</u> (No-28) Feb 9, 1978

4-J.Rivera et F.Ventura. European Symposium "Analysis of Organic Micropollutants in Water", Berlin 1979. Preprint. Volume 1 pag 105-110.

5-H.C.Hu and P.H.Weiner. J.of Chromat Science <u>18</u>, 333-342 (1980)

6-L.Smith, I.Cech. J.H.Brown and G,F.Bodgan. Env.Sci & technol. <u>14</u>,190-196
(1980).

ANALYSIS OF TRIHALOMETHANES FORMED DURING DRINKING WATER CHLORINATION

L. ŚCHOU and J. KRANE
University of Trondheim, Department of Chemistry, NLHT
N-7055 Dragvoll

H. ØDEGAARD
Div. of Hydraulic and Sanitary Engineering, University of
Trondheim, Norwegian Institute of Technology, N-7034 Trondheim

G.E. CARLBERG
Central Institute for Industrial Research, P.O. Box 350
Blindern, Oslo 3

Summary

Analyses of trihalomethanes (mainly chloroform) were performed in connection with chlorination of drinking water.
A typical Norwegian water source with relatively high content of humic constituents was treated to reduce the organics that cause trihalomethanes to be formed in the
chlorination process. A gas chromatograph with an electron capture detector (GC-ECD) was used in the analyses.
The workup of the trihalomethanes was done through a
liquid-liquid extraction of the water samples with n-pentane. Chlorination of the raw water with 2mg Cl_2/l gave
an average of 37.3µg $CHCl_3$/l. Raw water from the same
source was then treated in a filtration pilot plant. A
decrease in the chlorine demand of about 50% was observed.
Reducing the amount of chlorine accordingly, gave a decrease in $CHCl_3$ formation of 50% on the filtered water.
The effluent from the pilot plant chlorinated with a constant dose of chlorine showed a linear correlation between
the amount of $CHCl_3$ formed and UV-extinction. Experimental
results are also presented showing the influence of variation in pH, temperature, chlorine dosage and humic and
bromine content on the production of trihalomethanes.

1. INTRODUCTION

The formation of trihalomethanes (THM) from chlorination
of drinking water with a high content of humic constituents has
been found in several studies during recent years (1-5). The
possible risk of cancer from these components has caused great
concern, and has led to maximum concentraion limits of 25-100
µg/l in some countries.
Handling of the problem can be done in 3 different ways:
i)   limit the use of chlorine in treatment of drinking water
ii) remove THM after their formation
iii) reduce the content of precursors before chlorination
By using raw water of good quality except for high content
of humic constituents, prechlorination can be omitted, and
alternative iii) seems to be the best way to reduce the amount
of THM. In an ongoing research program at the Norwegian
Institute of Technology different methods for removal of humics

are being investigated. The method we report on here is on a pilot plant scale. The plant consists of one filtration unit with sand and anthracite as filter medium. The conventional sedimentation basin was omitted and the sedimentation as well as the filtration process take place in the filtration unit. The purification parameters studied were UV-extinction and turbidity of the effluent.

2. <u>EXPERIMENTAL</u>

The raw water was diluted to a colour content correspon-ding to 45mg Pt/l. Three sampling sites were chosen in the pilot plant, i.e. in the raw water reservoir, in the column above the filter and in the effluent. The samples buffered to pH 7, were chlorinated with NaOCl and the THM potential after 3 days contact time was registered.

The water samples (100 ml) were extracted with n-pentane (5 ml), and the n-pentane extracts were, without further concentration, injected splitless onto a fused silica capilla-ry column (SP-2100, 25m x 0.3mm). The temperature of the column was increased 10°C/min from 30°C to 100°C, and the injector and detector temperatures were 200°C and 250°C, res-pectively.

Four chloro- and bromo-trihalomethanes were used as an external standard mixture for quantification purposes. An internal standard (CBrCl$_3$) added to the n-pentane before the extraction eliminated the problem with non-reproducable injection volumes.

3. <u>RESULTS</u>

Chlorination of the drinking water used in our investi-gation gave mainly choroform (CHCl$_3$), with bromodichloro-methane (CHBrCl$_2$) as a minor product. The following discussion will mainly focus on the formation of CHCl$_3$. The analytical procedure gave a standard deviation of 1-3% for 10 parallel injections of a standard mixture, and 13% for the chlorination step (7 parallels).

We have investigated the influence of variation in pH, temperature, chlorine dosage, humic and bromine content on the production of THM.

i) An increase in pH from 7 to 11 using high (8mg Cl$_2$/l) chlorine dosage gave an increase in formation of CHCl$_3$, while the formation of CHBrCl$_2$ decreased. These variations are, however, dependent of the chlorine dosage. A chlorine dose of 1-2mg Cl$_2$/l for the same pH-window gave only minor variation in the CHCl$_3$ formed.

ii) The temperature effect is also dependent on the amount of chlorine used. At high chlorine dose, an increase in temperature results in a large increase in both CHCl$_3$ and CHBrCl$_2$ formed, while lower chlorine dose gave only minor variations in the products formed over the same temperature range.

iii) In Figure 1 is shown the amount of CHCl$_3$ formed for three chlorine doses as a function of humic concentration measured by the UV-extinction coefficient. Figure 1 also shows maximum production of CHCl$_3$ for certain ratios of chlorine/humic concentration. Obviously, there is no simple

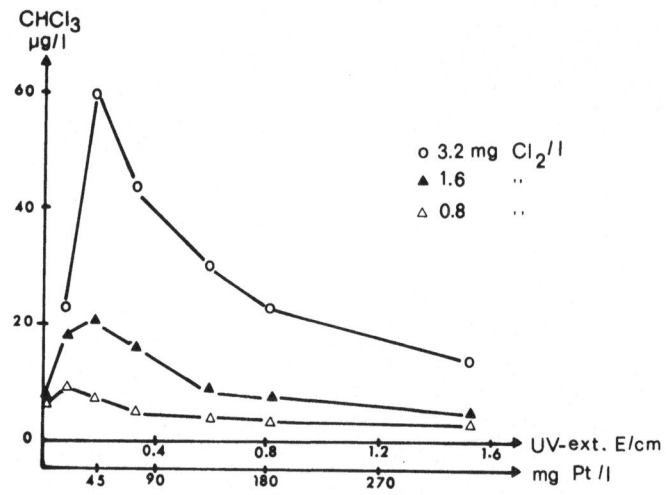

Figure 1.   Formation of CHCl$_3$ vs. humic concentration.
Contant time 3 days. pH 7. Temperature 20°C.

relation between the chlorine/humic concentration ratio and the
production of CHCl$_3$.   We suspect that chlorine is consumed in
competitive oxidation reactions which do not lead to CHCl$_3$.
     iv) The total amount of THM as well as the relative[3]
amounts of brominated compounds to CHCl$_3$ increase with increasing bromine content.
     Chlorination of the raw water to a constant amount of
0.5mg Cl$_2$/l after a contact period of 0.5 hr gave an average
of 37.3µg CHCl$_3$/l after 3 days.   Raw water from the same source
was then treated in a filtration pilot plant.   A decrease in
the chlorine demand of about 50% was observed.   Reducing the
amount of chlorine accordingly, gave a decrease in CHCl$_3$
formation of 50% on the filtered water.
     To investigate the dependence of variables such as
type and dosage of coagulants the effluent after filtration
was chlorinated with a constant dose of 2mg Cl$_2$/l.   The
results shown in Figure 2 indicates a linear correlation
between CHCl$_3$ and the UV-extinction coefficient at low humic
concentrations.

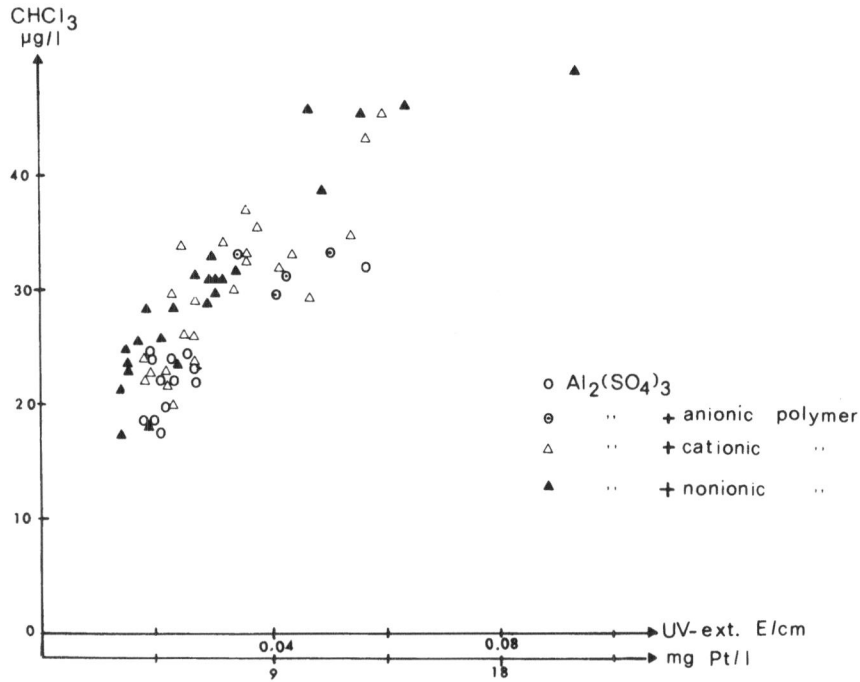

Figure 2. Formation of CHCl$_3$ after filtration vs. humic
concentration. Chlorine dose 2mg Cl$_2$/l.
Contact time 3 days. pH 7. Temperature 20°C.

4. REFERENCES

(1) J.J. Rook
J.Soc.Water Treatm. and Exam., 23:2, 234 (1974)

(2) T.A. Bellar, J.J. Lichtenberg and R.C. Kroner
JAWWA, 66:12, 703 (1974)

(3) J.M. Symons, T.A. Bellar, J.K. Carswell, J. Demarco, K.L.
Kropp, G.G. Robeck, D.R. Seeger, C.J. Slocum,B.L. Smith
and A.A. Stevens
JAWWA, 67:11,634 (1975)

(4) Canadian Environmental Health Directorate Nationale Survey
for halomethanes in drinking water
Report 77-EDH-9 (1977)

(5) W. Fonahn, L. Berglind and H. Drangsholt
"Halomethanes in Norwegian Drinking Water". Proceedings
from 17th Symposium on Water Research, Porsgrunn, Norway
(1981)

DETERMINATION OF PHENOLICS IN THE AQUATIC ENVIRONMENT

Lars Renberg
National Swedish Environment Protection Board
Special Analytical Laboratory
Wallenberg Laboratory
S-106 91 Stockholm, Sweden

SUMMARY

Different methods based on chromatography are summarized
for trace level determination of phenolics. Hydrophilic
phenols can be isolated from water samples with steam distillation, continuous liquid extraction, or on ion exchange
resins, or extracted as lipophilic ion pairs. Procedures,
based on *in situ* conversion of hydrophilic phenols into
hydrophobic derivatives in combination with an extraction
step, have been found to be useful. In addition, hydrophobic phenols are easily concentrated from water samples
by batch liquid/liquid extraction or reversed phase adsorption. Phenols, present in sediment and sewage sludge, can
be extracted either as undissociated phenols with organic
solvents or as phenolates with alkaline aqueous solutions.
The isolation of phenols from organism tissues are usually
carried out with a mixture of semipolar and unpolar organic solvents. Clean-up procedures, involving batch par-
titioning at different pH values, are usually sufficient.
Extracts with a high content of organic material require a
more thorough and selective clean-up. For the quantification of phenolic compounds, capillary gas chromatography is
a commonly used powerful tool. Conversion of phenols into
unpolar derivatives are often applied before the gas chromatographic determination. For many non-halogenated phenols,
the sensitivity can be drastically increased after the conversion into halogenated derivatives when using the electron capture detector. Other chromatographic methods have
been less used for trace levels determination of phenols,
mainly due to less sensitivity and selectivity compared to
gas chromatography. However, liquid chromatography in combination with electrochemical detection have been demonstrated to be a promising technique.

INTRODUCTION

Phenolic compounds may be defined as any aromatic nucleus bear-
ing one or several hydroxy groups (the word phenol is derived
from "phene" an old name for benzene). Several members of the
group are widely used in a number of applications, ranging from
pesticides to food additives, indicating the very broad span of
uses - but also of the toxicity of the different compounds. Besides being intentionally used, some phenolics are uninten-
tionally formed in industrial processes, e.g. chlorinated guaiacols and catechols have been identified in kraft pulp mill

bleachery effluents (Rodgers 1973, Lindström and Nordin 1976).

The interest in the presence of phenolics in the aquatic environment is due to the fact that several phenols have shown to be toxic against aquatic organisms at concentrations often below the mg/l level. Certain phenols have also the ability to impart taste and odour of drinking water supplies and edible aquatic animals at μg/l level. Bioaccumulating properties have also been established for some phenols - e.g. pentachlorophenol - which often results in considerably higher levels in the organisms compared to the surrounding water. Comprehensive surveys of toxicity data and evaluations for different phenols are given in the reviews carried out by Buikema et al (1979), Kozak et al (1979) and by US Environment Protection Agency (EPA 1980).

ANALYTICAL PROCEDURES

General aspects

As discussed above, many phenols may thus be potentially harmful against man and/or the environment. The US Environment Protection Agency has identified ten phenols to be of particular environmental concern (EPA 1977). These phenols, included in the priority pollutants list, together with the phenols on the list of consensus voluntary reference compounds originating from the American Chemical Society (Keith 1979) and the pesticides hexachlorophene, 5-chloro-2-(2,4-dichloro)phenoxyphenol and clopidol (3,5-dichloro-2,6-dimethyl-4-pyridinol) represent a wide range of chemical properties. They are highly suitable for testing and evaluating analytical procedures for trace level determination (see Table I).

Besides volatility of the undissociated compound and steric hindrance of the hydroxy group(s) an important parameter from an analytical point of view is the distribution ratio (D) between a hydrophobic solvent and a hydrophilic solvent (usually water). The distribution ratio is mainly regulated by the distribution coefficient K_D (also often called the partition coefficient) which reflect the hydrophobicity of the undissociated phenol and the acid strength, expressed as the corresponding K_a-value.

The correlation between the different parameters discussed for a monohydroxyphenol is illustrated in Figure 1.

The distribution coefficient, distribution ratio and the acid strength are important factors to be considered not only from an analytical point of view, e.g. when optimizing partition procedures but also from an ecotoxicological standpoint. The distribution ratio can thus be used to predict the bioaccumulation potential and the degree of adsorption to suspended matter. There are also indications that usually the toxicity increases with increasing acid strength.

The procedures for the determination of phenolics are similar to trace level determination of other organic environmental

TABLE I

Hydrocarbon based phenol	Halophenols
P phenol	P 2-chlorophenol
C guaiacol	PC 2,4-dichlorophenol
C catechol	P 2,4,6-trichlorophenol
P 2,4-dimethylphenol	2,3,4,6-tetrachlorophenol
C 2,6-di-*tert*-butyl-4-methylphenol	PC pentachlorophenol
C 4-phenylphenol	P 4-chloro-3-cresol
C 1-naphtol	C 4-chloro-1-naphtol
C bisphenol A	C tetrachlorocatechol
	C tetrabrombisphenol A

Nitrophenols	hexachlorophene (2,2'-methylene-bis) (3,4,6-trichlorophenol)
P 2-nitrophenol	5-chloro-2-(2,4-dichloro- phenoxyphenol
PC 4-nitrophenol	
P 2,4-dinitrophenol	clopidol (3,5-dichloro-2,6-dimethyl-
P 4,6-dinitro-2-cresol	-4-pyridinol

P = Priority pollutant phenol
C = Consensus voluntary reference phenol

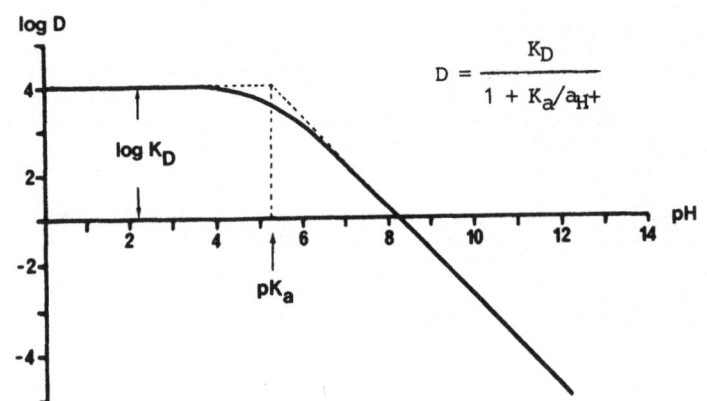

Figure 1. Log D = f(pH) for a monohydro phenolic compound with
        log D = 4 and pKa = 5.3

    D = distribution ratio K_D = distribution constant
    K_a = dissociation constant a_{H^+} = hydrogen ion activity

pollutants: After an isolation/concentration step to remove the phenols from the matrix, the extract is treated in a clean-up procedure to remove interfering substances and the purified extract is then analyzed by gas chromatography or liquid chromatography.

Prior to a gas chromatographic determination, the phenols are usually converted into less polar derivatives. In addition, methods have been developed for samples where the extraction and derivatisation steps are combined into one step. Such methods, based on extractive alkylation (e.g. Ehrsson 1971, Fogelqvist et al 1980) and extractive acylation (Coutts et al 1979, Renberg 1981a) are particularly useful for the determination of phenols with unfavourable partition coefficients.

Isolation/enrichment and clean-up methods

Although methods using direct injection of water samples have been described for both gas chromatographic determination (e.g. Baird et al 1974) and liquid chromatographic determination (e.g. Pinkerton 1981) such procedures have not been extensively used. Besides the obvious risk for contamination of the chromatographic system by injection of samples containing various types of dissolved and suspended matter, the detection levels can be considerably lowered if an enrichment step is used.

Many methods have successfully been used for the isolation and concentration of phenolic compounds from water samples. The by far most commonly used method is liquid/liquid extraction (partition), i.e. after acidification of the water sample to a pH less than the pK_a-values of the components to be analyzed, the sample is shaken one or several times with a suitable organic solvent, or mixtures thereof (e.g. EPA 1977), Renberg 1981b) and references cited herein).

Batch extractions are simple and straight forward procedures, but sometimes problems can be caused by particles and formation of emulsions which tend to obstruct the phase separation. The partition coefficient of the individual components must be carefully considered. Highly hydrophobic phenols - e.g. pentachlorophenol - do not usually create any problems of principal nature. The recovery of relatively polar phenols with less favourable partition coefficients can be increased by repeating the extraction steps with several portions of organic solvents. Also a change towards more polar solvents and using smaller organic solvent/water sample volume ratios will also increase the recovery. An alternative method to increase the recoveries of relatively polar phenols is the continuous extraction of a batch sample (Lindström and Nordin 1976, Carlberg et al 1980).

Column techniques have often been applied to the trace enrichment of phenols. Two principles have thereby been used, i.e. ion exchange and reversed phase adsorption. Ion exchange techniques are based on the fact that phenols are quantitatively converted into the corresponding phenolate ion at sufficiently high pH-values and therefore can be retained on an anion exchanger. With such a method, also relatively polar chlorophenols are easily isolated (Renberg 1974, Criswell et al 1975).

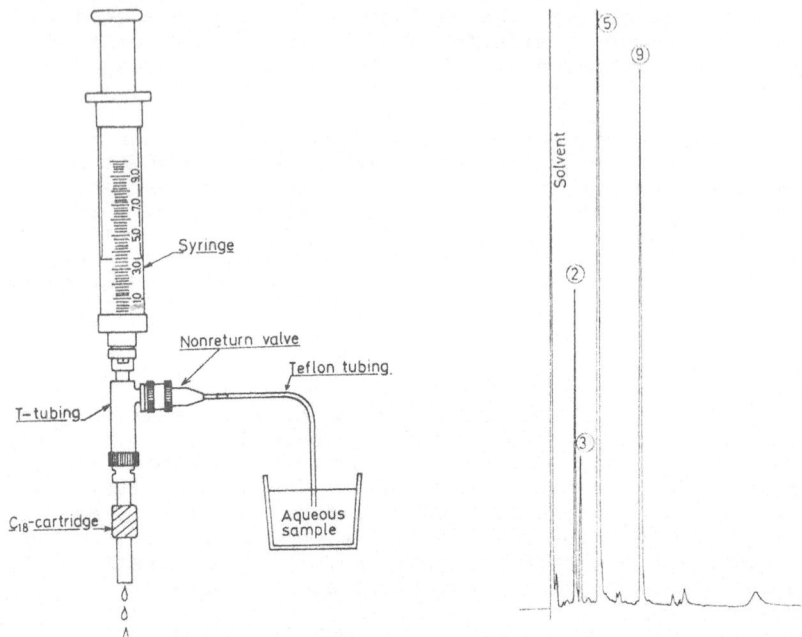

Figure 2. Enrichment of an aqueous sample on a C_{18} cartridge and a gas
chromatogram of a lake water sample down stream a saw mill,
the peaks represent acetylated 2,4,6-trichlorophenol (2),
2,3,4,6-tetrachlorophenol (5) and pentachlorophenol (9), corresponding to 0.5, 1.8 and 0.3 µg/l, respectively. Data from
Renberg and Lindström 1981.

Adsorption techniques have gained much popularity in residue
analysis and undissociated phenols easily adsorbed on polymers
such as polystyrene resins (Amberlite XAD-4 etc). Hydrophobic
phenols show an exceptionally strong retardation on such resins.
This behaviour, probably due to interaction between the
electrons of the aromatic ring systems of the phenol and the
polystyrene matrix, may be disadvantageous as large amounts of
the desorbing solvent is needed (often several times compared
to the volume of the resin).

Reversed phase chemically bonded stationary phases offer an
attractive alternative for the isolation and concentration of
organic substances. The development of such phases for high performance liquid chromatography has also resulted in commercially
available cartridges, packed with octadecyl modified silica gel.
Such cartridges have been applied to trace enrichment of chlorinated phenols, guaiacols and catechols (Renberg and Lindström
1981, see also Figure 2).

Generally, trace enrichment, using columns, are highly effici-

ent, however, particles present in the water sample tend to
clog the columns which can seriously disturb the procedure.
This can be overcome by a pretreatment step, e.g. centrifugation or filtration.

For hydrophobic phenols - e.g. chlorinated phenols - having a
tendency for adsorption and bioaccumulation from several analytical methods have been designed for sediments, sewage sludge
and organisms. From sediments and sewage sludge, the phenols
are extracted as undissociated compounds during acidic conditions (e.g. Buhler et al 1973, Jensen et al 1977, Pierce and
Victor 1978, Eder and Weber 1980) or as water soluble phenolates with an alkaline buffer (Stark 1969, Renberg 1974).

Aquatic organism samples are usually extracted under acidic conditions with a mixture of organic solvents during homogenisa-
tion (e.g. Rudling 1970, Renberg 1974, Hattula et al 1979).
Methods, in which the fat is not destroyed during the extraction should be preferred as such methods allow a gravimetric
determination of the fat content. Thus, the concentration of
the phenols can also be calculated on a fat weight basis, which
is valuable when considering the values from an ecological
point of view. Figure 3 shows the presence of chlorophenols in
fish after an accidental discharge from a saw mill.

In addition to the methodologies referred to above, steam distillation has proved to be very useful for the isolation and
clean-up of both water, sediment and organism samples (e.g.
Stark 1969, Kuehl and Dougherty 1980, Giger and Schaffner 1981,
Renberg 1981b). Compared to other isolation techniques, steam
distillation is relatively time consuming but will, on the
other hand, produce quite concentrated extracts which usually
do not require any further clean-up procedures.

Using other methodologies, a separate clean-up step is necessary. For extracts originating from samples with a low content of
organic material, a simple partition procedure is usually sufficient, e.g. the organic extract is shaken with an alkaline water solution, which after acidification is reextracted with an
organic solvent. Extracts with a high content of organic material - e.g. fat - require a more thorough clean-up based on e.g.
ion exchange (Renberg 1974), column chromatography (Faas and
Moore 1979) or thin layer chromatography (Hattula et al 1979).

Gas chromatography

In general, chromatography is superior to any other metodology
- e.g. colorimetric determination - depending on the separation
power in combination with sensitive detectors. Gas chromatography has become a widely used and sensitive tool for trace
level determination of phenolic compounds, usually in combination with flame ionization or electron capture detection. The
development of glass capillary columns has resulted in efficiencies superior to any other chromatographic system, irrespec-
tive of whether this parameter is measured as the number of
theoretical plates or effective plates, or their corresponding
values per unit time.

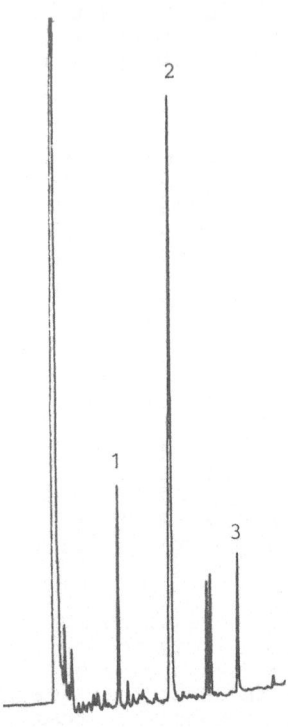

Figure 3.
Gas chromatogram obtained
from a fish (*Perca flu-*
viatilis) down stream a
saw mill. Peaks 1, 2 and
3 represent the acetates
of 2,4,6-tri-, 2,3,4,6-
-tetra- and pentachloro-
phenol, corresponding to
0.42, 0.61 and 0.11 mg/kg,
respectively (fresh weight
basis).

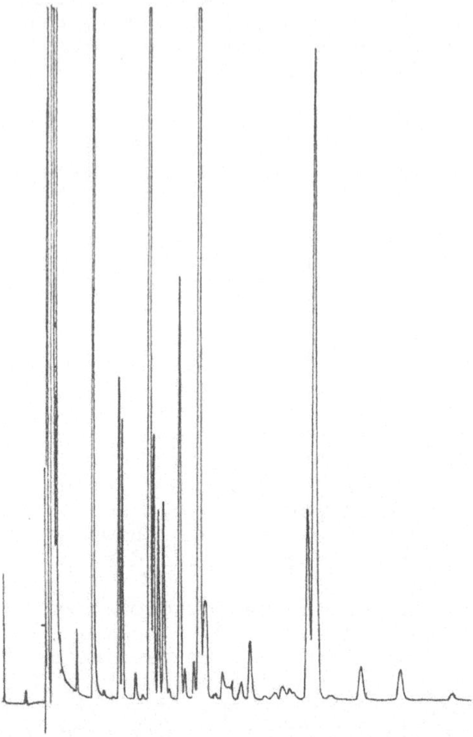

Figure 4.
Gas chromatogram of the pentafluoro-
benzoates of isomeric cresols, xylenols
isopropylphenol obtained after hydrolysis of a commercial tricresylphos-
phate product. (injected amount correspond to 0.5 ng triarylphosphate).

As the polar hydroxy group(s) tends to decrease the volatility
and efficiency of the gas chromatographic system, the phenols
are commonly converted into more volatile, unpolar and often
more stable derivatives. For many phenolics, the sensitivity
can - using suitable detectors - be drastically increased.

A number of derivatization procedures have been reported based
on alkylation, acylation or silylation techniques.

Alkylation has usually been carried out with a diazoalkane (e.g.
Stark 1969, Shafik et al 1973, Renberg 1974, Lindström and
Nordin 1976, Jensen et al 1977, Faas and Moore 1979) or with

pentafluorobenzyl bromide (e.g. Ehrsson 1971, Kawahara 1971, Fogelqvist et al 1980), the latter reagent giving derivatives with high electron capture detector (ECD) responses.

The commonly used acylation techniques involve the use of acetic anhydride (e.g. Rudling 1970, Chau and Coburn 1974, Coutts et al 1979, Renberg and Lindström 1981) and - to increase the ECD response - chloroacetic anhydride (Argauer 1968), perfluorinated aliphatic anhydrides (e.g. heptafluorobutyric anhydride, Ehrsson et al 1971) and pentafluorobenzoyl chloride (e.g. McCallum and Armstrong 1973, Renberg 1981). Halogenated aliphatic anhydrides seem, however, to give less stable derivatives than other acylation reagents.

Silylation is less used for quantification purposes and does not offer any substantial advantage compared to alkylation or acylation for trace level determination of phenols.

However, it should be pointed out that a derivatization procedure adds an extra step to the analysis, requiring time and labour. Disturbing artifacts from more or less toxic reagents are often produced and the yield may be influenced to various degrees by steric hindrance and the acid strength of the individual compound.

Good results have been reported on the determination of underivatized phenols using capillary columns, both with glass capillaries, e.g. coated with OV-73 (Giger and Schaffner 1981), SE_54 (Supelco 1980), or graphitized carbon black (Bacaloni et al 1979), and quartz capillary columns (personal observations). As pointed out by several authors (e.g. Grob et al 1979), thoroughly deactivated columns are absolutely essential, due to the polar character of the hydroxy group.

The difficulties with a capillary gas chromatographic determination increase with increasing acid strength and hydrophilicity of the phenol. Compounds with more than one phenolic hydroxy group - e.g. the readily oxidizable tetrachlorocatechol - are particularly difficult, or impossible, to determine underivatized, using gas chromatography.

Liquid chromatography

The rapidly growing interest for modern liquid chromatography has resulted in the development of several methods for phenolic compounds, based on photometric or electrochemical detection. Usually, the column packings consist of chemically bonded stationary phases. The commonly used UV-spectrophotometric detection has also been applied to phenols in water (e.g. Fass and Moore 1979, Realini 1981), but the relatively low sensitivity of such a chromatographic system indicates the need for a careful enrichment of the substances from the matrix before the liquid chromatographic determination.

Electrochemical detectors, in combination with liquid chromatographic systems, represent a more sensitive and selective system for readily oxidizable (or reducible) compounds, in-

cluding phenols (e.g. Armentrout et al 1979, King et al 1980).
The sensitivity is highly dependent on the particular phenols.

Liquid chromatographic methods seem not to have gained the same
popularity as the gas chromatographic methods for the determination of phenols in environmental samples. A main reason is
probably that a capillary gas chromatographic system, at the
present time, is more efficient - in terms of separation power -
compared to a high performance liquid chromatographic system.

CONCLUSIVE REMARKS

The most important conclusion to be drawn from the previous discussion is probably that there is no general method for the
trace level determination of all environmentally interesting
phenols. There are several factors influencing the choice of
method - or rather the combination of suitable enrichment,
clean-up and quantification procedures. Some of the most important factors to be considered in each specific case are:

Which phenolics are to be determined?	Structure depending on source.
Which matrix is to be analyzed?	Biota, sediment and/or water.
Need for specificity?	Depending on complexity and matrix.
What is the desirable detection limit?	Depending on expected concentration and/or toxicity.
Simplicity of techniques?	Decisive factors are time, available instruments, number of samples and the skill of the staff.
Reproducibility?	Depending on the specific investigation and the variation of the matrix.
Costs?	Consider all the above variables!

In addition, it should be pointed out that methods discussed
above can with - or without - minor modifications be applied
to the determination of phenolic metabolites (e.g. Shafik et al
1973). Also technical products, having a phenolic backbone,
can be characterized by similar techniques. As an example, in
Figure 4, the phenolics derived,after hydrolysis of a tricresyl-
phosphate product, are characterized as the corresponding penta-
fluorobenzoates.

REFERENCES

Argauer R J. Rapid procedure for the chloroacetylation of microgram quantities of phenols and detection by electron-capture
gas chromatography. *Anal Chem* 40 (1968) 122-124.

Armentrout D N, J D McLean, M W Long. Trace determination of
phenolic compounds in water by reversed phase liquid chromatography with electrochemical detection using a carbon-poly-
ethylene tubular anode. *Anal Chem* 51 (1979) 1039-1045.

Bacaloni A, G Goretti, A Laganà, B M Petronio. Graphitized capillary columns for the determination of chlorinated compounds. *J Chromatogr* 175 (1979) 163-173.

Baird R B, C L Kuo, J S Shapiro, W A Yanko. The fate of phenolics in waste-water-determination by direct-injection GLC
and Warburg respirometry. *Arch Environ Contam* 2 (1974)
165-178.

Buhler D R, M E Rasmusson, H S Nakaue. Occurrence of hexachlorophene and pentachlorophenol in sewage and water. *Environ
Sci Technol* 7 (1973) 929-934.

Buikema A L, M J McGinnis, J Cairns Jr. Phenolic in the aquatic
ecosystems. A selected review of recent literature. *Marine
Environ Res* 2 (1979) 87-181.

Carlberg G E, N Gjøs, M Møller, K O Gustavsen, G Tveten, L Renberg. Chemical charcterization and mutagenicity testing of
chlorinated trihydroxybenzenes identified in spent bleach
liquors from a sulphite plant. *Sci Total Environ* 15 (1980)
3-15.

Chau A S Y, J A Coburn. Determination of pentachlorophenol in
natural and waste waters. *JAOAC* 57 (1974) 389-393.

Chriswell C D, R C Chang, J S Fritz. Chromatographic determination of phenols in water. *Anal Chem* 47 (1975) 1325-1329.

Coutts R T, E E Hargesheimer, F M Pasutto. Gas chromatographic
analysis of trace phenols by direct acetylation in aqueous
solution. *J Chromatogr* 179 (1979) 291-299.

Eder G, K Weber. Chlorinated phenols in sediments and suspended
matter of the Weser estuary. *Chemosphere* 9 (1980) 111-118.

Ehrsson H. Quantitative gas chromatographic determination of
carboxylic acid and phenols after derivatization with pentafluorobenzyl bromide. Evaluation of reaction conditions.
Acta Pharm Suec 8 (1971) 113-118.

Ehrsson H, T Walle, H Brötell. Quantitative gas chromatographic
determination of picogram quantities of phenols. *Acta Pharm
Suec* 8 (1971) 319-328.

EPA (Environment Protection Agency, U S). Sampling and analysis
procedures for screening of industrial effluents for priority pollutants, April 1977.

EPA (Environment Protection Agency, U S). Ambient water quality criteria for chlorinated phenols (EPA 440/5-80-032),
for 2-chlorophenol (EPA 440/5-80-034), for 2,4-dichloro-
phenol (EPA 440/5-80-042), for pentachlorophenol (EPA 440/5-
-80-065, 1980a).

Faas L F, J C Moore. Determination of pentachlorophenol in marine biota and sea water by gas-liquid chromatography and high pressure liquid chromatography. *J Agric Food Chem* 27 (1979) 554-557.

Fogelqvist E, B Josefsson, C Roos. Determination of carboxylic acids and phenols in water by extractive alkylation using pentafluorobenzylation, glass capillary GC and electron capture detection. *J High Res Chromatogr* 3 (1980) 568-574.

Giger W, C Schaffner. Determination of phenolic water pollutants by glass capillary gas chromatography. In "Advances in the identification and analysis of organic pollutants in water", 1 H Keith (ed), *Ann Arbor Science Publ* in press (1981).

Grob K, G Grob, K Grob Jr. Deactivation of glass capillary columns by silylation. Part 1. Principles and basic techniques. *J HRC & CC* 2 (1979) 31-35.

Hattula M L, H Reunanen, V-M Wasenius, R Krees, A U Arstila. Toxicity of 4-chloro-o-cresol to fish. Light microscopy and chemical analysis of the tissue. *Bull Environ Contam Toxicol* 22 (1979) 508-511.

Jensen S, L Renberg, L Reutergårdh. Residue analysis of sediments and sewage sludge for organochlorines in the presence of elemental sulfur. *Anal Chem* 49 (1977) 316-318.

Kawahara F K. Gas chromatographic analysis of mercaptans, phenols and organic acids in surface waters with the use of pentafluorobenzyl derivatives. *Environ Sci Technol* 5 (1971) 235-239.

Keith L H. Analysis of organic water pollutants. *Environ Sci Technol* 13 (1979) 1469-1471.

King W P, K T Joseph, P T Kissinger. Liquid chromatography with amperometric detection for determining phenolic preservatives. *JAOAC* 63 (1980) 137-142.

Kozak V P, G V Simsiman, G Chesters, D Stensby, J Harkin. Reviews of the environmental effect of pollutants: XI. Chlorophenols. Oak Ridge Nat Lab ORNL/EIS-128, June 1979 (EPA/600/10).

Lindström K, J Nordin. Gas chromatography - mass spectrometry of chlorophenols in spent bleach liquors. *J Chromatogr* 128 (1976) 13-26.

Mc Callum N K, R J Armstrong. The derivatization of phenols for gas chromatography using electron capture detection. *J Chromatogr* 78 (1973) 303-307.

Pierce R H, D M Victor. The fate of pentachlorophenol in an aquatic ecosystem in "Pentachlorophenol", K Ranga Rao (ed), Plenum Publ Corp (1978) 41-51.

Pinkerton K A. Direct LC analysis of selected priority pollutants in water at ppb levels. *J of HRC & CC* 4 (1981) 33-34.

Realini P A. Determination of priority pollutant phenol in water by HPLC. *J Chromatogr Sci* 19 (1981) 124-129.

Renberg L. Ion exchange technique for the determination of chlorinated phenols and phenoxy acids in organic tissue, soil and water. *Anal Chem* <u>46</u> (1974) 459-461.

Renberg L. Gas chromatographic determination of phenols in water samples as their pentafluorobenzoyl derivatives. *Chemosphere* <u>10</u> (1981a) 767-773.

Renberg L. Gas chromatographic determination of chlorophenols in environmental samples. Natl Swedish Environment Protection Board, report SNV PM 1410 (1981b).

Renberg L, K Lindström. C_{18}-reversed phase trace enrichment of chlorinated phenols, guaiacols and catechols. *J Chromatogr* <u>214</u> (1981) 327-334.

Rodgers I H. Isolation and chemical identification of toxic components of kraft mill wastes. *Pulp Pap Mag Can* <u>74</u> (1973) T303-T308.

Rossum P van, R G Wegg. Isolation of organic water pollutants by XAD resins and carbon. *J Chromatogr* <u>150</u> (1978) 381-392.

Rudling L. Determination of pentachlorophenol in organic tissues and water. *Water Res* <u>4</u> (1979) 533-537.

Shafik T M, H C Sullivan, H R Enos. Multiresidue procedure for halo- and nitrophenols. Measurement of exposure to biodegradable pesticides yielding these compounds as metabolites. *J Agric Food Chem* <u>21</u> (1973) 295-298.

Stark A. Analysis of pentachlorophenol residues in soil, water, and fish. *J Agric Food Chem* <u>17</u> (1969) 871-873.

Supelco Inc, Bellefonte, Pennsylvania 16823, USA. A single capillary column for the separation of priority water pollutants. GC reporter V (September 1980).

ANALYSIS OF ALKYLPHENOLS IN AN AQUEOUS MATRIX

CONTAINING AROMATIC HYDROCARBONS

D. BOTTA, F. MORANDI and E. MANTICA

Politecnico - Istituto di Chimica Industriale "G.NATTA"
Piazza Leonardo da Vinci, 32 - 20133 MILANO

Summary

The problem of analysing alkylphenols in an aqueous matrix with a
more or less high content of aromatic hydrocarbons may be met when
performing the analysis of ground or surface waters in proximity
of factories where coal tar is produced or processed.
As concerns the analysis of phenols, a general method based on
liquid-liquid extraction at different pH values was proposed.
We applied this method, but with negative results: the extraction
yields were low, and great differences were evidenced for alkylphenols
with the alkyl substituents in different positions.
We have adopted a modified method which foresees the removal by the
purging of the aromatic hydrocarbons from the aqueous sample, the
acetylation of alkylphenols at controlled temperature and pH, the
adsorption on resin (Chromosorb 102), the subsequent elution with
carbon disulfide, the gas chromatographic analysis on a persilanized
glass capillary column.
The method may be applied to clear ground waters, whereas it may
involve difficulties or interferences as concerns industrial waste
waters or cloudy waters.

1. INTRODUCTION

The identification and quantitation of phenol and alkylphenols C_7-
C_9 (cresols, xylenols, ethylphenols, trimethylphenols, etc.) in the
ground or surface waters represent an analytical problem which can be
met in the proximity of many industrial plants. It is sufficient to
mention the coke ovens and the tar distilleries, the industries which
use alkylphenols as raw materials for chemical syntheses (production
of disinfectants, dyes, pharmaceuticals, pesticides, explosives, antioxi
dants, etc.), the petrochemical industries etc. Recently special attention
was paid to the alternative fuels (shale-oils and solvent-refined coal)
in which phenol concentrations considerably higher than those found in
petroleum can be present.
Alkylphenols are frequently present in traces in aqueous matrices
which contain more or less important quantities of interfering substances,
in particular of aromatic hydrocarbons (alkylbenzenes, alkenylbenzenes,

alkylindanes, alkylnaphthalenes, alkyldiphenyls and so on). In these conditions, their analysis involves serious difficulties since the methods commonly used may be exposed to interferences due to the main pollutants.

Phenols are compounds which have a particular influence on the delicate environmental equilibria for their well-known toxic effects on fish and other forms of aquatic life. We note that in the EPA list of the priority pollutants (1) only phenol and 2,4-dimethylphenol have been included. However, also the other alkylphenols may represent a danger to the public health especially after their conversion, always possible, into chlorophenols, a class of even more toxic compounds which give water a disagreeable smell and taste and prevent its use in many fields.

2. SELECTION OF THE ANALYTICAL METHOD

Many methods had been suggested in the specialized literature for the determination of the phenolic compounds and we have taken them into consideration with a view of choosing a method suitable for our purposes.

a - Colourimetric Methods.

They are based on the reaction between the phenolic compound and a reagent which develops a characteristic colour, thus allowing the determination of the concentration on the basis of an absorbance measure at a given wavelength. Among these reagents, the more widely known and used is 4-aminoantipyrine, also mentioned in the standard method of test D 1783-70 of ASTM (2). None of these methods was considered suitable for our purposes because of their many drawbacks. The colourimetric methods give total phenol concentration rather than showing the presence of the more toxic components, do not allow us to evaluate some para-substituted phenols and give results which are lower than the real concentration values. They give incomplete reactions as a function of the type of phenolic compound analyzed and reagent used, and finally they are not very accurate for spectrophotometric reasons (displacement of the maximum of the absorption band of derivatives of substituted phenols in respect to that of phenol derivative).

b - Spectrophotometric Methods.

They are based on ultraviolet spectrophotometric measurements of phenolic compounds in basic solution or on infra-red absorption of phenols brominated in ortho position. These methods too, were not considered suitable for our purposes since they involve certain difficulties. The analytical information is only general,difficulties may arise with the ionization of compounds when high steric hindrance substituting groups exist at positions 2 and 6 in respect to the phenolic hydroxyl (UV measurements), and with bromination in ortho position of phenols which have positions 2 and 6 occupied by other groups (IR measurements).

c - Gas Chromatographic Methods.
    They are based on gas chromatographic separation, on non-polar,
    polar or porous polymer (type Tenax GC) packed columns, of phenolic
    compounds which can be injected 1. directly into an aqueous solu-
    tion, 2· after pre-concentration in suitable solvents, 3· after
    derivatization with silylating, alkylating or acylating reagents.

    Direct aqueous injection procedure: this is the method described
    in the standard method of test D. 2580-68 of ASTM (3) and recommended
    for the determination of phenol, cresols, monochlorophenol, dichloro-
    phenols in water for concentrations higher than 1 mg/l. The gas chro-
    matographic analysis is carried out in a column packed with Carbowax
    20M with TPA or FFAP as stationary phase; however, Bartle et al. (4)
    recommended the use of Tenax GC as a more suitable phase. We discarded
    this method because of its insufficient sensitivity, of the difficul-
    ties arising from the large quantities of injected water, of the
    frequency of ghost peaks and of the presence of many interfering
    compounds.

    Injection of Solutions Obtained from Pre-concentration Treatments:
    The following different treatments of isolation and pre-concentration
    of the phenolic compounds from the aqueous matrix were suggested:

1. Discontinuous or Continuous Liquid-Liquid Extraction (5-8) with
    solvents like ether or methylene chloride, followed by the
    reduction of the extract to a small volume in such a way as to
    obtain enrichment factor ranging from 10^2 to 10^4.
    The extraction techniques are adversely affected by drawbacks
    such as the great differences in the values of the partition
    coefficients between the aqueous phase and the solvent with
    consequent different recoveries of the analyzed phenols, the
    presence of variable quantities of impurities in the solvent
    which, after concentration, may interfere with the gas chromato-
    graphic analysis of alkylphenols, the time-consuming operations
    of extraction and concentration.

2. Adsorption of Phenolic Derivatives on macro-reticular resins
    (9,10) followed by the elution with a solvent and the partial
    removal of the latter. By properly choosing the quantities of
    sample, resin and solvent, losses during distillation can be
    prevented. Also, when using these methods, difficulties are met
    with owing to the different values of the partition coefficients
    resin/aqueous matrix and resin/solvent for differently substituted
    phenolic compounds.

3. Ion Exchange on Anionic Resins (6, 11) followed by elution with
    aqueous hydrochloric acid, which re-transforms the phenolated
    ions into free phenols, and then with an acetone-water mixture.
    The eluates are extracted with methylene chloride and the organic
    phase is concentrated by distillation. This process is very
    laborious and requires additions of reagents and solvents which

introduce impurities into the sample with the possibility of
interference. The recovery of the phenolic compounds is a function
of the pH of the aqueous solution, but the addition of sodium
hydroxide to very hard waters to obtain pH values higher than 12,
brings about the precipitation of calcium and magnesium carbonates
which obstruct the column containing the resin or the filtering
elements thus extending the duration of the operation and favouring
the losses of phenols due to oxidation.

4. Steam distillation (6,12) followed by liquid-liquid extraction
   with chloroform or ether. This process can be useful to separate
   the phenolic compounds from non-volatile substances or inorganic
   salts but, however, it involves, in the course of the extraction,
   the same difficulties as those previously mentioned in 1.

Injection of Solutions Obtained from Derivatization Reactions.
The derivatization methods of the phenolic compounds were introduced
in order to obtain three different results:

1. To facilitate the gas chromatographic separation of this class of
   compounds which, owing to their high polarity, pose serious problems due to adsorption phenomena on the solid support or on the
   walls of the column and prevent, in practice, the use of non-
   polar phases.

2. To differentiate, through the formation of derivatives, the gas
   chromatographic behaviour of the phenolic compounds which have
   very similar retention parameters.

3. To increase the sensitivity of the method of analysis forming derivatives which can be detected with selective detectors   (type ECD).
The commonly used derivatization processes are silylation, alkylation and acylation.

Silylation (13, 14 and cited bibliography). The conversion of alky
alkylphenols into trimethylsilylethers is obtained through the action
of a silylating agent (such as hexamethyldisilazane, bis(trimethylsilyl)-
trifluoroacetamide and similar). It is rapid, sufficiently quantitative,
in spite of the steric influences of the alkyl groups in ortho position; the chromatographic properties of the derivatives are good.

Alkylation(5,15,16,17 and cited bibliography). The phenolic compounds
may react with alkylating agents, such as diazomethane, diazoethane,
pentafluorobenzylbromide, 1-fluoro-2,4 dinitrobenzene and similar,
thus bringing about ethers having a polarity lower than that of
phenols. Particularly interesting are the derivatives obtained from
the last two reagents which impart to the alkylated molecules a
response to the electron-capture detectors and allow better sensitivity and selectivity. It can be pointed out however that the alkylating
agents have different reaction times as a function of the nature,
number and position of the substituents, so that conversions are usually
not quantitative. Also the pH value of the processed mixture can
considerably influence the formation of the derivatives.

Acylation (18-21 and cited bibliography). The phenolic compounds
react with acylating agents such as aliphatic anhydrides (mainly
acetic),perfluorinated aliphatic anhydrides (trifluoroacetic, penta
fluoropropionic, heptafluorobutyric),heptafluorobutyrylimidazole
and similar, to form esters having a better gas chromatographic
behaviour. Acyl derivatives can be prepared under different conditions:
A - Acylation of phenolic compounds after extraction of the aqueous
    matrix with a solvent.
B - **Direct** acylation in aqueous phase followed by the extraction of
    the acylated derivatives obtained with a solvent or by the
    adsorption of the same in a column containing a layer of macro-
    reticular resin followed by the elution with a solvent.
Both these reaction paths have been considered in the papers cited.
In case A - the same drawbacks as those common to all the extraction
processes are met. Furthermore, there are the effects of the different
reactivity of the differently substituted phenolic compounds which
affect the conversion yields. In case B - the difficulties due to
the latter cause of error still remain and, furthermore, there are
drawbacks due to the values of the resin/aqueous matrix and resin/
solvent partition coefficients which, however, are more favourable
than in the case of phenols. The operating condition of the acylation
reaction (pH, temperature, ratios among reagents) must be strictly
controlled since each one of these factors remarkably affects the
conversion yields, especially in the case of polysubstituted
alkylphenols.

d - HPLC Chromatographic Methods.
Many reports were published on the use of these techniques in the
analysis of phenolic compounds (22 and cited bibliography). These
methods received no further consideration owing to the lack of
adequate equipment at our laboratory.
In view of the points briefly explained hereabove and of the
difficulties which arise while using the different methods, we
decided to re-examine more closely the methods of the liquid-liquid
extraction and of direct acetylation of phenols in aqueous phase.

3. LIQUID-LIQUID EXTRACTION

We applied to the mixtures of phenols prepared by us the liquid-
liquid extraction process proposed by EPA (1) for the quantitation of
the 11 phenols considered to be priority pollutants in the industrial
waste waters. This method proposes two different approaches:

1 - Extraction of neutral/basic compounds and then of acidic compounds
    with methylene chloride in basic (pH 12) and acidic(pH 2) solution.

2 - Extraction of acidic neutral compounds and then of basic compounds
    with methylene chloride in acidic (pH 2) and basic (pH 12) solution.
    To meet our requirements, we chose the first approach and operated
    on aqueous solutions containing approximately 20 ppb of each of the
    8 alkylphenols selected in such a way as to represent the behaviour

of compounds substituted by alkyl groups in different positions. From our experimental work some observations were made which can be summarized as follows:

a - The removal of neutral or basic compounds present in the sample is satisfactory and their interferences with the gas chromatographic analysis are reduced or eliminated.

b - The recovery of the alkylphenols is, in general, incomplete with a trend to a selectivity in the extraction which favours some phenolic compounds more than others. A portion of alkylphenols enters prevalently the fraction containing the neutral/basic compounds rather than the fraction of the acidic compounds, as shown in fig. 1 and 2.

c - The large quantities of solvent used for the extraction introduce interfering impurities which make the detectability limit of alkylphenols worse.

d - During the distillations made in order to reduce to a small volume the methylene chloride extracts, alkylphenol losses are noted.

First of all, the observation pointed out in b - is, in our opinion, of great importance. We do not have to date reliable quantitative data considering the remarkable differences in the values of the response factors of the different alkylphenols for the detector used (FID). Semiquantitative measurements, however, indicate that there is a special influence on the recoveries of the alkylphenols with at least one occupied ortho position, in particular 2,4-dimethylphenol, 2,4,6-trimethylphenol, 2,3,5-trimethylphenol. These are compounds whose pK (23) value is higher than that of the phenols which are prevalently found in the acidic fraction. This difficulty, however, does not arise in the more frequently studied analyses of chlorophenols having clearer acidic characteristics.

4. DIRECT ACETYLATION OF ALKYLPHENOLS IN AQUEOUS PHASE.

The method we used was derived from the Coutts, Hargesheimer and Pasutto papers (20, 21) with some modifications concerning the control of pH of the solution (buffered to 10 in order to better recover the less acidic phenols), of the reaction temperature (reactor thermostated to 18°C) and of ratios among reagents. Furthermore, we employed a different method for the recovery of the acetylated compounds from the aqueous phase, based on the adsorption of same on a macro-reticular resin bed. The compounds retained by the resin are then eluted with small quantities of a solvent (CS_2).

Using the operating procedure indicated above on a sample of water containing both alkylphenols and aromatic hydrocarbons (naphtha solvent) the chromatogram in Fig. 3 was obtained, where the peaks of unreacted phenols and acetylated compounds can be hardly distinguished from those of the alkylbenzenes which have similar retention times. To obviate this difficulty, on the basis of our previous experiences in the recovery of aromatic hydrocarbons from water by purging with an inert gas at 70°C, we removed the interfering compounds before acetylation.

By setting the nitrogen flow-rate between 50 and 500 ml/min as a function of the range of the boiling points of the polluting aromatic hydrocarbons, we substantially simplified the chromatogram as shown in Fig. 4.

In this figure are shown, as main peaks, those pertaining to the 8 acetylated alkylphenols and, as weak peaks, those due to the nonderivatized residual phenols. Only for 2,4,6-trimethylphenol, a compound remarkably inhibited owing to the presence of two methyl groups in the two ortho position, is the conversion broadly incomplete and the peak of the non-acetylated 2,4,6-trimethylphenol has an area which can be compared with that of the peak of the acetyl derivative. It is not possible to date to report reliable quantitative data allowing to evaluate the completeness of the conversion and recovery of the acetylated phenols. Indicative quantitative measurements showed that the response factors greatly differ according to the compound and that the disregarding of these differences involves errors higher than 20%. We are now preparing and purifying the different acetylated compounds in order to refine the results obtained and to improve the criteria of evaluation of the whole procedure.

5. EXPERIMENTAL

Apparatus:
Gas chromatograph, C. Erba Mod. 2150, equipped with Grob type split/splitless injector and flame ionization detector (FID).
Recorder, Leeds and Northrup Italiana, Mod. Speedomax XL 681 B (1 mV f.s.; chart speed 1 cm/min.).
Computing Integrator, Spectra-Physics Mod. 4100.
Capillary column, persilanized Pyrex glass column, coated with silicone rubber JXR (0.27 mm I.D., length 16 m, film thickness 2.7 μm).
Ultra-thermostat, NB Colora.
Glassware:
1 l gas-washing bottles (purging apparatus), equipped with a fritted disc (porosity 2).
Tubular traps, glass tubes (100 x 6 mm O.D. x 4 mm I.D.), containing 150 mg Chromosorb 102 (80-100 mesh).
1 l separatory funnels, with teflon stopcocks.
Graduated test tubes, 1 ml.
Kuderna-Danish evaporators.

Reagents:
The following reagents were used for the different operations:
Gas: hydrogen, nitrogen, air, pure for gas-chromatography
Phenols: phenol, o-cresol, m-cresol, 2,4-xylenol, 3,5-xylenol, p-ethylphenol, 2,3,5-trimethylphenol, 2,4,6-trimethylphenol, of purity higher than 98%, were contained in a kit supplied by Supelco and not submitted to further purification treatments.

Solvents: Methylene chloride (RS Erba for pesticide determination), sufficiently pure to obtain a blank compatible with the requested detectable concentration limit (1 ppb).
Carbon disulphide (RS Erba for spectrophotometry).

Acetone (Merck UVASOL for spectrophotometry).
Acetic anhydride: Merck, chemically pure pro analysi.
Sodium sulphate anhydrous: Merck, chemically pure pro analisi.
Sodium bicarbonate : RPE Erba.
Sodium hydroxide : pellets, RPE-ACS Erba.
Hydrochloric acid : Ultrex of J.T. Baker Chemical Co.
Water : free from phenols.
Chromosorb 102 : Johns Manville, 80-100 mesh, washed with acetone before use.

Preparation of Standard Solutions

Alkylphenols : In a 10 ml volumetric flask, approximately 10 mg of each of the different considered alkylphenols were weighed and acetone was added to make a total volume of 10 ml. A solution which is approximately 1‰ of each phenol was obtained and then used for the different tests.

Naphtha solvent : In a 10 ml volumetric flask, approximately 50 mg of naphtha solvent were weighed and the flask was filled to the mark with ethanol thus obtainng a solution at approx. 5‰ .

Extraction Procedure

The typical procedure used was the following: in a 1 1 separatory funnel 500 ml of phenol-free water and 10 µl of phenol standard solution were introduced by means of a 50 µl syringe, taking care of inserting the tip of the needle just below the surface of the liquid. The pH of the solution was brought to 12 by addition of NaOH 6N and then three extractions were made with methylene chloride (150 + 75 + 75 ml). The collected extracts, containing the neutral and basic compounds present in the sample, were dried on anhydrous sodium sulphate and reduced to a small volume (less than 1 ml) in a Kuderna-Danish evaporator. The residual aqueous phase was acidified to pH 2 with 6M hydrochloric acid and extracted three times with methylene chloride (100 + 50 + 50 ml). The collected extracts, containing the compounds having an acidic character, were dried on anhydrous sodium sulphate and reduced to less than 1 ml in a Kuderna-Danish evaporator. The concentrated liquids from the two extractions were transferred to 1 ml graduated test tubes and these were filled to the mark with methylene chloride.

Purge, Acetylation, Adsorption and Elution

The typical procedure used was the following: in a gas-washing bottle 500 ml of water, 10 µl of alkylphenols standard solution and 200 µl of naphtha solvent standard solution were introduced.The bottle was immersed in a water bath heated to 70°C.When temperature equilibrium was reached,the bottle was connected to the nitrogen supply and the gas was allowed to bub - ble with a flow-rate of 500 ml/min for 30 minutes in order to remove all the purgeable compounds present therein (aromatic hydrocarbons of naphtha solvent).The flow of gas was reduced to 20 ml/min and the gas-washing bottle was cooled and thermostated to 18°C. After about half an hour,the temperatu re of the sample stabilized to this value, 1.71 g of sodium bicarbonate and 0.5 g of sodium hydroxide were added.When the dissolution of the two salts was completed, the nitrogen flow was discontinued. Meanwhile, the trap containing the adsorbing resin, usually preserved in methylene chloride, was washed separately with carbon disulphide and dried in a nitrogen stream. The trap was then mounted on the nitrogen inlet pipe of the washing bottle. 400 µl of acetic anhydride were added to the water and, as soon as carbon dioxide evolution stopped, the bottle was pressurized with nitrogen (0.3 kg/cm2) in such a way as to force the liquid through the adsorbing trap. Under this pressure the flow-rate was 7-10 ml/min. As soon as the flow of the aqueous solution was completed the trap was connected directly to the nitrogen line and dried for 15 minutes under a gas flowrate of 500 ml/min. The adsorbed compounds were eluted with 1 ml of carbon disulphide and the solution was collected in a 1 ml gradua-ted test tube and filled to the mark with solvent.

Gas chromatographic conditions
The injection of the sample was made according to the Grob's splitless technique; 2 µl of sample solution were injected at a temperature of 275°C and a splitless period of 45 sec. was used. Hydrogen was the carrier gas (3 ml/min). The gases for the detector (FID) were hydrogen (30 ml/min) and air (300 ml/min). The temperature conditions of the oven were the following:
room temperature at injection,
to 70°C ballistically after 45 sec.
70°C isothermal for 2 minutes
70-250°C at 4°C/min
250°C final isotherm.

6. CONCLUSIONS

Considering the critical examination of the methods recommended in the literature and the results obtained, we have come to the conclusion that none of the available methods is able to give reliable results when phenol concentrations are as low as those considered by us and when rather considerable quantities of aromatic hydrocarbons are present. This is still more evident when we consider the real problems which arise when waters contain large quantities of inorganic salts or surfactants which, during the extractions, bring about the formation of foam or persistent emulsions. The continuous extraction, which can be of help on these occa-sions, is time-consuming and involves the risk of phenol losses by oxydation in basic medium. Other drawbacks associated with the extractive processes have been pointed out previously: the most serious of them is the incompleteness of the recovery which is especially bad at the lower concentrations. Even though the extraction with solvent at basic pH represents an efficient means for the removal of the interfering aromatics, this method cannot be considered satisfactory under the quantitative aspect.

As far as the analytical problem under examination is concerned, better results are obtained with acetylation in aqueous solution if the precaution of previously eliminating the interfering compounds by purge with an inert gas is used. The conversions of the various phenols are

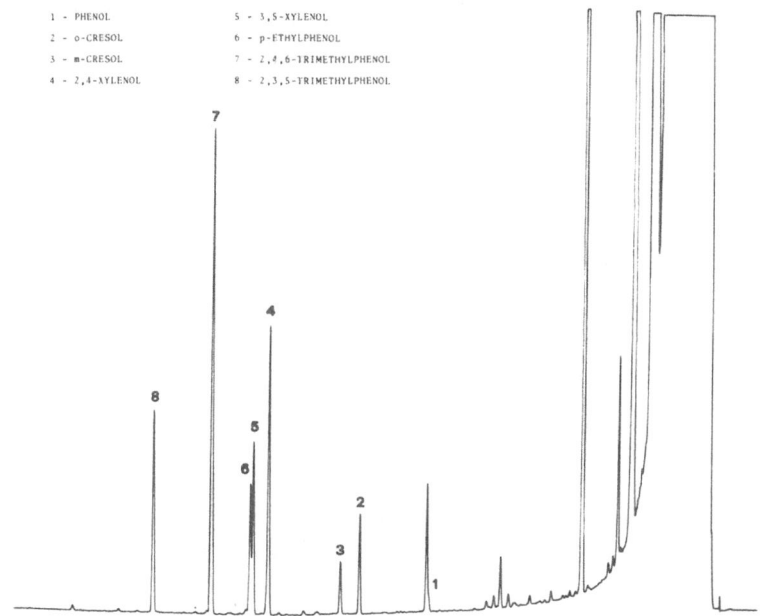

1 - PHENOL
2 - o-CRESOL
3 - m-CRESOL
4 - 2,4-XYLENOL
5 - 3,5-XYLENOL
6 - p-ETHYLPHENOL
7 - 2,4,6-TRIMETHYLPHENOL
8 - 2,3,5-TRIMETHYLPHENOL

Fig. 1 – CHROMATOGRAM OF ALKYLPHENOLS EXTRACTED AT pH 12

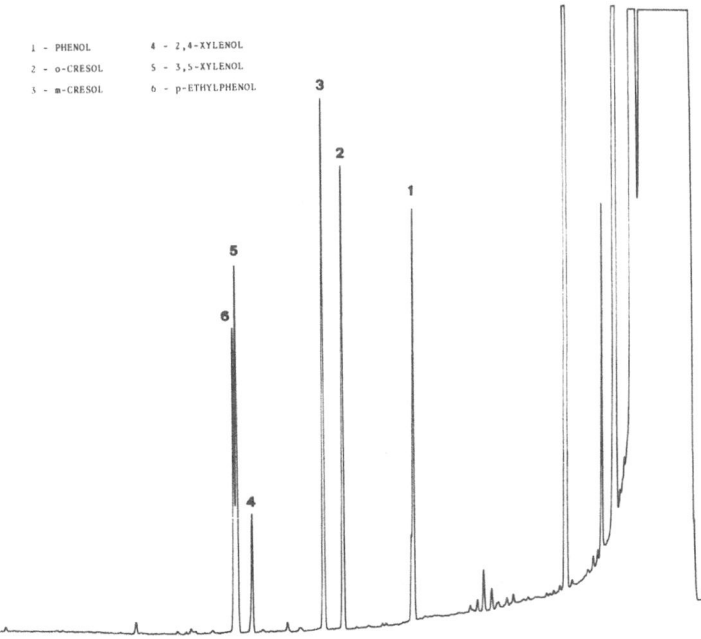

1 - PHENOL
2 - o-CRESOL
3 - m-CRESOL
4 - 2,4-XYLENOL
5 - 3,5-XYLENOL
6 - p-ETHYLPHENOL

Fig. 2 – CHROMATOGRAM OF ALKYLPHENOLS EXTRACTED AT pH 2

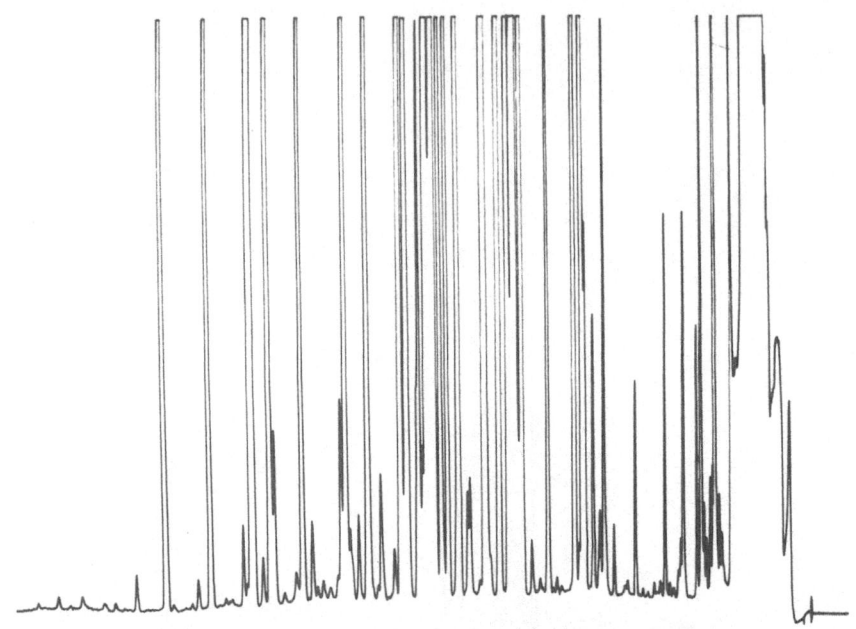

Fig. 3 — CHROMATOGRAM OF CARBON DISULFIDE ELUATE FROM THE
RESIN TRAP, ACETYLATION WITHOUT PURGE PRETREATMENT

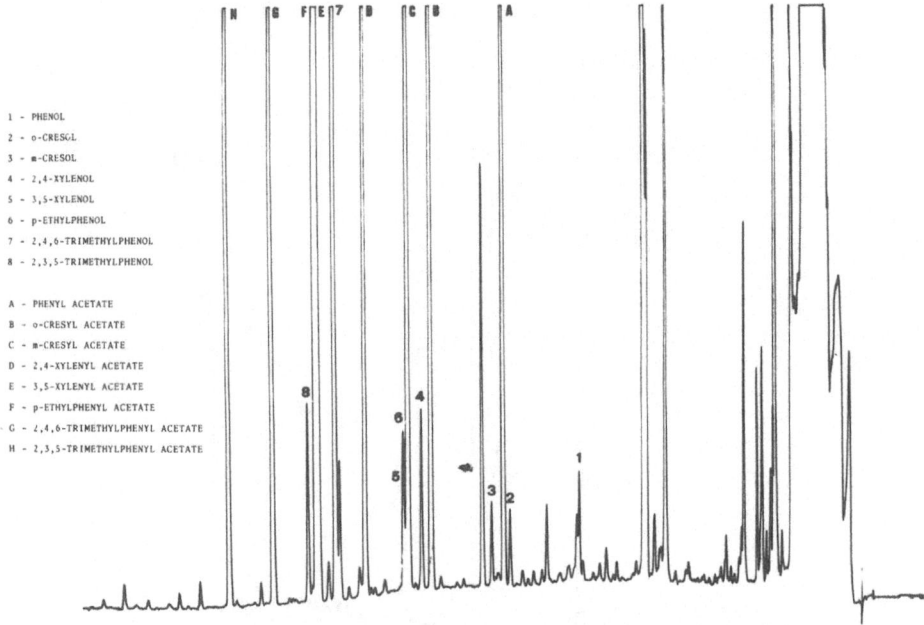

1 - PHENOL
2 - o-CRESOL
3 - m-CRESOL
4 - 2,4-XYLENOL
5 - 3,5-XYLENOL
6 - p-ETHYLPHENOL
7 - 2,4,6-TRIMETHYLPHENOL
8 - 2,3,5-TRIMETHYLPHENOL

A - PHENYL ACETATE
B - o-CRESYL ACETATE
C - m-CRESYL ACETATE
D - 2,4-XYLENYL ACETATE
E - 3,5-XYLENYL ACETATE
F - p-ETHYLPHENYL ACETATE
G - 2,4,6-TRIMETHYLPHENYL ACETATE
H - 2,3,5-TRIMETHYLPHENYL ACETATE

Fig. 4 — CHROMATOGRAM OF CARBON DISULFIDE ELUATE FROM THE
RESIN TRAP, ACETYLATION AFTER PURGE PRETREATMENT

never complete, but, in many cases, they are higher than 95% even when concentrations are very low. Regarding less reactive compounds, the close control on the acetylation conditions may allow us to obtain reproducible conversions and the correction by calibration processes of the deviations observed. The recovery of acetyl derivatives by adsorption on resin causes some difficulties when waters are cloudy or hard. In the latter situation, operating at basic pH, it is possible to obtain the precipitation of the dissolved calcium or magnesium salts which tend to obstruct the filter or the resin trap and prolong the percolation times.

BIBLIOGRAPHY

1. Guidelines establishing test procedures for the analysis of pollutants; proposed regulations. Federal Register 44, 69464 (1979)

2. ASTM, 1978 Annual Book of ASTM Standards. Pt. 31, pp.650-655, American Society for Testing and Materials, Philadelphia, 1978.

3. ASTM, 1978 Annual Book of ASTM Standards. Pt. 31, pp.656-663, American Society for Testing and Materials, Philadelphia, 1978.

4. Bartle K.D., Elstub J., Novotny M. e Robinson R.J., J.Chromatogr. 135, 351-358 (1977).

5. Shackelford W.W. e Webb R.G., "Survey analysis of phenolic compounds in industrial effluents by gas chromatography-mass spectrometry", in Measurement of organic pollutants in water and wastewaters, ASTM STP 686, C.E. Van Hall, Ed., American Society for Testing and Materials, Philadelphia, 1979, pp.191-205.

6. Mousa J.J. e Whitlock S.A., "Analysis of phenols in some industrial wastewaters", in Measurement of organic pollutants in water and wastewaters, ASTM STP 686, C.E. Van Hall, Ed., American Society for Testing and Materials, Philadelphia, 1979, pp.206-220.

7. Hertz H.S., Brown J.M., Chessler S.N., Guenther F.R., Hilpert L.R., May W.E., Parrish R.M., Wise S.A., Anal. Chem. 52, 1650-1657 (1980).

8. Guenther F.R., Parrish R.M., Chessler S.N. e Hilpert R.L., J.Chromatogr. 207, 256-261 (1981).

9. Junk G.A., Richard J.J., Grieser M.D., Witiak D., Witiak J.L., Arguello M.D., Vick R., Svec H.J., Fritz J.S., e Calder G.V., J. Chromatogr. 99, 745-762 (1974).

10. Dressler M., Chromatogr. Reviews 23, 167-206 (1979).

11. Chriswell C.D., Chang R.C. e Fritz J.S., Anal.Chem. 47, 1325-1329 (1975).

12. APHA, Standard methods for the examination of water and wastewaters, 14th ed., American Public Health Association, Washington, D.C., 1975 pp.576-577.

13. Rossemyr L.I., J. Chromatogr., 170, 463-467 (1979).

14. Tullberg L., Peetre I.B. e Smith B.E.F., J.Chromatogr. 120, 103-113 (1976).

15. Davis B., Anal.Chem. 49, 832-834 (1977).

16. Rosenfeld J.M. e Crocco J.L., Anal.Chem. 50, 701-704 (1978)

17. Lehtonen M., J. Chromatogr., 202, 413-421 (1980).

18. Shulgin A.T., Anal.Chem. 36, 920-921 (1964).

19. Lamparski L.L. e Nestrick T.J., J.Chromatogr., 156, 143-151 (1978).

20. Coutts R.T., Hargesheimer E.E. e Pasutto F.M., J.Chromatogr., 179, 291-299 (1979).

21. Coutts R.T., Hargesheimer E.E. e Pasutto F.M., J. Chromatogr., 195, 105-112 (1980).

22. Realini P.A., J.Chromatogr.Sci. 19, 124-129 (1981).

23. Kortüm G., Vogel W. e Andrussow K., Dissociation constants of organic acids in aqueous solution. IUPAC Commission on Electrochemical Data, Butterworths, London, 1961, pp. 428-437.

ANALYSIS OF NITROGENOUS ORGANIC SUBSTANCES IN WATER

M.ELMGHARI-TABIB, C.LE CLOIREC, J.MORVAN, G.MARTIN
Ecole Nationale Supérieure de Chimie de Rennes
Avenue du Général Leclerc
35000 RENNES
FRANCE

Summary

Elaboration of an analytical protocol which pats up with the analysis
of most nitrogenous substances which might exist in the surface or
drinking waters : amino-acids, herbicides and nitro-derivatives. Gas
and liquid chromatography, with electron capture detectors for the
analysis of herbicides and nitro-derivatives, and by fluorescence for
the amino-acids'one, permits to reach high sensitiveness. Concentration
of products is necessary either on the ions exchange resin, or by
extraction with an organic solvent.

1. INTRODUCTION

Les méthodes classiques de dosage de l'azote organique permettent
d'obtenir une valeur mesurant la charge d'une eau, mais elles ne fournissent
aucun renseignement sur la nature des molécules et des fonctions présentes.

(Nous rappelons que certaines fonctions azotées comme les groupements
nitro- ne sont pas pris en compte dans le dosage NTK). Les eaux de notre
région sont tributaires de rejets urbains, agricoles et d'industries agro-
alimentaires. Les composés azotés les plus probables sont donc les protéïnes et leurs produits de dégradations ainsi que les biocides. Le premier groupe représente de l'ordre de 70% de l'azote organique dans les
eaux de surface (1) .
Les quelques études consacrées à l'identification des produits organiques azotés dans les eaux de surface sont fragmentaires. STANDER (2)
indique qu'un faible pourcentage de ces molécules a été identifié dans les
eaux potables : environ 10% aux USA et 7% en Europe. Nous avons cherché
à mettre au point une méthodologie pour approfondir le problème.
Les études analytiques réalisées à ce jour portent généralement sur
un produit ou une famille de produits. Les principales difficultés viennent
de la variété des produits présents (3) et des faibles concentrations
rencontrées, souvent inférieures aux limites de détection.
Nous pouvons faire une distinction entre les formes macromoléculaires
facilement éliminées par floculation ou ultrafiltration et les formes
solubles plus résistantes à ces traitements.
L'analyse des acides aminés provenant de la dégradation des protéïnes
a suscité de nombreux travaux. Les différents auteurs semblent d'accord
pour effectuer une concentration sur résines échangeuses d'ions, mais
pour la séparation ils utilisent des techniques chromatographiques diverses:

estérification et analyse par CPG (4 et 5) avec possibilité de détecter
les produits séparésen spectrométrie de masse, ou chromatographie liquide
et détection par fluorescence(6).

Pour ce qui concerne les autres formes d'azote organique, la diver-
sité est encore plus grande, et pour les méthodes de concentration le
choix est donné entre un passage sur résines macromoléculaires et une
extraction liquide-liquide. De même la partie séparation et identification
utilise aussi bien la chromatographie (7) sous toutes ses formes que la
colorimétrie (8). Toutefois il ressort de toutes ces études ponctuelles
que le dichlorométhane apparaît comme un solvant préférable pour extraire
les composés organiques azotés des solutions aqueuses puisqu'il a été
utilisé pour la concentration des amines aromatiques (8), des pesticides
azotés (9) (10) (7), les carbamates (12), les urées substituées (11) et les
nitrosamines (13).

2. MISE AU POINT D'UN PROTOCOLE D'ANALYSE

Les données bibliographiques montrent que les deux problèmes majeurs
posés par l'analyse de la micropollution en général et azotée en particulier sont d'une part la concentration et d'autre part la séparation et la
détection. Ces deux contraintes dépendent principalement des caractéristiques
physico-chimiques des substances analysées.

2.1. Concentration

2.1.1. Acides aminés

Ces produits sont très solubles dans l'eau et ont des températures
de fusion assez élevées ; de plus, leur caractère ionique est bien connu :

$$R-C\begin{smallmatrix}COO^-\\NH_2\end{smallmatrix} \xleftarrow[pK]{pK_2\atop -H^+} R-C\begin{smallmatrix}COO^-\\NH_3^+\end{smallmatrix} \xrightarrow[+H^+]{pK_1} R-C\begin{smallmatrix}COOH\\NH_3^+\end{smallmatrix}$$

forme anionique   point isoionique   forme cationique

$$pH_i = 1/2 \; (pK_1 + pK_2)$$

Selon le pH du milieu et le pKa de l'acide aminé , nous nous trouvons
en présence soit de la forme anionique (pH > pK$_2$) soit de la forme cationique (pH < pK$_1$) . Pour un pH très acide (< 2) tous les acides aminés sont de
forme cationique, cette propriété nous a permis la concentration sur résine
échangeuse de cations forte IRC 120. C'est une Amberlite sulfonique et qui
retient les acides aminés selon la réaction :

$$Amb-SO_3^- \; Na^+ \; + \; R-NH_3^+ \longrightarrow Amb-SO_3^- \; NH_3^+ -R \; + \; Na^+$$

Les produits retenus sont élués ensuite en percolant une solution
basique :

$$Amb-SO_3^- \; NH_3^+ -R \; + \; Na^+ OH^- \longrightarrow Amb-SO_3^- \; Na^+ \; + \; R-NH_2 \; + H_2O$$

Pour éviter l'élution des métaux, nous avons utilisé l'ammoniaque (N).
Après évaporation à sec le résidu est repris avec une solution tampon.
Nous pouvons ainsi obtenir des coefficients de concentration allant de 200
à 1000 .

Nous avons testé plusieurs solutions d'acides aminés à 10 mg/l, seuls
ou en mélange . Les résultats obtenus montrent que l'efficacité de la

de la résine est comprise entre 80 et 100% . (Figure I et Tableau I)

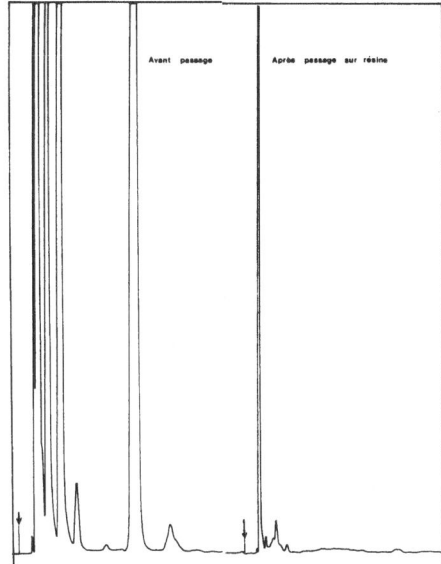

Fig. I — Mélange d'acides aminés
avant et après passage
sur résine

Acides aminés	efficacité %
Thréonine	86
Asparagine	91
Alanine	95
Phénylalanine	99

Tableau I : Efficacité de la résine.

2.1.2. Herbicides. Amines. Dérivés nitrés et nitrosés.

Ces composés sont moins solubles que les acides aminés dans l'eau, ils
le sont beaucoup plus dans les solvants organiques. L'extraction liquide-
liquide nous a semblé être la technique la plus adaptée à leur concentration.

Nous avons choisi le dichlorométhane comme solvant d'extraction,
compte tenu qu'il a été utilisé pour extraire toutes les familles de composés azotés. Le pH auquel l'extraction est effectuée peut favoriser celle de
telle ou telle famille de produits. L'extrait après séchage et évaporation
à l'évaporateur rotatif est repris avec du méthanol, le facteur de concentration peut aller de 500 à 1000.

Nous avons effectué les extractions sur des mélanges contenant chacun
la même famille de produits à des pH en fonction de leur caractère acide
ou basique. Le tableau II donne les coefficients de réponse en ng/cm^2 pour
chacun des produits testés compte tenu du rendement d'extraction.

Produit	Nitrobenzène	Monolinuron	Linuron	Monuron
Coefficient de réponse en ng/cm^2	1,5	2	7	7
	Linuron	Ortho-nitrophénol	Dichloro-2,3 aniline	Paranitro-aniline
	7	20	2,4	0,3
	Chloramphé-nicol	Trichloro-2,3,4 aniline	Dinitro-2,4 toluène	Triazine
	0,8	0,8	0,2	60
	Dinoterbe	Chloroxuron	Dinosebe	Metobromuron
	6	6	6	18

Tableau II : Coefficients de réponse compte tenu du
rendement d'extraction (sur OV 17)

Une fois les produits concentrés nous procédons à l'analyse proprement dite.

2.2. Séparation et détection

2.2.1. Acides aminés

Pour une analyse HPLC les acides aminés peuvent être détectés en UV
aux basses longueurs d'onde après séparation sur une colonne échangeuse
d'ions faible (sphérisorb NH$_2$). Cette méthode est limitée lorsqu'il s'agit
de séparer les constituants d'un mélange, étant donné que tous les acides
aminés ont des temps de rétention assez voisins. Nous avons fait appel à
une méthode moins directe et qui nécessite la dérivation des acides aminés
avec un dérivé fluorescent.

En effet, l'orthophtaldialdéhyde réagit avec les acides aminés selon
la réaction : (14)

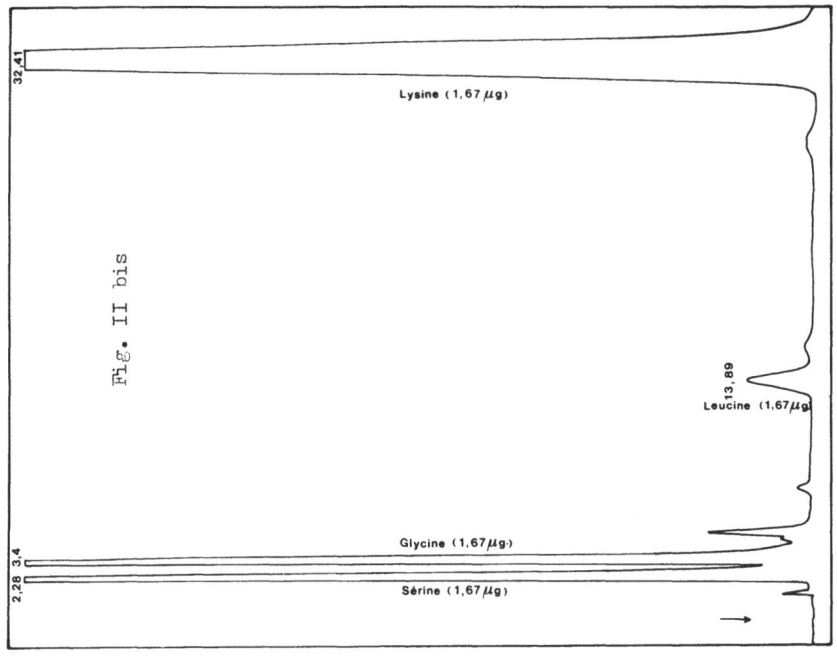

Fig. II bis

Lysine (1,67 μg)

32,41

13,89

Leucine (1,67 μg)

3,4

Glycine (1,67 μg.)

2,28

Sérine (1,67 μg)

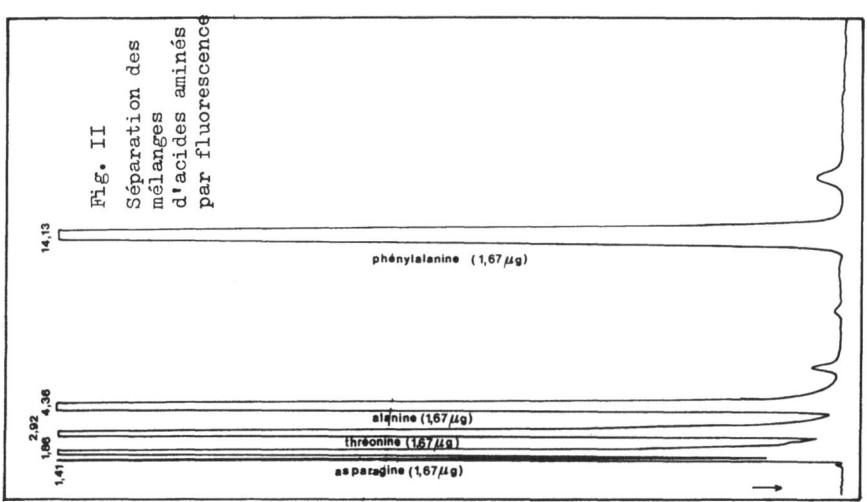

Fig. II

Séparation des
mélanges
d'acides aminés
par fluorescence

14,13

phénylalanine (1,67 μg)

2,92
4,36
1,86
1,41

alanine (1,67 μg)

thréonine (1,67 μg)

asparagine (1,67 μg)

Acide aminé	Formule	Temps de rétention en mn
Asparagine	$H_2N - C - CH_2 - CH \overset{COOH}{\underset{NH_2}{<}}$, $\overset{\shortparallel}{O}$	$1,48 \pm 0,19$ $1,92 \pm 0,1$
Acide aspartique	$HOOC - CH_2 - CH \overset{COOH}{\underset{NH_2}{<}}$	$1,65 \pm 0,1$
Acide glutamique	$HOOC - (CH_2)_2 - CH \overset{COOH}{\underset{NH_2}{<}}$	$1,76 \pm 0,1$
Glutamine	$H_2N - \underset{O}{\overset{\shortparallel}{C}} - (CH_2)_2 - CH \overset{COOH}{\underset{NH_2}{<}}$	$2,22 \pm 0,1$
Sérine	$HO - CH_2 - CH \overset{COOH}{\underset{NH_2}{<}}$	$2,33 \pm 0,14$
Histidine	$\underset{HC \diagdown \underset{H}{N} \diagup CH}{\overset{N - C - CH_2 - CH \overset{COOH}{\underset{NH_2}{<}}}{}}$	$2,54 \pm 0,2$
Thréonine	$HO - CH - CH \overset{NH_2}{\underset{COOH}{<}}$, $\underset{CH_3}{\shortmid}$	$3,45 \pm 0,3$
Glycine ou Glycocolle	$H - CH \overset{NH_2}{\underset{COOH}{<}}$	$3,47 \pm 0,1$
Arginine	$\underset{H_2N}{\overset{HN}{>}} C - NH - (CH_2)_3 - CH \overset{NH_2}{\underset{COOH}{<}}$	$3,5 \pm 0,18$
Tyrosine	$HO - \bigcirc - CH_2 - CH \overset{NH_2}{\underset{COOH}{<}}$	$3,72 \pm 0,18$
Alanine	$CH_3 - CH \overset{NH_2}{\underset{COOH}{<}}$	$4,68 \pm 0,8$
Tryptophane	$\underset{H}{indole} - CH_2 - CH \overset{NH_2}{\underset{COOH}{<}}$	$9,83 \pm 0,1$
Valine	$\underset{CH_3}{\overset{CH_3}{>}} CH - CH \overset{NH_2}{\underset{COOH}{<}}$	$10,37 \pm 0,2$
Méthionine	$CH_3 - S - CH_2 - CH_2 - CH \overset{NH_2}{\underset{COOH}{<}}$	$10,72 \pm 0,1$
Phénylalanine	$\bigcirc - CH_2 - CH \overset{NH_2}{\underset{COOH}{<}}$	$16,07 \pm 2,35$
Isoleucine	$CH_3 - CH_2 - CH - CH \overset{NH_2}{\underset{COOH}{<}}$, $\underset{CH_3}{\shortmid}$	$18,03 \pm 0,2$
Lysine	$H_2N - (CH_2)_4 - CH \overset{NH_2}{\underset{COOH}{<}}$	$30,31 \pm 2,1$

Tableau III – Séparation des acides aminés

conduisant ainsi à des dérivés qui sont bien séparés sur une colonne C_{18} à compression radiale et détectés par fluorescence (λ_{ex} = 390 nm et λ_{em} = 455 nm). Le système à compression radiale permet une bonne reproductibilité des résultats.

Nous pouvons constater que seuls deux acides aminés échappent à l'analyse : la proline et la cystéine.; dans le premier le groupement aminé est engagé dans une structure cyclique et ne peut pas réagir avec le réactif fluorescent. Pour le second, c'est certainement la présence du groupement H-S libre qui peut changer le site réactionnel.

Tous les autres acides aminés répondent dans un intervalle de 30 minutes (Tableau III et Figure II). L'éluant étant un mélange eau-méthanol à 50% 0,1 M de NaH_2PO_4 .

Les quantités minimales détectées sont de l'ordre de 20 ng par acide aminé.

2.2.2. Herbicides. Amines. Dérivés nitrés et nitrosés

Ces composés peuvent être analysés soit par chromatographie gazeuse, soit par chromatographie liquide.

- La chromatographie en phase gazeuse :

Nous avons testé différents produits représentatifs de chaque famille en utilisant deux types de phase stationnaire : l'une polaire et l'autre moyennement polaire (respectivement Carbowax 20 M et OV 17). Le détecteur à capture d'électron étant insensible aux hydrocarbures est particulièrement adapté à la détection des composés azotés surtout les triazines, les dérivés nitrés et nitrosés, les chloroanilines et les urées substituées.

Nous pouvons constater (Tableau IV et Figure III) que les mélanges sont assez bien séparés sauf pour les urées substituées (Monuron, Métobromuron, Linuron) . Les quantités minimales détectées varient entre 1 et 70 ng.

- La chromatographie liquide à haute performance :

Tous les composés ayant des caractéristiques leur permettant d'absorber en UV à254 nm (essentiellement les aromatiques) répondent avec des intensités plus ou moins importantes pour des concentrations plus élevées qu'en CPG. Les séparations sont moins nettes. Son seul avantage est la possibilité de détecter les amines aromatiques non détectées en CPG en plus des dérivés nitrés et des chloranilines. Les triazines et les urées substituées sont également détectées à condition d'effectuer des gradients d'élution (méthanol dans l'eau de 50 à 100% à 2ml/mn).

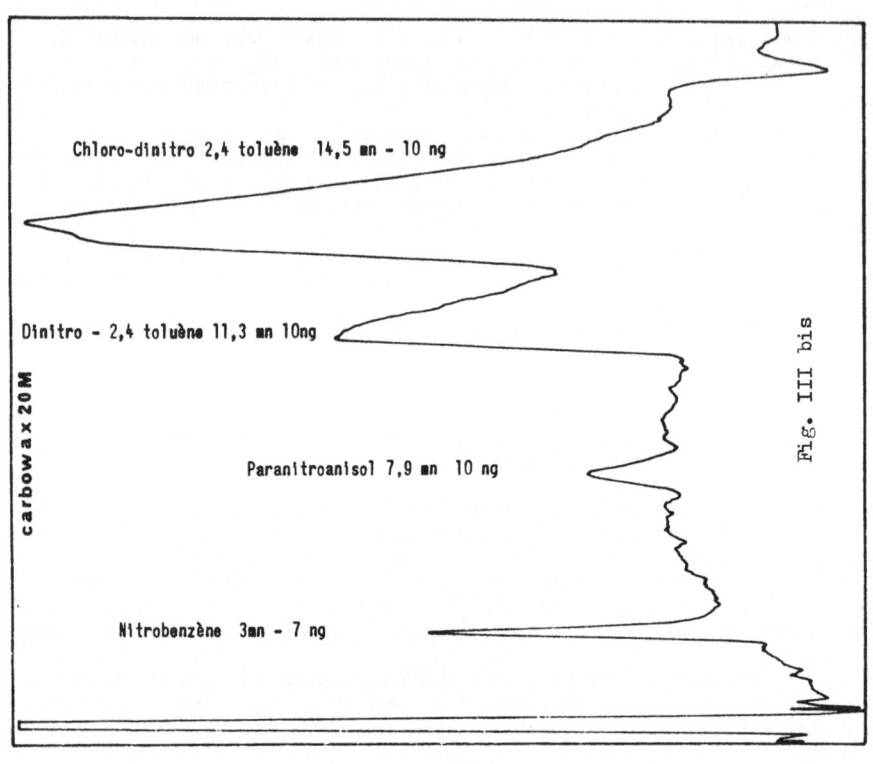

carbowax 20 M

Chloro-dinitro 2,4 toluène 14,5 mn - 10 ng

Dinitro - 2,4 toluène 11,3 mn 10ng

Paranitroanisol 7,9 mn 10 ng

Nitrobenzène 3mn - 7 ng

Fig. III bis

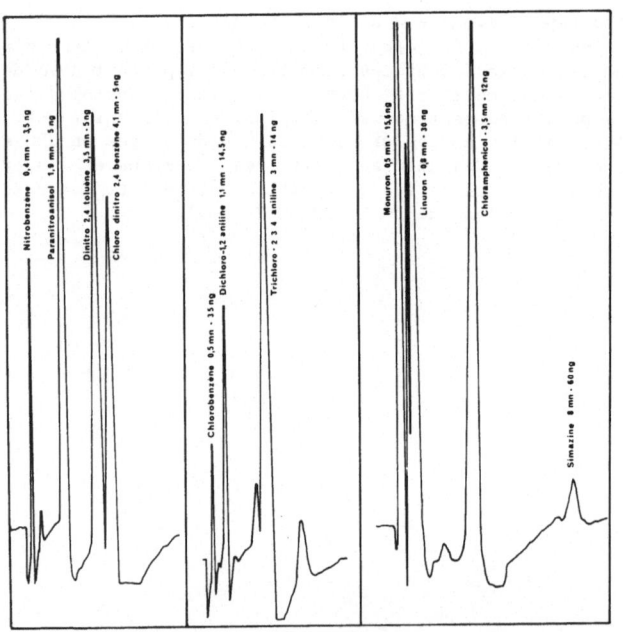

Nitrobenzène 6,4 mn - 15 ng
Paranitroanisol 1,9 mn - 5 ng
Dinitro 2,4 toluène 3,5 mn - 5 ng
Chloro dinitro 2,4 benzène 4,1 mn - 5 ng

Chlorobenzène 0,5 mn - 15 ng
Dichloro-3,2 aniline 1,1 mn - 14,5 ng
Trichloro - 2 3 4 aniline 3 mn - 14 ng

Monuron 0,5 mn - 15,6 ng
Linuron - 0,8 mn - 30 ng
Chloramphenicol - 3,5 mn - 12ng
Simazine 8 mn - 60 ng

Fig. III - Séparation des mélanges de composés azotés sur OV17 et Carbowax 20M avec détecteur à C.E.

- 318 -

Produit testé	OV$_{17}$ t_R mn	Carbowax 20 M t_R mn	Quantité décelable ng	Conditions d'injection
Nitrobenzène	0,4	2	1	
Menuron				carbo. 20 M
Métobromuron	0,6 à 0,8	18 à 26	50	70 < T$_f$ < 150 à 5°C
Linuron				
o - nitrophénol	1	4,5	70	OV$_{17}$ 150 < T$_f$ < 250
Dichloro - 2,3 aniline	1,1	8	5	Qt injectee 5 µl S = 1/16
p - nitroanisol	1,9	8,2	1	
Chloramphénicol	2	11,5	1	
Trichloro 2,3,4 aniline	3	14,5	1	
Dinitro 2,4 toluène	3,5	14,5	1	
Chloro, dinitro - 2,4 benzène	4,1	14,5	1	
Simazine	7,5	18,3	50	
Dinoterbe	10	24,5	20	
Chloroxuron	17,5	17	20	
Dinosebe	20	22	20	
Nitrophénide	25 à 27		20	

Tableau IV - Séparation par chromatographie en phase gazeuse
avec détecteur à Capture d'Electrons.

En complément, nous avons effectué des essais sur une eau de surface, ce qui nous a conduit à définir un protocole analytique (Figure IV) permettant l'analyse de la plupart des composés azotés susceptibles d'être rencontrés dans les eaux de surface et (ou) potables.

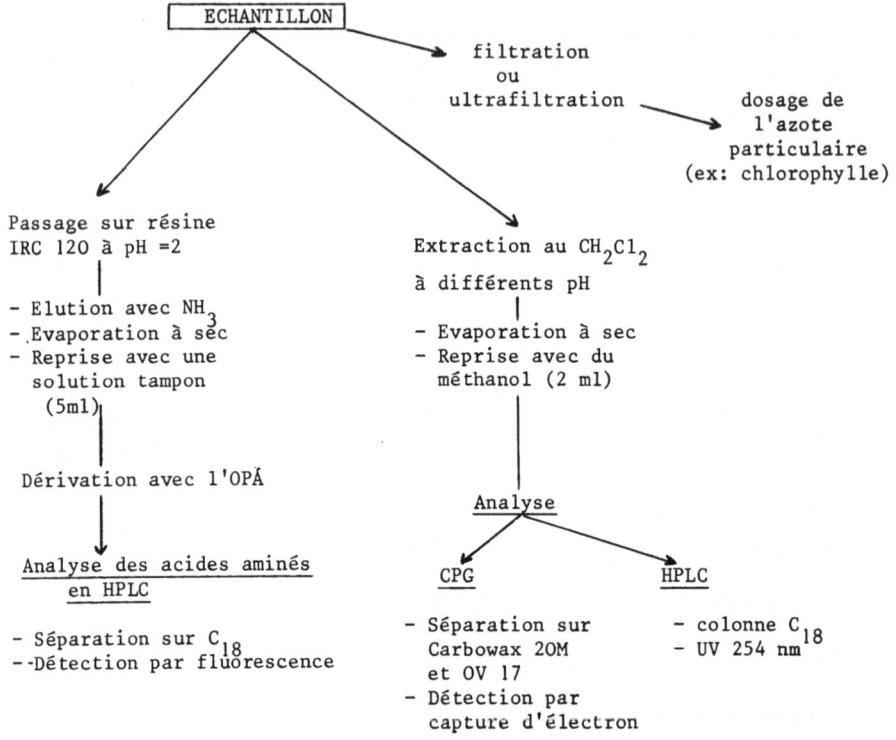

Figure IV : Protocole analytique des substances azotées.

Ce protocole ne concerne que les substances solubles. La partie macromoléculaire peut être analysée par HPLC en perméation de gel ; la méthode n'est pas tout à fait au point mais les premiers résultats montrent que la part de l'azote macromoléculaire représente 16 à 30% de l'azote total.

3. CONCLUSION

Nous avons appliqué ce protocole au suivi des micropolluants azotés dans des stations de traitement d'eau potable, ce qui permet de juger de l'efficacité de chaque étape de traitement (15). Les points de prélèvement sont situés à l'entrée de l'eau brute B , après la floculo-décantation D, la filtration F, et la désinfection O ; un dernier contrôle est effectué sur l'eau refoulée R. Nous avons ainsi mis en évidence l'augmentation du taux de certains acides aminés après filtration et après ozonation (Figure V). Pour les substances extractibles au dichlorométhane, nous avons pu constater l'augmentation des dérivés nitrés après l'ozonation, celle des chloroanilines après chloration. L'action de l'ozone sur des édifices macromoléculaires peut libérer des herbicides complexés d'où l'augmentation dans certains cas de ces derniers.

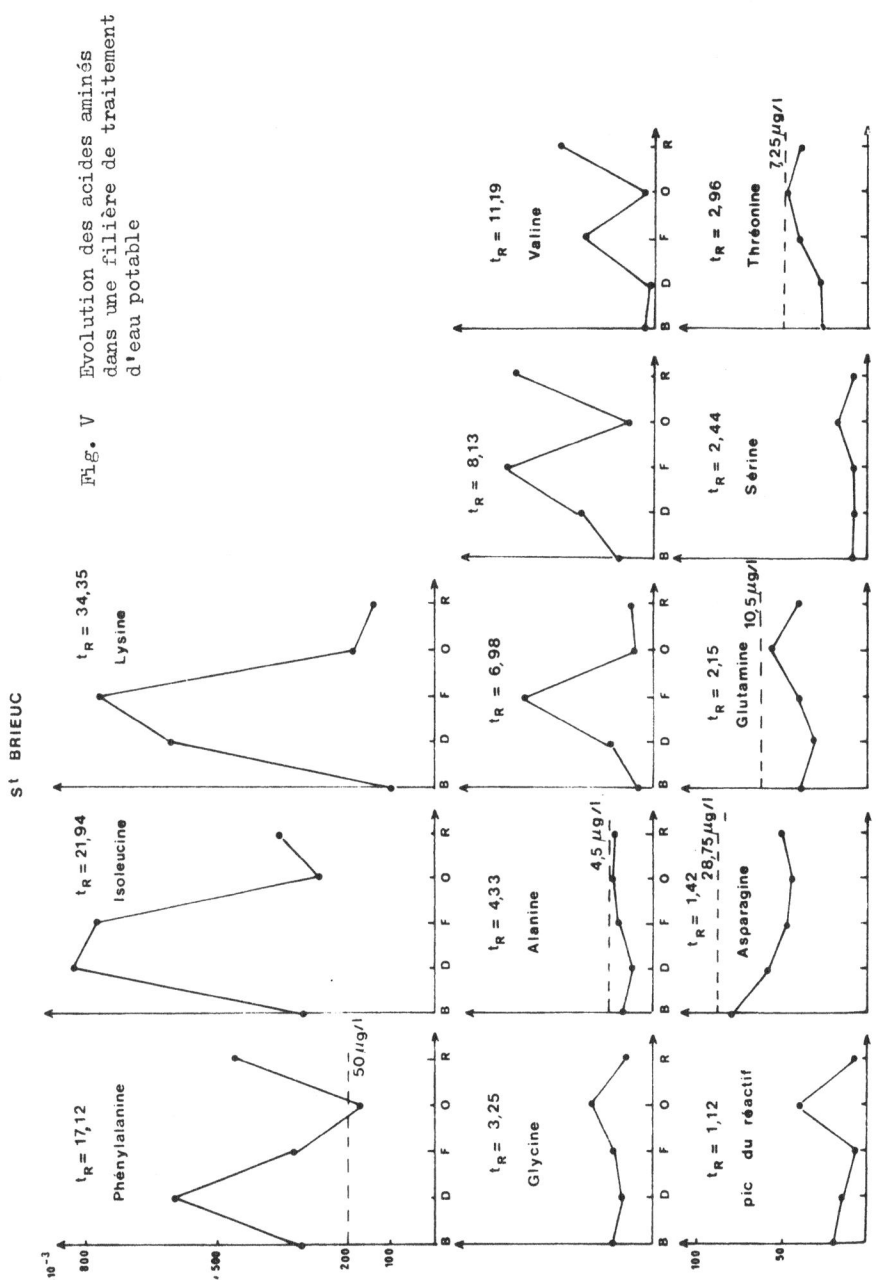

St BRIEUC

Fig. V Evolution des acides aminés dans une filière de traitement d'eau potable

BIBLIOGRAPHIE

(1) G.MARTIN , 1979 , L'azote dans les eaux , Lavoisier , Paris

(2) G.J.STANDER , 1980 , Water SA , 6 , 1 , 1-14

(3) M.FIELDING and R.F.PACKHAM , 1977 , Journal Institution Water Engineers and Scientists , 31 , 353-367

(4) J.L. BURLESON , G.R. PEYTON and W.H.GLAZE , 1980 , Environmental Science and Technology , 14 , 11 , 1354-1354

(5) C.R.LYTLE and E.M.PERDUE , 1981 , Environmental Science and Technology , 15 , 2 , 224-228

(6) D.KASISKE, K.D.KLINKMULLER, M.SONNEBORN , 1978 , Journal of Chromatographie , 149 , 703-710

(7) M.J.AARONSON,, K.W.KIRBY , J.D.TESSARI ,1980 , Bull. Environm. Contam. Toxicol. , 25 , 492-497

(8) R.C.C.WEGMAN , G.A.L. DE KORTE , 1981 , Water Research , 15 ,391-394

(9) J.T.HURLEY , 1974 , J.A.W.W.A. , 66 , 1 , 27-31

(10) F.ERB , J.DEQUIDT , A.DOURLENS , J.POMMERY et P.COLIN , 1978 , Journal Français d'hydrologie , 9 , 1 , 25 , 9-18

(11) D.S.FARRINGTON , R.G.HOPKINS , J.H.A.RUZICKA , 1977 , Analyst , 102 , 377-381

(12) G.R.PIEPER , 1979 , Bull. Environm. Contam. Toxicol. 22 , 167-171

(13) M.M.N.KAIDO , D. DEAN-RAYMOND , A.J.FRANCIS , M.ALEXANDER , 1977 , WATER Research , 1085-1087

(14) P.LINDROTH , K.MOPPER , 1979 , Analytical chemistry , 51 , 11 , 1667-1674

(15) M.ELMGHARI-TABIB , 1981 , Thèse 3ème cycle , E.N.S.C.RENNES

THE ELECTROANALYSIS OF ORGANIC POLLUTANTS IN AQUATIC MATRICES

W.F. Smyth and M.R. Smyth
Department of Chemistry, University College Cork, Cork, Ireland and
National Institute for Higher Education, Ballymun, Dublin 9

Summary

This paper "Electroanalysis of Organic Pollutants in Aquatic Matrices" will comment on the applicability of electroanalytical methods such as DC and pulse polarography and voltammetry, anodic and cathodic stripping voltammetry and on-line electrochemical detection to the determination of selected organic molecules and their metabolites of significance in the analysis of aquatic matrices. Where possible, a critical comparison will be made between them and alternative analytical methods based primarily on chromatographic and spectroscopic techniques.

1. INTRODUCTION

Voltammetric methods, based on the use of the techniques such as DC and pulse polarography and voltammetry, anodic and cathodic stripping voltammetry and on-line electrochemical detection have seen extensive exploitation and resulting diverse analytical application in recent years for the determination of trace concentrations of many organic molecules of significance in environmental chemistry (1), (2), (3). The process of electron exchange for many organic molecules, occurring at a mercury or solid electrode-aqueous solution interface is governed not only by the rate of mass transport to the electrode surface but also by the rate of electron transfer itself. Factors such as the stereochemistry of interaction of the molecule at the electrode surface and the protonation and adsorption reactions it and/or its electrochemical reaction product(s) undergo at this interface will tend to make the overall electrochemical reaction "irreversible" and thus reduce the selectivity of polarographic and voltammetric analysis in that broad, ill defined curves/peaks result. It is for this reason that separation techniques such as solvent extraction, t.l.c. and h.p.l.c. are frequently a necessary pre-requisite for the polarographic or voltammetric determination of trace concentrations of these organic molecules; this also decreases the effect of electro-active interferences in these aquatic matrices. Solvent extraction techniques can have the added advantage of pre-concentration of the organic molecules under investigation. Adsorption of the molecule and/or its electrochemical reaction product(s) can give rise to post and pre waves which complicate DC and to a lesser extent NPP polarography where the potential is only applied to the electrode surface during the pulsing period. The phenomenon of adsorption can be used to analytical advantage in stripping voltammetry where the product of the electrochemical reaction is adsorbed on the electrode surface by imposition of a suitable applied positive (c.s.v.) or negative (a.s.v) potential for a selected time interval prior to stripping of the adsorbed species from the electrode by imposition of a potential scan in an opposite direction to the original applied potential. The sensitivity and limit of detection for a limited number of organic molecules can be improved by several orders of magnitude over the corresponding analytical parameters for differential pulse polarographic analysis (d.p.p.) of these molecules (4). On-line electrochemical

detection with solid electrodes such as glassy carbon is inherently more
sensitive than polarographic and voltammetric analysis in quiescent
solution due to the convective element in addition to that of diffusion in
the overall mass transport to the electrode surface.

It is the purpose of this paper to comment on the applicability of the
aforementioned electroanalytical methods to selected organic molecules (and
their metabolites) of significance in the analysis of aquatic matrices.
Where possible, a critical comparison will be made between them and
alternative analytical methods based primarily on chromatographic and
spectroscopic techniques.

2. CARBONYL COMPOUNDS

Although the carbonyl group is electro-reducible, the little work that
has been carried out on carbonyl compounds by direct polarographic methods
has mainly been confined to aldehydes such as formaldehyde, acetaldehyde,
butyraldehyde and furfural (5-10). Limits of detection in the low $\mu g\ mL^{-1}$
region in waste waters have been reported.

The use of derivatization of the carbonyl group has been exploited by
Afghan et al. (11) in a systematic study of these compounds in various
supporting electrolytes, using twin cell potential sweep voltammetry. The
formation of the semicarbazone was found to be most satisfactory. The
formation of the semicarbazone was optimised at pH 4-6 and excess of the
electro-inactive semicarbazide was added to ensure complete reaction.

Using a citrate buffer with EDTA added to complex interfering heavy
metals, a limit of detection of 0.25 ng mL^{-1} was claimed for the
determination of carbonyl compounds present in natural waters and
industrial effluents without any separation or preconcentration of the
sample.

3. SIMPLE AROMATIC COMPOUNDS

(i) Nitrocompounds

Direct polarographic methods for nitro-compounds in waste water and
effluents have been developed. For example, nitrobenzene in industrial
wastes was determined by Dyatlovitskaya et al. (12) after separating
nitrophenols by distillation. The supporting electrolyte consisted of
aqueous ethanol containing hydrochloric acid. Fleszar (13) has determined
0.005 $\mu g\ mL^{-1}$ of nitrochlorobenzenes in water following extraction with
activated charcoal and polarography of the eluted acetone solution in a
pyridinium hydrochloride supporting electrolyte. Nitrophenols, nitro-
cresols and nitrotoluenes, in addition to nitrochlorobenzenes present in
waste water, have been polarographically analysed by Zaitsev and
Dichenskii (14). The supporting electrolyte used was 0.25 M NaOH in 3:1
$CH_3OH:H_2O$, and $E_{\frac{1}{2}}$ values in the region -0.70 to -1.20 V were obtained. A
very sensitive method for dinitro-o-cresol in water has been described by
Supin et al. (15). Extraction of the acidified water sample with petroleum
ether was followed by linear sweep voltammetry (l.s.v.) in an aqueous borate
buffer. This method offered a sensitivity of 0.004 $\mu g\ mL^{-1}$.

(ii) Phenolic Compounds

Phenols are not reducible at the d.m.e. but do exhibit anodic waves at
a variety of solid electrodes (16). Of particular electroanalytical
significance is the work of Kissinger and co-workers who have been able to
determine trace concentrations of biologically important compounds such as
neurotransmitters in body fluids following h.p.l.c. separation and electro-

chemical detection at a semi-micro carbon paste electrode. This work has been reviewed in a recent article (17, 18). Until these developments in electrochemical detection in conjunction with h.p.l.c., most polarographic methods for the determination of phenolic compounds were based on prior derivitisation procedures. In particular, the use of nitration has found many applications for the determination of these compounds in biological fluids, e.g., phenol in water (19) has been determined by this procedure using pulse polarography with a quoted limit of detection of 1 ng mL^{-1}.

Armentrout et al. (20) have selectively detected individual phenolic compounds, including the more toxic halogenated ones, at low ppb levels using h.p.l.c. with a polymeric cation exchange resin column, acidic acetonitrile/water eluent and an electrochemical detector containing a unique carbon-black/polyethylene tubular anode.

(iii) Polychlorinated Compounds

Organochlorine pesticides in aquatic matrices are usually determined by glass capillary g.l.c. where the electron capture and coulometric detectors offer sensitivity and selectivity not usually obtained by other methods involving spectrophotometry or polarography/voltammetry.

4. CARBOXYLIC ACIDS

Materials of the amino polycarboxylic acid type can be determined polarographically, by observing the reduction of a heavy metal-ligand complex, in the presence of excess metal ions. The reduction potential of this complex is shifted to more negative values, and the limiting current of this reduction is used for the quantitative determination of the ligand. Haberman (21) demonstrated how nitrilotriacetic acid (NTA), which is a detergent builder with the formula $N(CH_3COOH)_3$, could be determined using In(III). After an excess of In(III) had been added to the NTA in the aqueous solution, the free metal gave a polarographic wave and the limiting current of the complexed In(III) could be found by difference. Studies on metal-NTA complexes have shown that the optimum pH for trivalent metal ion complexation is pH 2 and for divalent metal ions such as Cd(II) and Pb(II), pH 7. Using an anion exchange column to concentrate NTA, Haberman (21) was able to determine 0.025 µg mL^{-1} when In(III) was used as the complexing metal. Afghan et al. (22) has improved the limit of detection of this method to 0.01 µg mL^{-1} using a Bi complex and has automated the procedure (15 samples hour^{-1}).

5. ALKYL SULPHONATES AND ALKYL BENZENE SULPHONATES

Polarographic methods such as depression of maxima methods and tensammetry have been developed for anionic, cationic and neutral surfactants (23) but, in general, they are not specific to surfactant type and an effective separation is a prerequisite to determination in real situations. Ion exchange is the most generally used surfactant separation method, either anionic or cationic type being separated from nonionic by suitable choice of resin.

Hart et al. (24) have determined the linear alkyl benzene sulphonate content of sewage and tap water samples by an indirect polarographic method based on nitration of the aromatic ring. The method was proved reliable when concentrations were of the order of 0.5 µg mL^{-1} or greater and it was found more selective than the colorimetric methylene blue method which gives a measure of total anionic surfactant present in the sample.

6. ORGANOPHOSPHORUS COMPOUNDS

G.l.c. with flame photometric or thermionic detection is commonly employed for the determination of organophosphorus compounds in samples of environmental significance. However, "cold" methods such as h.p.l.c., polarography/voltammetry are increasingly being investigated as alternative methods due to the instability of these insecticides under the conditions employed in g.l.c. analysis.

(i)   Determination of $\overset{\backslash}{\underset{S}{\overset{\|}{P}}}$ *– and* $\overset{\backslash}{\underset{\|}{P}}$ *– S – containing compounds*

Nangniot (25, 26) has studied the polarographic behaviour of a wide range of organophosphorus pesticides. Although phosphoric acid esters cannot be reduced at the d.m.e., compounds containing the above functional groups can produce sharp adsorption peaks using fast linear sweep voltammetric techniques (250 mV s^{-1}). This method could determine concentrations of selected organophosphorus compounds at the μM level. Cathodic stripping voltammetry (c.s.v.), following alkaline hydrolysis of these thiophosphates to release sulphur containing molecules capable of forming insoluble mercury salts, has been applied (27) to the determination of the insecticides, phthalophos and benzophosphate in apples with a detection limit of 0.2 μg kg^{-1}.

(ii)   Determination of –NO$_2$ containing compounds

Perhaps most applications of polarography for the determination of nitro-containing agrochemicals have come from investigations of nitrophenyl esters, e.g. parathion, fenitrothion, etc.

Smyth and Osteryoung (28) have made a detailed pulse polarographic study of parathion, methyl parathion, paraoxon and other structurally related organophosphorus agrochemicals and their metabolites but were unable to use the differences observed in their behaviour for quantitative purposes. Structurally related nitrophenyl esters should therefore be separated by a chromatographic procedure prior to analysis. This is exemplified by the work of Koen and Huber (29) who determined parathion and methyl parathion in lettuce following a liquid chromatographic separation.

(iii) Enzymopolarographic Determination

Davidek and Seifert (30) have developed an enzymopolarographic method for the determination of the organophosphorus pesticide intration. The method was based on the inhibition of anticholinesterase activity by the organophosphorus moiety (a reaction which parallels the *in vivo* biological activity of these compounds). Unreacted enzyme is then incubated with β-naphthyl acetate and the β-naphthol liberated measured by polarography following nitrosation. The method was applied to the analysis of intration in lettuce, cabbage, cherries and tomatoes. Naturally occurring enzymes which would be capable of hydrolysing β-naththyl acetate were removed by precipitation with C$_2$H$_5$OH and subsequent centrifugation. No interference was observed in the presence of carotenes, xanthophyll, chlorophyll or anthocyanidines and the method was found to be relatively simple and rapid to perform.

(iv)   Determination by Derivitisation

The herbicide glyphosate (N-phosphonomethyl-glycine) (I) was found to

be more conveniently analysed by polarography than by g.l.c. (31).
(Bronstad and Friestad 1976). For the latter method a lengthy four-stage
clean-up and two-stage derivatization procedure was necessary, whereas ion
exchange followed by nitrosation only was required for polarography. The
eluate from ion exchange was treated with 50% sulphuric acid and potassium
bromide and sodium nitrite solutions. After fifteen minutes, ammonium
sulphamate was added to destroy excess nitrite and polarography was carried
out after de-aerating with nitrogen.

$$OH - \overset{\overset{O}{\|}}{C} - CH_2 - \underset{\underset{H}{|}}{N} - CH_2 - \overset{\overset{O}{\|}}{\underset{\underset{OH}{|}}{P}} - OH$$

$$OH - \overset{\overset{O}{\|}}{C} - CH_2 - \underset{\underset{NO}{|}}{N} - CH_2 - \overset{\overset{O}{\|}}{\underset{\underset{OH}{|}}{P}} - OH$$

(I)   (II)

The N-nitroso derivative (II) gave a reduction peak at -0.78 V, which could
be used to monitor between 35 and 210 ng mL^{-1} of glyphosate, in natural
waters. It was suggested that one analyst could analyse twenty samples a
day, using the above procedure.

7. COMPOUNDS CONTAINING ENDOCYCLIC AND EXOCYCLIC $\overset{}{>}C \equiv N^{\diagup}$ GROUPS

Polarographic methods of analysis have found widespread application
for the determination of many drug substances containing the azomethine
group e.g., 1,4-benzodiazepines (32, 33), antidiabetic compounds (34) and
benzhydrylpiperazine derivatives (35). Separation procedures involving
solvent extraction, t.l.c. and h.p.l.c. are increasingly a necessary pre-
requisite for the simultaneous determination of mixtures of trace amounts
of the parent compound and its metabolites particularly in complex matrices
such as blood (36).
    The triazine pesticides, terbutryne, ametryne and atrazine, which
contain reducible endocyclic $>C = N^{\diagup}$ bonds, have been determined (37) in
pond and canal water with a limit of detection of 5 ng mL^{-1} following
extraction with dichloromethane, evaporation of the solvent, dissolution of
the residue in 50% methanol/0.01 N sulphuric acid and application of linear
sweep voltammetry. The polarographic method matched a g.l.c. procedure in
terms of time of analysis but was inferior with respect to sensitivity and
selectivity.
    Smyth et al. (38) have investigated the polarographic behaviour of
agrochemicals, cytrolane, cyolane, chlordimeform and drazoxolon, all of
which contain exocyclic $>C = N^{\diagup}$ groups, and recommended optimum conditions
for their determination by pulse polarography.

8. COMPOUNDS CONTAINING —SH GROUPS

Compounds containing sulphur in the —SH form are particularly amenable
to c.s.v. analysis due to their ability to form partially insoluble
complexes with mercury. The application of c.s.v. to the determination of
some thiourea-containing agrochemicals has been investigated by Smyth and
Osteryoung (39). Limits of detection at low ng mL^{-1} levels were found
using the differential pulse mode and this method was found to be more
sensitive than polarographic methods based on anodic waves observed at the
d.m.e.
    Thiourea can also be determined polarographically following
complexation with Cu(II) ions (40) or by liberation of the S atom and
subsequent determination of H_2S (41). This latter procedure has also been

used for the determination of other S-containing pesticides, e.g. diazinon, rogor and phenkapton and involved reduction of the pesticide by Al in HCl solutions in the presence of Ni. The H_2S evolved is then determined by monitoring the decrease of Pb(II) concentration in the Pb(OAc)$_2$ trapping solution. The method could determine down to 0.25 µg mL^{-1} in pure solution.

Brand and Fleet (42) have investigated the application of c.s.v. to the determination of the fungicide, tetramethylthiuram disulphide (thiram) in aqueous solutions. They reported that the best results were obtained using a mercury plated platinum electrode in a solution of thiram containing an excess of ascorbic acid. This addition had the effect of chemically reducing the disulphide moiety in thiram to form free —SH groups which were then amenable to c.s.v. analysis. Using this method, they were able to determine thiram down to 10^{-8} M in pure solution. This method offers a much greater sensitivity over the d.c., linear sweep or a.c. techniques which have limits of detection for this compound of 6 x 10^{-6}, 3.5 x 10^{-6} and 1 x 10^{-6} M respectively.

9. CARBAMATES

Their reduction at negative potentials has been investigated by polarography (43, 44) after derivitisation via nitration or nitrosation and recently, Anderson et al. (45) have applied their little investigated oxidation reactions to a reverse phase liquid chromatographic method with thin layer Kel-F-graphite electrochemical detection operated in the constant potential amperometric mode at +1.1 V (vs. Ag/AgCl). Calibration curves were linear over at least three orders of magnitude with relative standard deviations at 1–2%. Detection limits in the range 40–150 pg, which correspond to sample concentrations of 2–7 ng mL^{-1}, were obtained and the method compared favourably with g.l.c. with electron capture detection following hydrolysis to the corresponding phenols or amines and reaction with halogen rich reagents. The electrochemical detector, in this case, was found more sensitive than both the u.v. detector, operated at 190–210 nm, and the fluorescence detector in which dansyl derivatives were formed prior to injection or post column derivitisation carried out with o-phthaldehyde.

Chloroanilines can be formed as hydrolytic metabolites of some carbamates e.g., by microbial action in the soil. These can also be determined at low to subnanogram levels by h.p.l.c.-electrochemical detection (46, 47).

10. BENZIDINES

Benzidine and 3,3'-dichlorobenzidine, EPA priority pollutants, have been detected by reversed phase h.p.l.c.-electrochemical detection (48) in waste water at sub–ppb levels in a 50 µL injection. An oxidising potential of +0.7 V (vs. Ag/AgCl) was applied to a three electrode electrochemical cell containing a tubular carbon–black/polyethylene working electrode and the detection method was found to be 50 times more sensitive than UV detection at 280 nm.

REFERENCES

1. W. Franklin Smyth, Editor, *"Polarography of Molecules of Biological Significance"*, Academic Press 1979.
2. W. Franklin Smyth, Editor, *"Electroanalysis in Hygiene, Environmental, Clinical and Pharmaceutical Chemistry"*, Elsevier 1980.

3. M.R. Smyth and W. Franklin Smyth, Analyst., 103, 529(1978).
4. I.E. Davidson, Ph.D. thesis, University of London, 1979.
5. N.A. Kuchumova, C.J. Bepuzo and Mamomova T. Vses. Nauchn. - Issled.
    Inst. Po Pererobotre I Ispol'z. Topliva 12, 237(1963).
6. F.G. Dyatlovitskaya and F.J. Berezouskii Gig. I. Sanit., 27, 50(1962).
7. V.I. Bodyn and Ya. S. Feldman, Gidrolizn I. Lesokhim. Prom., 16(7),
    11(1963).
8. I. Melcer and A. Melcerova, Drev. Vyst., 16(1-2), 59(1971).
9. Yu. P. Ponomarev, O.I. Glazyrina, T.V. Kassai, Fiz. Khim. Metody.
    Ochistki. Anal. Stochnykh Vod. Prom. Predpr., 91(1974).
10. B.P. Zhantalai, A.S. Sergeeva and L.R. Kalichuk, Zn. Khim. Abst.,
    221, 298(1976).
11. B.K. Afghan, A.V. Kulkarni and J.F. Ryan, Analyt. Chem., 47(3),
    488(1975).
12. F.G. Dyatlovitskaya, F.I. Berezonskii and S.K. Potemkina, Gig. I.
    Sanit., 28, 38(1963).
13. B. Fleszar, Chem. Anal. (Warsaw), 9, 1075(1964).
14. Z.P.M. Zaitsev and V.I. Dichenskii, Zavod Lab., 32(7), 800(1966).
15. G.S. Supin, F.F. Vaintraub, C.V. Makarova, Gig. Sanit, 5, 61(1971).
16. V.D. Parker in M.M. Baizer, Editor, "Organic Electrochemistry",
    Marcel Dekker, New York, 531(1973).
17. P.T. Kissinger, Analyt. Chem., 49, 447A(1977).
18. W.R. Heineman and P.T. Kissinger, Anal. Chem., 50, 166R(1978).
19. Y. Audonard, A. Suzanne, O. Vittori and M. Porthaut, Bull. Soc. Chim.
    Tr., 130(1975).
20. D.N. Armentrout, J.D. McLean, and N.W. Long, Analyt. Chem., 51,
    1039(1979).
21. J.P. Haberman, Analyt. Chem., 43, 63(1971).
22. B.K. Afghan, P.D. Goulden and J.F. Ryan, Analyt. Chem., 44, 354(1972).
23. H. Jehring, Tenside, 3(6), 187(1966).
24. J.P. Hart, W. Franklin Smyth and B.J. Birch, Analyst., 104, 853(1979).
25. P. Nangniot, Anal. Chem. Acta, 31, 166(1964).
26. P. Nangniot, In "La Polarographie en Agronomie et Biologie", Duculot,
    Gembloux.
27. G.S. Supie and G.K. Budnikov, Zh. Anal. Khim., 28, 1459(1973).
28. M.R. Smyth and J.G. Osteryoung, Anal. Chim. Acta, 96, 335(1978).
29. J.G. Koen and J.F.K. Huber, Anal. Chim. Acta, 51, 303(1970).
30. J. Davidek and J. Seifert, Die Nahrung, 15, 691(1971).
31. J.O. Bronstad and H.O. Friestad, Analyst., 101, 820(1976).
32. J.M. Clifford and W. Franklin Smyth, Analyst., 99, 241(1974).
33. M.A. Brooks and J.A.F. de Silva, Talanta 22, 849(1975).
34. W.V. Malik and R.N. Goyal, Talanta, 23, 705(1976).
35. M.R. Smyth, W. Franklin Smyth and J.M. Clifford, Anal. Chim. Acta,
    94, 119(1977).
36. W. Franklin Smyth, Anal. Proceedings, RSC.
37. C.E. McKone, T.H. Byast and R.J. Hance, Analyst., 97, 653(1972).
38. M.R. Smyth and J.G. Osteryoung, Analyt. Chem., 50, 1632(1978).
39. M.R. Smyth and J.G. Osteryoung, Analyt. Chem., 49, 2310(1977).
40. H. Sohr and K. Wienhold, Anal. Chim. Acta, 83, 415(1976).
41. E.S. Kosmatyi and V.N. Kavetskii, Zav. Lab., 41, 286(1975).
42. M.J.D. Brand and B. Fleet, Analyst., 95, 1023(1970).
43. R.J. Gajan, W.R. Benson, J.M. Finnochiaro, J. Assoc. Off. Agric. Chem.,
    48, 958(1965).
44. D.O. Eberle, E.A. Gunther, J. Assoc. Off. Agric. Chem., 48, 927(1965).
45. J.L. Anderson and D.J. Chesney, Analyt. Chem., 52, 2156(1980).
46. J.P. Hart, M.R. Smyth and W. Franklin Smyth, Analyst., 106, 146(1981).
47. E.M. Lores, D.W. Bristol, R.F. Moseman, J. Chromat. Sci., 16, 358(1978).
48. D.N. Armentrout and S.S. Cutie, J. Chromatogr. Sci., 18, 370(1980).

DETERMINATION OF NONYLPHENOLS AND NONYLPHENOLETHOXYLATES IN
SECONDARY SEWAGE EFFLUENTS

C. Schaffner, E. Stephanou and W. Giger
Swiss Federal Institute for Water Resources and Water Pollution Control
(EAWAG), CH-8600 Dübendorf / Switzerland

SUMMARY

Nonylphenols and nonylphenolethoxylates with one, two, and three
oxyethylene groups were identified in effluents from activated sludge
sewage treatment plants. A steam-distillation and solvent-extraction
procedure was used to isolate the organic compounds. The complex
mixtures of isomers were analyzed by glass capillary gas chromato-
graphy/mass spectrometry. Concentrations of 100 to 300 µg of total
4-nonylphenol derivatives per liter effluent were determined in the
effluents of three out of six activated sludge treatment plants.
Nonionic detergents of the nonylphenolethoxylate type are considered
to be precursor chemicals of these refractory compounds.

1. INTRODUCTION

Effluents from sewage treatment plants contain complex mixtures of
organic chemicals covering a broad range of compound classes [1]. Qualitative and quantitative determinations of individual compounds should be
performed to identify the origins of these substances, to follow their
behaviour during sewage treatment, and to assess their impact on the quality
of the receiving waters. The application of gas chromatography/mass spectrometry (GC/MS) has provided much molecular information on such persistent
pollutants [e.g., 1,2]. We wish to report here identifications and quantitative determinations of nonylphenols (NP) and nonylphenolethoxylates (NP1,
NP2 and NP3; for structure formulae, see legend to Fig. 1) in effluents
from activated sludge treatment plants. These chemicals were discovered as
major constituents in a fraction obtained after steam-distillation and
solvent-extraction.

2. EXPERIMENTAL

Materials and samples. 4-Nonylphenol (technical grade, Fluka AG, Buchs,
Switzerland) and Marlophen 83 (Chemische Werke Hüls AG, Marl, Germany) were
analyzed as received from the supplier. 4-Nonylphenol is a mixture of
isomers with differently branched nonyl side chains. Marlophen 83 is a
nonionic detergent produced by the addition of ethylene oxide to 4-nonyl-
phenol. Samples from various sewage treatment plants in the Zurich area and
a small number of samples from rivers and groundwaters were analyzed.

Isolation. A closed-loop apparatus for steam-distillation and solvent-
extraction was applied to isolate the organic compounds from the aqueous
matrix of the samples [3,4]. One advantage of this procedure is that no
additional clean-up steps are necessary. The concentrate contains mostly
compounds which are well amenable to gas chromatography and few interfering
chemicals are coextracted.

Samples of 1.5 ℓ, after adding 30 g of sodium chloride, were kept under reflux for three hours, and 2 mℓ of cyclohexane was used as organic solvent. The extracts were dried over sodium sulfate and directly analyzed by GC and GC/MS.

Gas chromatography / mass spectrometry (GC/MS). A Finnigan mass spectrometer (Model 4021C) with an INCOS 2000 data system was used for mass spectrometric identifications. A Carlo Erba gas chromatograph (Model 4160) equipped with a Grob-type split-splitless injector was connected to the mass spectrometer. The glass capillary GC column was directly coupled to the ion source by a fused silica capillary.

Gas chromatographic separations were performed on glass capillary columns (20 m x 0.3 mm) kindly supplied by K. & G. Grob. The columns had been deactivated by persilylation and coated with OV-1 as an immobilized stationary phase [5].

3. RESULTS AND DISCUSSION

Identifications. Trace A of Fig. 1 shows the total ion current chromatogram which was obtained by GC/MS analysis of an extract of a 24 h-composite sample from the secondary effluent of a sewage treatment plant in Zurich. The three major peak groups, assigned NP, NP1, and NP2 contain compounds with molecular ions at m/z 220, 264, and 308 respectively, corresponding to the molecular masses of nonylphenols, nonylphenolmono-ethoxylates, and nonylphenoldiethoxylates.

The total ion chromatogram B of Fig. 1 was recorded by analyzing a reference mixture of the technical products 4-nonylphenol and Marlophen 83. The peak identifications NP, NP1, NP2, NP3, and NP4 were obtained from the molecular ions of the corresponding mass spectra. The nonylphenols and their polyoxyethylene derivatives can be identified by their fragmentation patterns. Six selected mass spectra are presented in Fig. 2. A detailed discussion of these mass spectra has been presented elsewhere [6].

Occurrence and origin. We have detected the persistent phenolic compounds described in this paper in the effluents of three out of the six activated sludge treatment plants which we have studied. In addition, NP, NP1, and NP2 were detected in water samples from the river Glatt which flows through a densely populated area east of Zurich, and also in a groundwater which is fed by infiltration from the river Glatt. Quantitative results are summarized in Table I.

NP, NP1, and NP2 are not widely used commercial products, and they are therefore not expected to occur ubiquitously in sewage effluents. Marlophen 83 is a nonionic detergent of minor importance because of its low water solubility. The most popular nonionic detergents of the alkylphenolpoly-ethoxylate type have longer oxyethylene side chains containing from approximately five to twenty oxyethylene groups. Investigations of the biological degradation of such nonionic detergents have shown that these molecules are altered by shortening of the oxyethylene side chains. Metabo-lites containing one or two residual oxyethylene groups were found in the effluents of laboratory-scale activated sludge systems [7-9]. Therefore, the occurrence of NP1 and NP2 in secondary sewage effluents can be explained by the higher persistency of these intermediates. Similarly NP is also believed to be a refractory metabolite of 4-nonylphenolpolyethoxylate detergents.

To the best of our knowledge, NP, NP1, NP2, and NP3 have not yet been determined in effluents of real treatment plants. Jones and Nickless [10] have reported short chain alkylphenolethoxylates in sewage treatment effluents. But they could not provide precise structural informations.

		Concentrations, µg/ℓ			
Sample		NP	NP1	NP2	NP3
Secondary sewage effluents	Dübendorf	8-35	50-130	13-70	n.d.
	Zürich	8-20	24-52	3-50	n.d.
	Kloten	0-16	70-130	30-70	n.d.
River Glatt		3	n.d.	25	9
Groundwater		2	2	4	n.d.

n.d.: not determinable

Table I   Concentration ranges of NP, NP1,2,3 in secondary sewage effluents, in a river and in a groundwater.

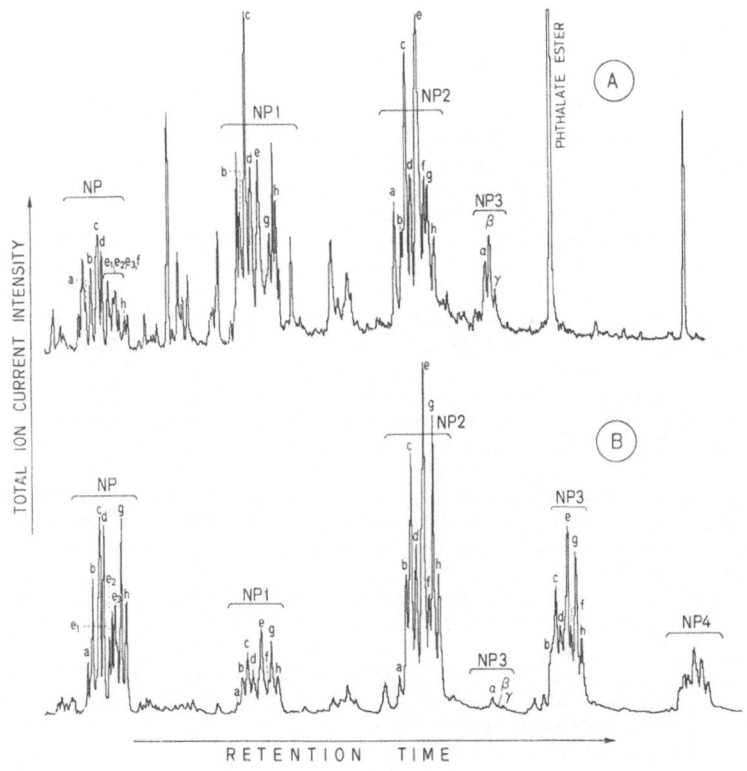

Figure 1   Total ion current chromatograms.
(A)   Extract of a secondary sewage effluent
(B)   Reference mixture of 4-nonylphenol and Marlophen 83

NP: $H_{19}C_9$-⬡-OH

NP1, NP2, NP3, NP4: $H_{19}C_9$-⬡-$(O-CH_2-CH_2)_n$-OH   n: 1,2,3,4

a-h: para-substituted isomers

α,β,γ: ortho-substituted isomers (tentatively)

- 332 -

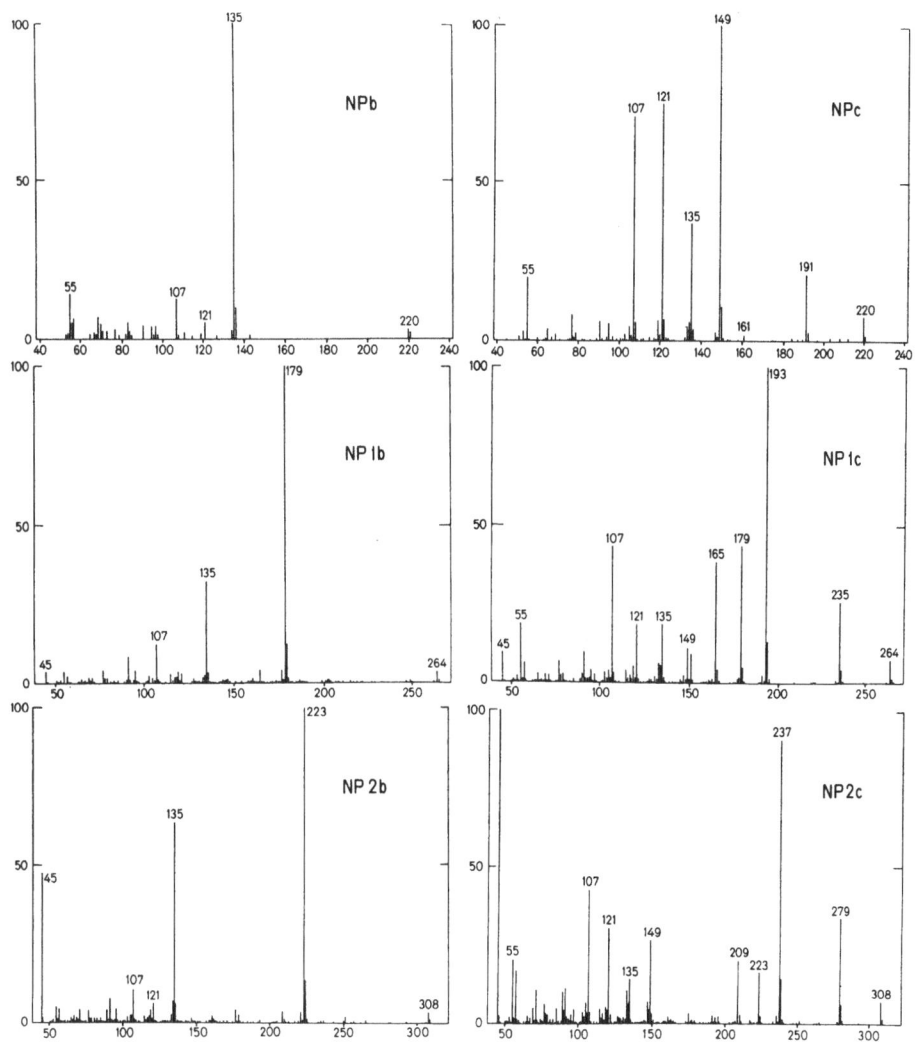

Figure 2

Six selected mass spectra of 4-nonylphenols and 4-nonylphenol-
ethoxylates extracted from secondary sewage effluent. Compound
assignments refer to Fig. 1.

Alkylphenolpolyethoxycarboxylates have been identified in effluents of an advanced waste water treatment plant [11]. Sheldon and Hites [12,13] in their study on organic compounds in the Delaware river, have achieved GC/MS identifications of nonylphenols, 4-(1,1,3,3-tetramethylbutyl)phenol, and 4-(1,1,3,3-tetramethylbutyl)phenolmono-, di-, and triethoxylates. Our results suggest that the primary sources of these phenolic water pollutants are effluents of domestic sewage treatment plants.

In conclusion, this investigation has shown the occurrence of NP, NP1, NP2, and NP3 in secondary sewage effluents and in receiving waters. These chemicals are considered to be persistent metabolites derived from widely used nonionic detergents. It should be noted that nonylphenols and nonyl-phenolethoxylates with short oxyethylene side chains are not detectable by the standard methods used for the collective determination of nonionic detergents. Gas chromatography / mass spectrometry accomplishes the identification of these more persistent metabolite which are overlooked in studies using methods for the determination of their precursor compounds. These results are of particular importance because of the high toxicity of the alkylphenols. Nonylphenols have LC 50 values between 130 and 300 µg/ℓ for shrimp and salmon, as recently reported by McLeese et al. [14].

ACKNOWLEDGMENT

This work was supported in part by the Swiss Department of Commerce (Project COST 64b). One author (E.S.) was recipient of a fellowship from the H. Schmid-Foundation. We thank K. & G. Grob for supplying capillary columns and very valuable advice, C. Jaques for technical assistance, and T. Conrad, K. Stadler, and R. Geiser for very helpful discussions.

REFERENCES

[1] Giger, W. and Roberts, P.V., In "Water Pollution Microbiology", Vol. 2, Mitchell, R., ed., Wiley, New York, 1978, pp. 135-175.
[2] Van Hall, C.E., ASTM Special Techn. Publ. 686 (1979).
[3] Veith, G.D. and Kiwus, L.M., Bull. Environ. Contam. Toxicol., 17, 631-636 (1977).
[4] Peters, T.L., Anal. Chem., 52, 211-213 (1979).
[5] Grob, K. and Grob, G., J. Chromatogr., 213, 211-221 (1981).
[6] Giger, W., Stephanou, E. and Schaffner, C., Chemosphere, 10, 1981, in press.
[7] Rudling, L. and Solyolm, P., Water Research, 8, 115-119 (1974).
[8] Baggi, G., Beretta, L., Galli, E., Scolastico, C. and Freccani, V., "The Oil Industry and Microbial Ecosystems", K.W.A. Chater and H.J. Somerville, eds., Heydon, London, 1978, pp. 129-136.
[9] Geiser, R., Ph.D. thesis, Nr. 6678, ETH Zurich, 1980.
[10] Jones, P. and Nickless, G., J. Chromatogr., 156, 99-110 (1978).
[11] Reinhard, M., private communication, 1981.
[12] Sheldon, L.S. and Hites, R.A., Sci. Total Environ., 11, 279-286 (1979).
[13] Sheldon, L.S. and Hites, R.A., Environ. Sci. Technol., 12, 1188-1194 (1978).
[14] McLeese, D.W., Zitko, V., Sergeant, D.B. and Metcalfe, C.D., Chemosphere, 10, 723-730 (1981).

DETERMINATION OF PHENOLS IN WATER BY HPLC

M. SONNEBORN, E. PABEL, R. SCHWABE

Institute of Water-, Soil and Air Hygiene
Federal Health Office
Berlin, Federal Republic of Germany

1 Principle of method

Determination of phenols by High Pressure Liquid Chromatography (HPLC) with a reversed phase column (RP-18) and application of an Electro Chemical Detector System (ELCD). Enrichment of the phenols on suitable resins and column material.

1a

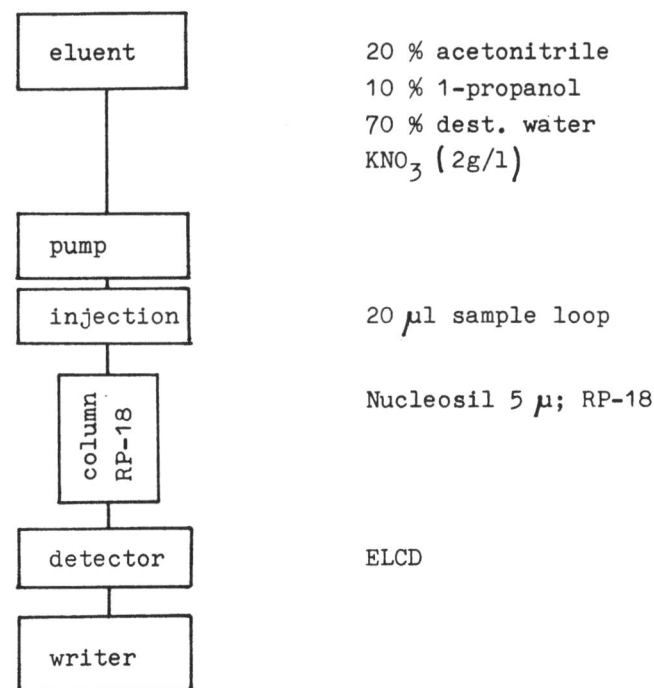

eluent	20 % acetonitrile
	10 % 1-propanol
	70 % dest. water
	KNO_3 (2g/l)
pump	
injection	20 μl sample loop
column RP-18	Nucleosil 5 μ; RP-18
detector	ELCD
writer	

1b Standard solution : phenols and indole compound
(addition of an indole compound be-
cause of similar chromatographic behaviour)

Peak No.	name	formula	concentration [ppm]
1	phenol		4.5
2	4-fluorophenol		2.4
3	4-cresol		2.5
4	2-chlorophenol		3.8
5	2,3-xylenol		2.7
6	2-ethylphenol		4.8
7	3-methyl-indole ("skatole")		2.3
8	4-phenylphenol		5.4
9	2-phenylphenol		10.6
10	2,3,6-trichloro-phenol		8.0
11	thymol		9.6
12	4-tert.-butyl-2-methylphenol		12.7
13	4-tert-butyl-2-chlorophenol		12.7

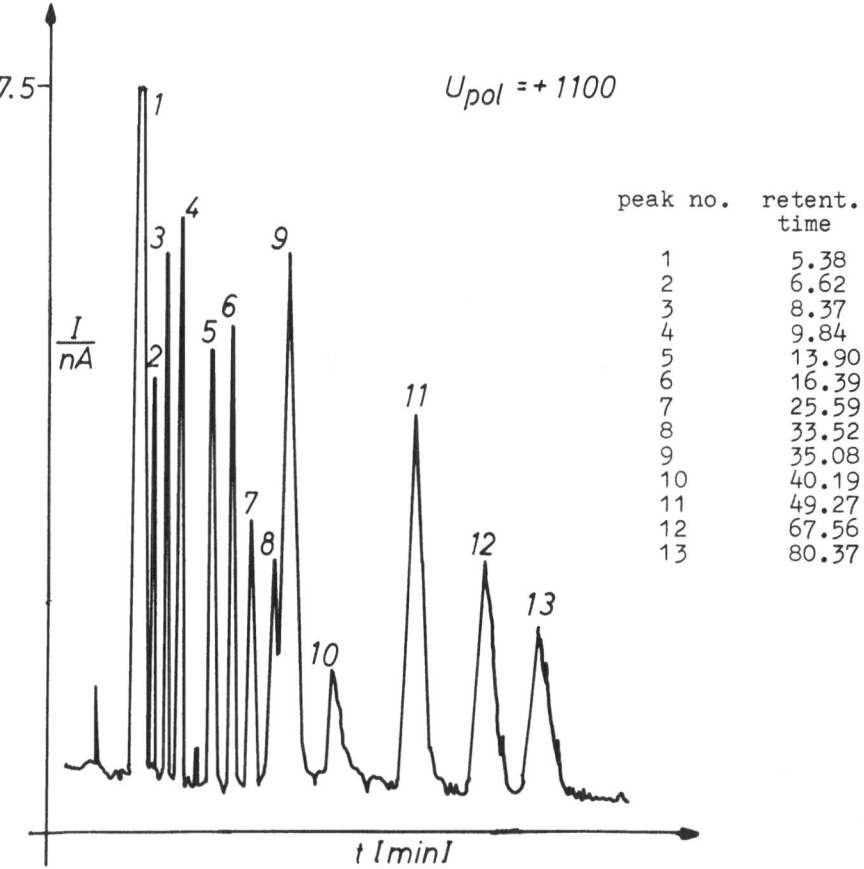

$U_{pol} = + 1100$

peak no.	retent. time
1	5.38
2	6.62
3	8.37
4	9.84
5	13.90
6	16.39
7	25.59
8	33.52
9	35.08
10	40.19
11	49.27
12	67.56
13	80.37

Chromatogram of the standard solution

2 Enrichment of phenols

An enrichment of phenols before injection on the analytical
RP-18 column is necessary as the concentration of the phe=
nols occuring in the different kinds of water is not sufficient enough for detection.

Column material: XAD-2, XAD-4 (resins)
                   Sep Pak C 18 (reversed phase material)

Preparation of XAD-2,XAD-4 and C 18:

a) Washing of the material with methanol and dest. water
   before use; filling of XAD-2 and XAD-4 in glass columns
b) Elution of the compounds after use with 20 ml methanol

3 Experiments for the retention of phenols in water
   a) Phenol standard solution (2 ml) in 1 l dest. water
   b) waste water (1 l)
   c) drinking water (1 l)

The solutions were given on   XAD-2, XAD-4 and C 18. The
compounds were eluted with 20 ml methanol.

4 RESULTS

Peak No.	XAD-2	XAD-4	C 18
1	65	100	-
2	88	97	-
3	97	122	-
4	102	88	-
5	111	91	10
6	127	105	11
7	99	39	13
8	70	12	69
9	95	44	88
10	25	-	30
11	104	91	99
12	109	74	95
13	119	74	114
the numbers indicate percentage of retention			

Conclusion:
The use of the tested materials for enrichment of phenols
enables the detection of small amounts (> 0.1 µg/l) of
these compounds by application of HPLC/ELCD.

SURFACTANTS SURVEY IN THE SURFACE WATERS
IN THE PARIS-AREA

C. HENNEQUIN, M. ROGGER and N. BOURGEOIS
Institut National de Recherche Chimique Appliquée
France

A survey has been done in the Paris Area rivers concerning anionic, cationic and in some cases non ionic surfactants. The sampling points are those of the National Surface Waters Survey (Map I).

We applied the following analytical procedures :
- Anionics : MBAS technique, results are indicated in dodecylbenzene sulfonate ($\mu g/l$)
- Non ionics : They are separated by a gas-stripping system and measured by the Wickbold method, results are in $\mu g/l$ NP 10
- Cationics : After a concentration step, cationics species are separated from anionics by ion exchange. Then, they are measured by the WATERS and KUPFER Technique (Disulfine blue VN.150). The original detection limit ($\simeq 0,1$ mg/l DMDSAC) has been lowered up to 0,01 mg/l by a careful analytical procedure.
At this level the recovery rate in distillated water as well in river water is generally higher than 95%.
About 40 ponctual samples has been done. A special treatment is applied to the "cationic samples" : we added 10 mg/l anionic sufactant to avoid the adsorption of eventually free cationics.

RESULTS

The whole results are given in the poster.

- Anionics : all the samples are between 20 and 230 $\mu g/l$ the maximal frequency lies in the 50-70 $\mu g/l$ range. They correspond to the known values in this area.

- Non ionics : generally not signicative values, close by the detection limit.

- Cationics : always below the detection limit, except

for one point (code 50000) where the water is heavy polluted.

Analytical problems hinder the measurement of cationics in
the bottom sludges where cationics are adsorbed.
And we need some analytical improvements to estimate the
cationics contents.

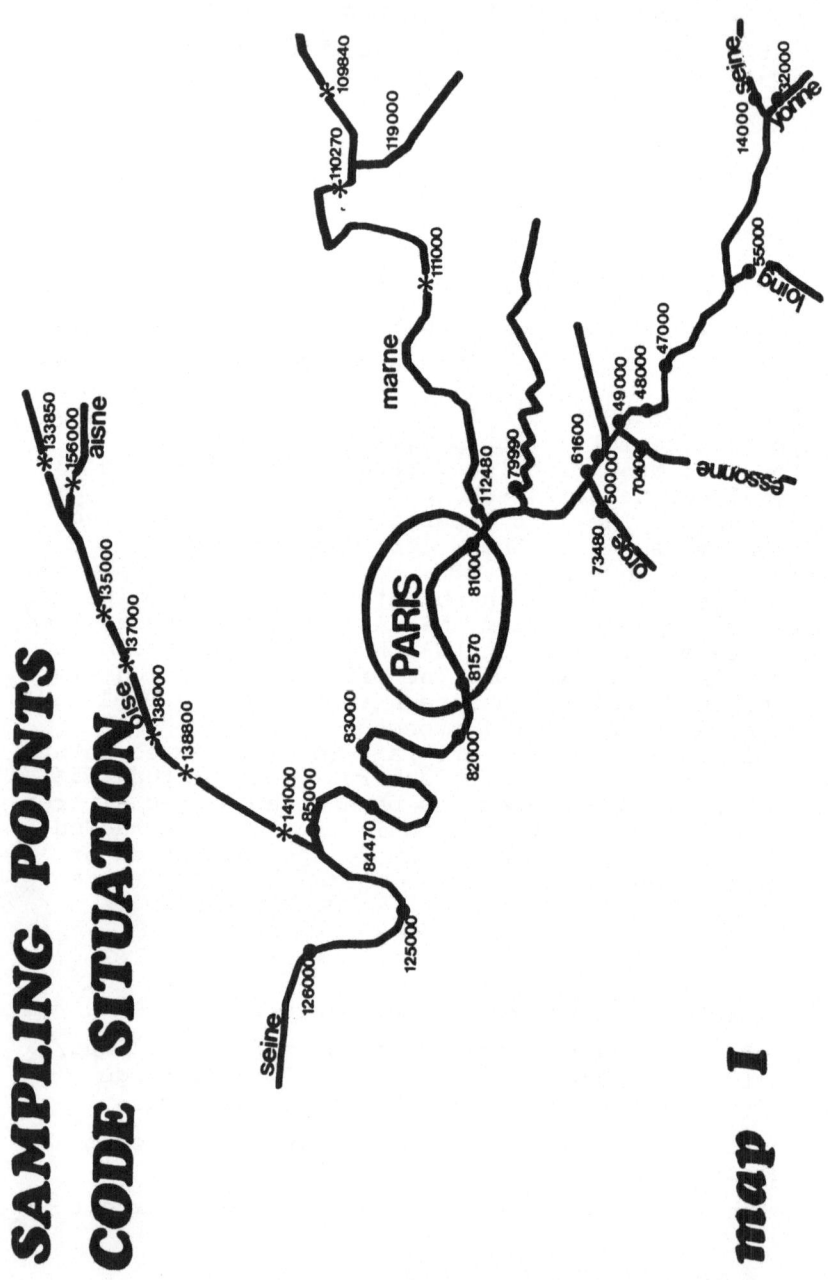

C O N C L U S I O N S

by the

Coordinators of the Working Parties

- SAMPLING AND SAMPLE TREATMENT
    F.J.J. Brinkmann, National Institute for Water Supply, Voorburg,
    The Netherlands

- GAS-CHROMATOGRAPHY
    R. Schwarzenbach, EAWAG, Dübendorf, Switzerland

- SEPARATION AND ANALYSIS OF NON-VOLATILE COMPOUNDS
    M. Fielding, Water Research Centre, Medmenham Laboratory, Marlow,
    Bucks. , United Kingdom

- MASS-SPECTROMETRY
    R. Ferrand, CERCHAR, Verneuil, France

- DATA PROCESSING
    P. Groll, Kernforschungszentrum Karlsruhe, Federal Republic of
    Germany

- SPECIFIC ANALYTICAL PROBLEMS
  A. - Organic halogens
  B. - Phenolic and other compounds
      D. Quaghebeur, Instituut voor Hygiëne en Epidemiologie, Brussels,
      Belgium

SUMMARY OF SESSION ON SAMPLING

F.J.J. BRINKMANN
National Institute for Water Supply, Voorburg, The Netherlands

Main lecture:

"Recent Concepts in Sampling Methodology", by Dr. Björn Josefsson.

Short lectures:

a) "A new method for the quantitative analysis of organochlorine pesticides and polychlorinated biphenyls", by Dr. P. Sandra

b) "Concentration and identification of the main organic micropollutants classes in waters", by Dr. C. Hennequin.

The discussion on the main lecture concentrated on questions on technical aspects (Dr. Wegman and Dr. Hennequin).
An animated discussion on the possibilities to distinguish between particle bound compounds and dissolved compounds in order to obtain information on the availability of these compounds in a sample was initiated by Dr. Schwarzenbach.

When questioned, Dr. Sandra and Dr. Hennequin elucidated some details.

The chairman expressed the conformity of Dr. Josefsson's lecture to the new approach of Working Group I. In contradiction to the past, in which concentrations and clean up-techniques were the topics, in this new approach sampling by itself will be the main field of interest.
The chairman expressed also his appreciation to Dr. Josefsson for his courage to accept the challenge to act as an expert in the field which is only poorly understood by the environmental chemist but is belonging to his task.

SUMMARY OF SESSION ON GAS CHROMATOGRAPHY

R. Schwarzenbach
EAWAG, 8600 Dübendorf
Switzerland

The papers presented in this session as well as the general discussions held have clearly demonstrated that, since the Berlin Symposium in
1979, further significant progress has been made in the field of capillary
gas chromatography. Capillary gas chromatography has become a reliable routine method for the separation, identification and quantification of many
important organic polluttants contained in complex mixtures as encountered
in water analysis. Furthermore, as demonstrated in Figure 1, the range of
compounds amenable to gas chromatography is being extended continuously.
However, in viewing the state of the art of this very important technique,
it has to be reminded that the gas chromatographic analysis is usually only
the last (and presently in most cases the least critical) step in the determination of organic compounds in water, and that the quality of the results
is, of cource, dependent on the performance of the various other steps involved in the analysis (i.e., sampling, enrichment).

In the papers in this session, practical aspects of contemporary capillary gas chromatography are discussed. A critical evaluation of injection
techniques, of selective detectors, and of the state of the art of column
technology is given. The aim of these papers is to provide the reader with
an overview over the most recent developments in this field, and to give
him a key to the pertinent recent literature. Together with the workshops,
the courses, and the information exchange ("GC-letters") organized in the
framework of working-party 2, this series of papers should also aid laboratories to acquire the know-how necessary for a proper application of gas
chromatographic techniques to water analyses.

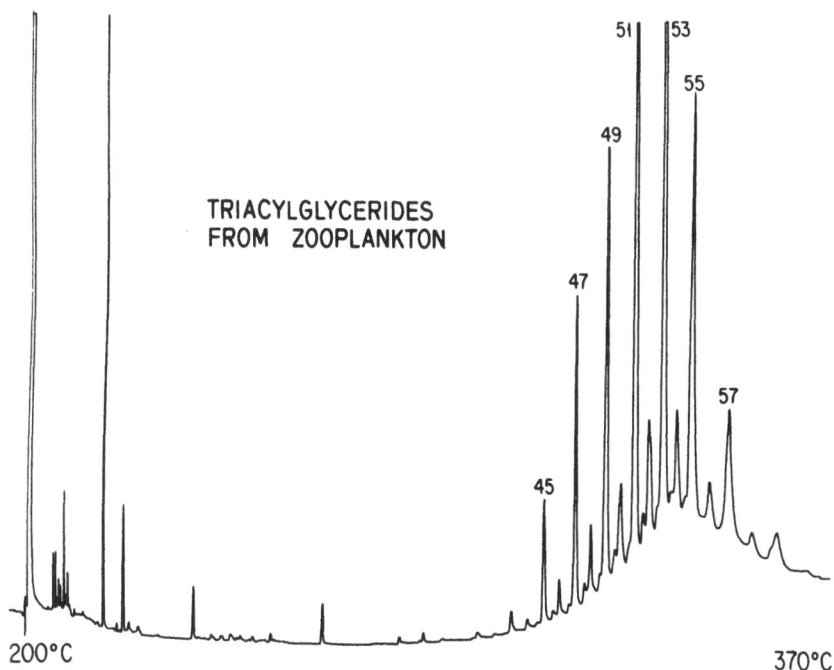

TRIACYLGLYCERIDES
FROM ZOOPLANKTON

200°C   370°C

Figure 1. Gas chromatogram of triacylglycerides extracted from
           mixed zooplankton.
           Numbers refer to total carbon numbers.

           Column: SE-52, persilylated, 25 x 0.3 mm i.d.
           Carrier gas: hydrogen, 1.6 atm
           Injection: on-column, 150 °C
           Detector: FID
           Oven: 4°/min, 200-370 °C.

           (From Wakeham et al., submitted to Anal. Chem.)

SUMMARY OF THE TECHNICAL SESSION ON SEPARATION AND ANALYSIS
OF NON-VOLATILE COMPOUNDS

M. FIELDING
PILOT, COST 64B bis WORKING GROUP 3

The session on separation and analysis of non-volatiles dealt with the
important problem of detection and identification of components separated
by HPLC. Three expert review papers were presented.

Dr Poppe reviewed recent developments in selective detectors for HPLC.
He covered various types of detection and made many important comments.
For instance, he reminded us that a detector of high selectivity also needs
to be very sensitive for handling complex mixtures, such as organics in
water, in order to avoid column overloading. He clearly felt that post-
column reaction detection offers good sensitivity, high selectivity and
wide applicability. Not many laboratories appeared to use post-column
reaction techniques, but as suitable commercial equipment is becoming
available, analysts can now apply these techniques to their current
problems.

The other two reviews dealt with methods of identification based upon
on-line and off-line HPLC. Professor Levsen covered the field of coupled
LC-MS, concentrating on the moving belt interface and direct liquid
introduction. He gave to us a very clear explanation of the basic
requirements for LC-MS coupling. The advantages and disadvantages of the
latest developments in coupling HPLC with MS were outlined and the use of
'soft' ionisation methods with LC-MS was discussed. Again many important
points were raised but one general point needs reiterating. Very rapid
developments are now taking place in the coupling of mass spectrometers
with HPLC, especially HPLC with micro-columns. It would be foolish not
to keep abreast of these developments, for in other fields, such as
biological and pharmaceutical analysis, they may become relatively common
tools.

Dr Crathorne dealt with an overall scheme for the identification of
non-volatile organics using off-line techniques. Firstly, he gave a clear
definition of the term 'non-volatile' and then outlined several important
reasons why techniques for their identification are needed. After review-
ing methods for isolation and separation of non-volatiles, he explained the
use of field desorption (FD) techniques in combination with collisional
activation (CA) and linked-scanning methods. He showed FD-MS, and FD-CA
spectra obtained from drinking water samples and used these to emphasize
two important points. Firstly, many non-volatile compounds are readily
detected in drinking water. Secondly, although the techniques described
are in their 'kindergarten' stage, they can provide valuable structural
data on non-volatile pollutants.

The discussions on the review papers were particularly useful. One
suggestion seems on reflection to be particularly attractive. The identification of non-volatiles in the complex mixtures encountered is difficult
and slow at the present time. However, it was suggested that one could
speed up things greatly by searching data, such as FD-MS data, for known
classes or types of compounds. Perhaps more realistic would be searches
for 'priority' non-volatiles, which have been selected according to some
criteria. Identification of non-volatiles, in this sense, may be
relatively quick.

A paper on a potential universal detector emphasised the widespread
need for such a detector with adequate sensitivity. With modern
techniques and developments it seems strange that such a detector is not
available.

In the session, the use of micro-HPLC columns was mentioned several times, and aspects of resolution and suitable detection systems discussed. Perhaps this is one specific area for future collaboration within the working group. In the introduction to the technical session the general objective of the working group was outlined. It is basically to review the development and application of techniques and to recommend aspects of problem areas requiring more research, and to promote collaboration. In this symposium several points have been raised which are worthy of future collaboration within the working group. However, it is important to finish the ongoing activities before commencing new work. Two ongoing activities need to be mentioned; the compilation of HPLC-based methods and collaborative qualitative analysis of a freeze-dried water sample. The compilation of HPLC-based methods is aimed at collecting brief, concise details of methods used by participating laboratories. Such a compilation should be particularly valuable. Information and a simple questionnaire was circulated about six months ago but to date only three replies have been received. Much more information is essential to make such a compilation successful. At a recent workshop meeting of the group at the ISPRA establishment, portions of freeze-dried material were distributed for subsequent qualitative analysis for non-volatile organics. It was hoped to have some results in time for this symposium. However, to date results from only one laboratory have been made available. Other laboratories are urged to provide data on analyses of the freeze-dried sample so that a workshop meeting can be arranged in 1982. This will enable participants to discuss identification of non-volatiles generally, with reference to the freeze-dried sample, and to discuss future collaboration.

Special thanks must go to the authors of the three excellent review papers, to participants who provided contributed papers and posters, and to those who took part in the useful and interesting discussions.

SUMMARY OF SESSION ON MASS SPECTROMETRY

R. FERRAND

CERCHAR - France

1 - SYMPOSIUM MS SECTION -

Monsieur CORNU presented a paper on MS-MS techniques which were thought as a potentially important field of activity in the near future for the participating laboratories.

These techniques, which are still at the stage of development, were mainly used, up to now, for fundamental research topics, such as structure elucidation and ion physics. From the discussions, it could be foreseen that, from their flexibility as regards the investigation of the fragmentation of separated molecular ions, they could be applied in the field of analysis of complex mixtures such as "Total extracts" of pollutants, that :

- With less stringent demands on previous separation steps such as L C or GC, up to molecular weights higher than those we can reach up to now.

- Even whithout chromatographic separation steps.

It is felt that these methods could apply

- Either to panoramic approaches (broad survey or multide-tection) by the investigation of every detected molecular ion.

- Or to specific approaches (target compounds or target groups of compounds) by screening of molecular ions giving specific daughter ions or neutral losses.

2 - ON GOING WORK -

a) A round robin test has been initiated quite recently, based on the examination of an actual extract spiked at different concentration levels with the so called "Base neutrals" priority pollutants of US EPA.

The participants are asked :

- To identify as exhaustively as possible, from multidetection approach, the compounds in the actual original
extract.

- To investigate the possibility of detecting and measuring
the "Base neutrals" adducts at the different concentration
levels and therefrom estimating the detection limit of
these compounds.

b) A questionnaire was also issued to inventorise the nowadays
state of the art of the MS equipment and MS methods in the working party.

3 - FUTURE WORK -

A meeting of the working party is planned for march or april
1982, possibly 2 days :

- For discussing the results of the round robin test.

- For discussing the trends in the participating laboratories
as regards new techniques such as MS-MS, FT-MS, different
ionisation techniques....

COMPUTERISIERTE DATENSAMMLUNG, HANDHABUNG UND INTERPRETATION

ZUSAMMENFASSUNG

P. Groll

Kernforschungszentrum Karlsruhe

Institut für heisse Chemie

D - 7500 Karlsruhe, BRD

Die Beiträge zu diesem Symposium auf dem Gebiet der computeriserten Datensammlung, Handhabung und Interpretation beschäftigten sich mit:

1. Sammlung von Massen-Spektren organischer Wasserverunreinigungen.
2. Software-Systemen für die Massenspektrometrie.
3. Sammlung von Daten, die das Auftreten identifisierter organischer Verunreinigungen in Oberflächengewässern charakterisieren.

Zu 1: Die Sammlung von Massen-Spektren wurde sowohl auf die manuelle Anwendung als auch auf den Einsatz von Computersystemen abgestimmt. Die Sammlung in Listenform enthält 1889 Massenspektren, die Sammlung auf Magnetband 1383 Massenspektren verschiedener Substanzen. Sowohl von den Leitern beider Arbeitsgruppen als auch von dem eingeladenen Referenten wurde die Wichtigkeit betont diese Sammlung fortzusetzen, wobei besonderes Augenmerk auf die Qualität der Spektren zu legen ist. Dies wird auch durch das Interesse der Arbeitsgruppe von Prof. MacLafferty, USA, an dieser Sammlung unterstrichen. Dies bedeutet auch, dass in Zukunft der Austausch von Massenspektren, die in den Labors gewonnen wurden, intensiviert werden sollte. Für den manuellen Zugriff sind entsprechend aufbereitete Listen dieser Sammlung in Vorbereitung. Eine zukünftige Sammlung von Massenspektren, die durch Stossionisation erhalten wurden, wurde angeregt.

Zu 2: Dr. Henneberg ist in seinem Vortrag auf die wesentlichen Aspekte des Einsatzes der Datenverarbeitung in der Massenspektrometrie eingegangen. Um hier Fortschritte zu erzielen, ist eine Optimierung aller verwendeten Software-Systeme nötig. Was die Dynamik betrifft, so sei hier ein Bereich von 100 000 ohne Änderung eines Sensitivitätsparameters erreichbar. Bei der Auswahl der Spektren der mittels GG getrennten Komponenten zeigte er eine Reihe von Fallstricken auf, die kommerzielle Systeme heute enthalten. Ein anderes Problem ist das Gebiet der Untergrundkorrektur. Das in den meisten kommerziellen Systemen enthaltene Softwaresystem, das auf Arbeiten von Biller und Biemann zurückgeht, kann bei nicht vollständiger Auftrennung der Komponenten im Gasschromatographen zu Ergebnissen führen, die eine anschliessende Substanzidentifikationen erschweren oder auch zu falschen Ergebnissen führen. Der von Dromey et al. vorgeschlagene Algorithmus scheitert daran, dass die GC-Peaks zu schmal sind, d.h. dass zu wenige Massenspektren während einer GC-Peaks gemessen werden. Das von Dr. Henneberg eingesetzte Programm würde diese Nachteile vermeiden. Anschliessend zeigte er an Beispielen wie die verschiedenen Suchstrategien, INCOS, PBM, STIRS und SISCOM, mit Problemen unreiner Spektren der zu identizierenden Substanzen und verfälschter Bibliotheksspektren fertig werden. Wenn das gesuchte Spektrum nicht in der Bibliothek enthalten ist, sind Hinweise auf Spektren ähnlicher Substanzen wünschenswert. Eine weitere Absicherung des Ergebnisses könnte erhalten werden, wenn eine Sammlung von Massenspektren GC-Retentionsindices als zusätzliche Information, nicht aber als Suchkriterium enthielte. Dies ist heute nicht der Fall.

Ein Vergleich verschiedener Suchstrategien wird im Moment von Arbeitsgruppe 6 der konzertierten Aktion der EG versucht. Brauchbare Ergebnisse sind aber nur bei längerfristigem Einsatz der verschiedenen Softwaresysteme unter Verwendung der gleichen Bibliothek an praktischen GC-MC-Messdaten möglich. Für die Zukunft wurde erneut eine Sammlung von Massenspektren angeregt, die bisher keinen Substanzen zugeordnet werden können. Eine derartige Sammlung muss sorgfältig geplant werden, und ich möchte diese Gelegenheit ergreifen, Beiträge einzuladen, die eine Abschätzung des Nutzens einen derartigen Sammlung ermöglichen.

Zu 3: Dr. Waggott zeigte die rasche Zunahme von Informationen über organische Verunreinigungen, die in Oberflächengewässern identifiziert wurden. Sie werden sowohl die Arbeiten der Praktiker als auch der Forscher unterstützen. Die geschaffene Datensammlung CICLOPS wird gegenwärtig durch ein Programmpaket CROSSBOW ergänzt, das die Suche nach chemischen Strukturen ermöglicht. Um CICLOPS den teilnehmenden Laboratorien leichter zugänglich zu machen, wird es in ECDIN integriert, wobei mittels ADABAS ein Verbindungsfile, ein bibliographisches File und ein geographisches File geschaffen wird. Die zukünftige Arbeit wird sich auf folgende zwei Schwerpunkte konzentrieren.

1. Erweiterung der Sammlung, wozu die teilnehmenden Laboratorien vor allem durch das Einbringen nicht veröffentlichter Daten beitragen sollen.
2. Realisierung der CROSSBOW und ECDIN-Files.

SUMMARY OF SESSION ON SPECIFIC ANALYTICAL PROBLEMS

D. QUAGHEBEUR
Instituut voor Hygiëne en Epidemiologie, Brussel

A. Organic Halogens

The organic halogens, representing a very wide range of compounds, can hardly be determined entirely. A partial answer to the problem can be provided by determining e.g. some of the non-volatiles or some of the volatiles by adequate techniques. Otherwise group or sum parameters could be useful at least as a screening tool, and by preference also an ecological or toxicological indication of the water quality. For this it must be possible to come to a common method.
The introductory lecture by R.C.C. Wegman, entitled *"Determination of organic halogens ; a critical review of sum parameters"* reviewed the analytical methods which are used for the determination of the sum of halogenated organic compounds in water samples, to conclude that :

- it is not possible to determine the sum of halogenated organic compounds in water samples with any of the methods described in the literature ;

- a reliable method for the determination of halogenated organic compounds in water samples seemed to be a method with
a) a purge step
b) an adsorption step with carbon as sorbent
c) a combustion step
d) conductance measurement after ion-chromatographic separation of each of the halogens.

- in future work attention must be given to particles in the water samples. Many compounds are adsorbed to the particles ;

- the number of abbreviations, sometimes covering quite divergent compounds, must be kept to a minimum.

B. Discussion

Comment Zürcher (Chairman)

In contrast to e.g. GC-analysis enabling an unambiguous ecological and toxicological approach for the individual compounds, the use of sum parameters causes a lot of interpretative problems, mainly due to the diversity of the compounds involved.
It is indeed not possible to determine the real total organic halogens, but only a part of them (volatiles, solubles...), depending on the determination method.
An evaluation and classification of these group parameters is necessary to provide for reliable ecological information. Group parameters can be used as a reference for the amount of individual products but cannot be related directly to toxicity (other criteria are also needed, e.g. structure...).
F. Zürcher presents also a scheme for the evaluation and classification of organo-chlorine water constituents (see annex).

Comment Kühn

Does not agree with Wegman's conclusions. - Using solvent extraction (e.g. hexane) a distinction can be made between dissolved and total or-

ganic halogen. - The toxicological information is very limited (**except**
when TOX = 0) ; anyway an analytical method does not give toxicological
information. Toxicology is out of the analysts domain.

Answer Wegman

Objection against the term "total" because not <u>all</u> organic halogen is
measured. One should look for other and a strictly limited number of
definitions.

Answer Zürcher

Does not see why to measure a "total sum", because different groups
have different physico-chemical properties. The proposed classification (see annex) has a well considered basis, giving indications of
ecological behaviour (but not directly toxicological). This selection
is important.

Comment Schwarzenbach

The analyst should care about toxicology, physico-chemical behaviour
is related to toxicology. TOX is not a desirable parameter. One
should not think only in analytical terms, one should decide case by
case.

Answer Kühn

A misunderstanding ! Would only like to stress the danger to correla-
te directly the physico-chemical behaviour with the ecological or
toxicological behaviour.

Comment Büchert

TOX is what the water really contains. Purgeable, non-purgeable, volatile... organic halogen is related to methods.

Answer Zürcher

A parameter like TOC does not allow an interpretation ; when an interpretation is needed one has to make a selection, so the analysis
should serve the information.

Answer Wegman

The determination of sum parameters is not complementary, but should
also be considered as a screening method to indicate the direction for
more specific analysis.

Comment Brinkmann

The term POX should be replaced in order to avoid confusion with POC
(particulate organic carbon).

Answer Zürcher

One should agree on terms and nomenclature, the selection steps should
be improved.

C. Other contributions on organic halogens

*Volatile halogenated hydrocarbons in river water, ground water, drin-
king water and swimming-pool water in the Federal Republic of Germany.
M. Sonneborn, S. Gerdes, R. Schwabe.*

Water samples of the rivers Weser, Lippe and Main were analyzed for the content of volatile halogenated hydrocarbons with the aim to get a profile of these compounds along the river.

The concentrations in the river Main are shown to be clearly higher. For checking the quality of drinking water 85 samples from the F.R.G. were investigated ; additionally chlorinated swimming-pool water samples were taken from public indoor swimming-pools ; all concentrations are clearly higher than the concentrations found in the mains drinking water of the corresponding city districts.
Examinations of ground water showed that even in this water the determination of halogenated hydrocarbons is possible, which can in particular circumstances affect the preparation of drinking water.
For the determination of the halogenated hydrocarbons, representing a widespread range of concentrations, GC and GS/MS were applied.

Analysis of trihalomethanes formed during drinking water chlorination. L. Schou, J. Krane, H. Ødegaard, G.E. Carlberg.

A typical Norwegian water source with relatively high content of humic constituents was treated to reduce the organics which cause trihalomethanes to be formed in the chlorination process. Chlorination of the raw water with 2 mg Cl_2/l gave an average of 37,3 µg $CHCl_3$/l. Raw water from the same source was then treated in a filtration pilot plant. A decrease in the chlorine demand of about 50% was observed. Reducing the amount of chlorine accordingly, gave a decrease in $CHCl_3$ formation of 50% on the filtered water. The effluent from the pilot plant chlorinated with a constant dose of chlorine showed a linear correlation between the amount $CHCl_3$ formed and UV-extinction. Experimental results are also presented showing the influence of variation in pH, temperature, chlorine dosage and humic and bromine content on the production of trihalomethanes.

Simultaneous determination of total volatile organochlorine, -bromine and- fluorine compounds in water by ion-chromatography. F. Zürcher.

A method for the determination of total volatile organohalogen compounds in water is described. Volatile halogenated organics are purged from water samples with oxygen and continuously combusted at 950°C. The resulting halogenides are trapped and quantified separately by ion-chromatography. The method gives the total amount of each group of purgeable organohalides and can be applied to a wide range of water samples. The detection limit of the described method is found to be below 0.1 µg per liter for individual groups of volatile organohalogens.

D. Phenolic compounds

The introductory lecture by L. Renberg, entitled *"Determination of phenolics in the aquatic environment"*, reviewed as well the different isolation, enrichment and clean-up methods as the methods based on chromatography for trace level determination of phenolics.
The most important conclusion is probably that there is no general

method for the trace level determination of all environmentally interesting phenols : there are several factors influencing the choice
of method- or rather the combination of suitable enrichment, clean-up
and quantification procedures.

E. Discussion on phenolic compounds

Question Schwarzenbach : 1) what time is needed for extractive alkylation ?

2) are the more acidic phenols the more toxic ?

Answer Renberg : 1) about 1 minute

2) the most toxic are the nitro- and chlorophenols.
The more lipophilic the less resistant they are against biodegradation.

Question Sdika : Acid chloride is used as a derivatisation reagent on
phenols in aqueous solution. Does the acid chloride not react with
the water at first ?

Answer Renberg : no alifatic but aromatic acid chlorides are used, the
latter being more stable against hydrolysis. It is also a question of
competition, but a great excess of acid chloride is added.

Question Sdika : is this a quantitative reaction ?

Answer Renberg : This aspect is not studied extensively but the reaction seems to be quantitative because a linear relationship and reproducibility is observed.

F. Other contributions (phenolic compounds and others)

Analysis of alkylphenols in an aqueous matrix containing aromatic
hydrocarbons.
D. Botta, F. Morandi and E. Mantica.

The problem of analysing alkylphenols in an aqueous matrix with a more
or less high content of aromatic hydrocarbons may be met when performing the analysis of ground or surface waters in proximity of factories
where coal tar is produced or processed.
Different analytical methods have been taken into consideration, such
as colorimetry, spectrophotometry, gaschromatography with direct
aqueous injection, injection of solutions from pre-concentration
treatments or injection of solutions from derivatization reactions,
HPL-chromatography.
Different conditions for liquid-liquid extraction were examined and
also the direct acetylation of alkylphenols in the aqueous phase was
considered.
The most suitable procedure was based on the removal of the aromatic
hydrocarbons from the aqueous sample by purging, the acetylation of
alkylphenols at controlled temperature and pH, the adsorption on resin, the subsequent elution with carbon disulfide followed by gaschro-
matographic analysis on a persilanized glass capillary column.

Determination of nonylphenols and nonylphenolethoxylates in se-
condary sewage effluents.
C. Schaffner, E. Stephanou and W. Giger.

Nonylphenols and nonylphenolethoxylates with one, two, and three oxyethylene groups were identified in effluents from activated sludge
sewage treatment plants. A steam-distillation and solvent-extraction
procedure was used to isolate the organic compounds. The complex
mixture of isomers were analyzed by glass capillary gaschromatography/
mass spectrometry. Nonionic detergents of the nonylphenolethoxylate
type are considered to be precursor chemicals of these refractory compounds.

Surfactants survey in the surface waters in the Paris-area.
C. Hennequin, M. Rogger and N. Bourgeois.

Anionic surfactants, determined by the MBAS technique, ranged between
20 and 230 µg/l (maximal frequency 50-70 µg/l). Nonionic surfactants,
measured by the Wickbold method, were close to the detection limit.
Cationic detergents, separated from anionics by ion exchange and measured by the Waters and Kupfer technique were always below the detection limit of 0.01 mg/l. The measurement of cationics in bottom sludges was not possible.

Analysis of nitrogenous organic substances in water.
M. Elmghari-Tabib, C. Le Cloirec, J. Morvan, G. Martin.

An analytical protocol is elaborated which pats up with the analysis
of most nitrogenous substances which might exist in the surface or
drinking waters : amino-acids, herbicides, nitro-and nitroso-derivati-
ves. After a suitable extraction and concentration technique the amino acids are determined by HPLC and fluorescence detection, the herbicides and nitro-derivatives by gaschromatography and election capture
detection.

The electroanalysis of organic pollutants in aquatic matrices.
W.F. Smyth and M.R. Smyth.

The applicability of electroanalytical methods such as DC and pulse
polarography and voltammetry, anodic and cathodic stripping voltammetry and on-line electrochemical detection is commented to the determination of selected organic molecules and their metabolites. The application of these techniques is described and critically comparized
with alternative analytical methods for the following groups of compounds :

- carbonyl compounds
- simple aromatic compounds (nitro-, phenolic-, and polychlorinated
  compounds)
- carboxylic acids
- alkyl sulphonates and alkyl benzene sulphonates
- organophosphorus compounds
- compounds containing endocyclic and exocyclic $>C = N^-$ groups
- compounds containing - SH groups
- carbamates
- benzidines.

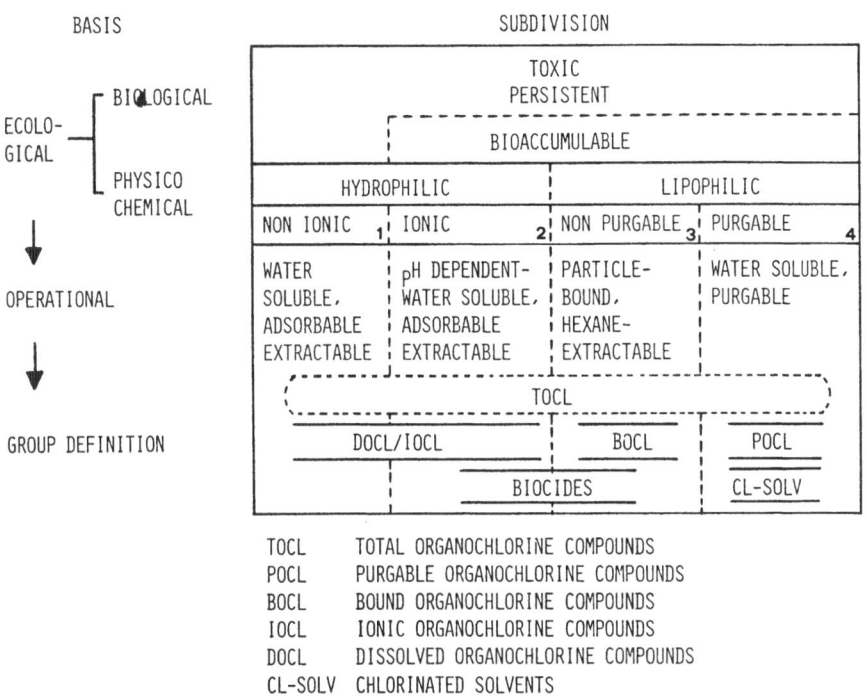

| EVALUATION AND CLASSIFICATION OF ORGANO-CHLORINE WATER CONSTITUENTS |

BASIS   SUBDIVISION

ECOLO- ┌ BIOLOGICAL   TOXIC
GICAL ─┤   PERSISTENT
        └ PHYSICO   BIOACCUMULABLE
          CHEMICAL

↓   HYDROPHILIC | LIPOPHILIC

                    NON IONIC | IONIC | NON PURGABLE | PURGABLE
                            1 | 2 | 3 | 4

OPERATIONAL   WATER | pH DEPENDENT- | PARTICLE- | WATER SOLUBLE,
                    SOLUBLE, | WATER SOLUBLE,| BOUND, | PURGABLE
                    ADSORBABLE| ADSORBABLE | HEXANE- |
                    EXTRACTABLE| EXTRACTABLE | EXTRACTABLE |

↓   TOCL

GROUP DEFINITION   DOCL/IOCL | BOCL | POCL

                                  BIOCIDES | CL-SOLV

TOCL   TOTAL ORGANOCHLORINE COMPOUNDS
POCL   PURGABLE ORGANOCHLORINE COMPOUNDS
BOCL   BOUND ORGANOCHLORINE COMPOUNDS
IOCL   IONIC ORGANOCHLORINE COMPOUNDS
DOCL   DISSOLVED ORGANOCHLORINE COMPOUNDS
CL-SOLV CHLORINATED SOLVENTS

PROCEDURE FOR THE GROUP-DIFFERENTIATION OF ORGANO-CHLORINATED COMPOUNDS *

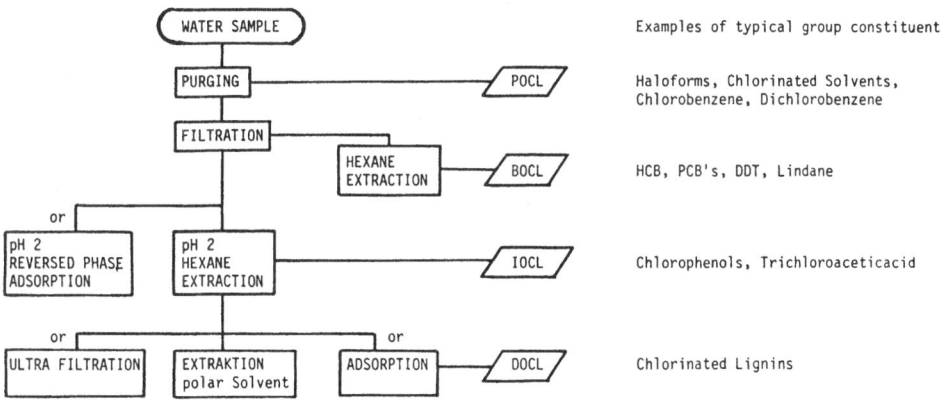

Examples of typical group constituent

Haloforms, Chlorinated Solvents,
Chlorobenzene, Dichlorobenzene

HCB, PCB's, DDT, Lindane

Chlorophenols, Trichloroaceticacid

Chlorinated Lignins

*) The procedure is also suitable for fluorinated, brominated and mixed halogenated derivatives.
   This implies detectors to discriminate between halogens.

LIST OF PARTICIPANTS

BELGIUM

QUAGHEBEUR, D.
  Instituut voor Hygiëne en
  Epidemiologie
  Ministry of Public Health
  Juliette Wytsmanstraat 14
  B - 1050 BRUSSELS

SANDRA, P.
  University of Ghent
  Laboratory of Organic Chemistry
  Krygslaan 271 S4
  B - 9000 GHENT

SAVOIR, R.
  Laboratories CIBE
  764, Chaussée de Waterloo
  B - 1050 BRUSSELS

TERMONIA, M.
  IRC-ISO
  Ministry of Agriculture
  Museumlaan 5
  B - 1980 TERVUREN

DENMARK

BUECHERT, A.
  National Food Institute
  19 Moerkhoej Bygade
  DK - 2860 SOEBORG

FOLKE, J.
  Water Quality Institute
  11 Agern Alle
  DK - 2970 HORSHOLM

HANSEN, N.
  Water Quality Institute
  11, Agern Alle
  DK - 2970 HORSHOLM

SORENSEN, A.K.
  Danish Civil Defence Analytical
  Chemical Laboratory
  Universitetsparken 2
  DK - 2100 COPENHAGEN

FEDERAL REPUBLIC OF GERMANY

GROLL, P.
  Nuclear Research Center Karlsruhe
  Institute for Hot Chemistry
  Postfach 3640
  D - 7500 KARLSRUHE

HENNEBERG, D.
  Max-Planck-Institut für Kohlen-
  forschung
  Postfach
  D - 6433 MUELHEIM/Ruhr

KARRENBROCK, F.
  ESWE Institut für Wasserforschung
  und Wassertechnologie GmbH
  Söhnleinstr. 158
  D - 6200 WIESBADEN

KUHN, W.
  DVGW-Forschungsstelle am
  Engler-Bunte-Institut
  Wasserchemie
  Richard-Willstätter-Allee 5
  D - 7500 KARLSRUHE 1

LEVSEN, K.
  Institut für Physikalische
  Chemie
  Wegelerstr. 12
  D - 5300 BONN 1

SCHWABE, R.
Institute for Water, Soil and
Air Hygiene
Federal Health Office
Correnzplatz 1
D - 1000 BERLIN 33

FRANCE

BRENER, L.
DEGREMONT
183, ave du 18 Juin 1940
F - 92500 RUEIL MALMAISON

CORNU, A.
Commissariat à l'Energie
Atomique
Centre d'Etudes Nucléaires
de Grenoble
Ave des Martyrs
F - 38041 GRENOBLE CEDEX

ELMGHARI-TABIB, M.
Laboratoire de Chimie des Eaux
et de l'Environnement
E.N.S.C. Rennes
Ave du Général Leclerc
F - 35000 RENNES-BEAULIEU

FERRAND, R.
CERCHAR
B.P. no 2
F - 60550 VERNEUIL

HENNEQUIN, C.
Institut de Recherche
Chimique appliquée
Centre de Vert le Petit
F - 91710 VERT LE PETIT

PHILIP, P.
Ministère de l'Environnement
Direction de la Prévention
des Pollutions
14, bvd du Général Leclerc
F - 92524 NEUILLY-sur-SEINE Cedex

SDIKA, A.
Institut National de Recherches
Chimiques appliquées - IRCHA
B.P. no 1
F - 91710 VERT LE PETIT

GREECE

MIMICOS, N.
N.R.C. Democritos
Chemistry Division
Aghia Pataskevi Attikis
GR - ATHENS

IRELAND

ARMONICI, E.
An Foras Forbartha
St. Martin's House
Waterloo Road
IRL - DUBLIN 4

BEHAN, J.J.
Southern Health Board
Public Analysts Laboratory
Douglas Road
IRL - CORK

CURRAN, F.
An Foras Forbartha
St. Martin's House
Waterloo Road
IRL - DUBLIN 4

DADGAR, D.
University College Cork
Western Road
IRL - CORK

DOWNEY, W.
An Foras Forbartha
St. Martin's House
Waterloo Road
IRL - DUBLIN 4

DYNES, K.
Institute for Industrial
Research and Standards
Ballymun Road
IRL - DUBLIN 9

EADES, J.
The Agricultural Institute
Oakpark Research Centre
IRL - CARLOW

FLANAGAN, P.
An Foras Forbartha
St. Martin's House
Waterloo Road
IRL - DUBLIN 4

GAULT, J.
Sligo Regional Technical
College
Ballinode
IRL - SLIGO

HORAN, H.T.
An Foras Forbartha
St. Martin's House
Waterloo Road
IRL - DUBLIN 4

LACEY, E.
An Foras Forbartha
St. Martin's House
Waterloo Road
IRL - DUBLIN 4

LYNCH, P.J.
National Board for
Science and Technology
Shelbourne House
Shelbourne Road
IRL - DUBLIN 4

MACLEAN, I.
Cork Country Council
42, Glen Heights Park
IRL - CORK

MEGARTY, C.
Cork Corporation
City Mall
IRL .- CORK

O'BRIEN, S.
Regional Technical College
Dublin Road
IRL - DUNDALK, Co. Louth

O'DONNEL, C.
State Laboratory
Upper Merrion Street
IRL - DUBLIN 2

O'GORMAN, V.
National Board for Science
and Technology
Shelbourne Road
IRL - DUBLIN 4

O'SULLIVAN, M.
Department of Fisheries
and Forestry
Fisheries Research Centre
IRL - ABBOTSTOWN, Co. Dublin

RYAN, J.
National Board for Science
and Technology
Shelbourne Road
IRL - DUBLIN 4

SMYTH, M.
National Institute of
Higher Education
School of Chemical Sciences
Ballymun Road
IRL - DUBLIN 9

SMYTH, W.F.
University College Cork
Western Road
IRL - CORK

WALSH, S.
Regional Technical College
Cork Road
IRL - WATERFORD

ITALY

BOTTA, D.
Politecnico - Istituto di
Chimica Industriale
Piazza Leonardo da Vinci 32
I - 2013 MILANO

GRIFFINI, O.
Laboratorio Chimico-
Biologico-Acquedotto
Vial Villamagna 31
I - FIRENZE

LIBERATORI, A.
Istituto Ricerca Acque
Via Reno, 1
I - 00198 ROMA

MANTICA, E.
Politecnico - Istituto di
Chimica Industriale
Piazza Leonardo da Vinci, 32
I - 20133 MILANO

THE NETHERLANDS

BRINKMANN, F.J.
  Rijksinstituut voor de
  Drinkwatervoorziening
  Gebouw Damsigt
  Nieuwe Havenstraat 6
  NL - LEIDSCHENDAM

NOORDSIJ, A.
  The Netherlands Waterworks'
  Testing and Research Institute
  Sir W. Churchilllaan 273
  P.O. Box 70
  NL - 2280 AB RIJSWIJK

POPPE, H.
  University of Amsterdam
  Nieuwe Achtergracht 166
  NL - 1018 WV AMSTERDAM

VAN DEN HOED, N.
  Koninklijke/ShellLaboratorium
  Postbus 3003
  NL - 1003 AA AMSTERDAM

VEENENDAAL, G.
  The Netherlands Waterworks'
  Testing and Research Institute
  Sir W. Churchilllaan 273
  P.O.Box 70
  NL - 2280 AB RIJSWIJK

WEGMAN, R.
  National Institute of
  Public Health
  P.O. Box 1
  NL - 3720 BA BILTHOVEN

VENEMA, A.
  AKZO-ARLA
  Velperweg 76
  ARNHEM
  THE NETHERLANDS

NORWAY

BJØRSETH, A.
  Central Institute for
  Industrial Research
  P.B. 350
  Blindern
  N - OSLO 3

CARLBERG, G.E.
  Central Institute for
  Industrial Research
  P.B. 350
  Blindern
  N - OSLO 3

KLUNGSOYR, J.
  Institute of Marine Research
  Directorate of Fisheries
  Havforskningsinst.
  P.O. Box 1870
  N - 5011 NORDNES-BERGEN

SCHOU, L.
  University of Trondheim
  Department of Chemistry
  NLHT
  N - 7055 DRAGVOLL

PORTUGAL

BISCAYA, Y.
  Instituto Hidrografico
  Rua das Trinas 49
  1296 - LISBOA codex

FERREIRA, A.M.
  Instituto Nacional de
  Investicacao das Pescas
  Avenida Brasilia
  Alges-Praia
  P - 1400 LISBOA

GIL DE CASTRO, M.O.
  Instituto Nacional de
  Investigacao das Pescas
  Avenida Brasilia
  Ales-Praia
  P - 1400 LISBOA

SPAIN

GONZALEZ-NICOLAS PEREZ, J.
  Centro de Estudios Hidrograficos
  Paseo Bajo Virgen del Puerto $N^{o}3$
  E - MADRID 5

RIVERA ARANDA, J.
  Instituto QUIMICA Bio-Organica
  CSIC
  c/Jorge Girona Salgado S/N
  E - BARCELONA 34

JOSEFSSON, B.
  Department of Analytical
  and Marine Chemistry
  Chalmers University of Technology
  Gothenburg University
  S - 41296 GOTHENBURG

LINDSTROM, K.
  Swedish Forest Products
  Research Laboratory
  Box 5604
  S - 11486 STOCKHOLM

RENBERG, L.
  National Swedish Environment
  Protection Board
  Special Analytical Laboratory
  Wallenberg Laboratory
  University of Stockholm
  S - 109 61 STOCKHOLM

SUNDSTROM, G.
  National Swedish Environmental
  Protection Board
  Special Analytical Laboratory
  Wallenberg Laboratory
  S - 10691 STOCKHOLM

SWITZERLAND

GROB K., Jr.
  Kantonales Labor
  P.O. Box
  CH - 8030 ZUERICH

SCHAFFNER, CH.
  Swiss Federal Institute for
  Water Resources and Water
  Pollution Control - EAWAG
  CH - 8600 DUEBENDORF

SCHWARZENBACH, R.P.
  Swiss Federal Institute for
  Water Resources and Water
  Pollution Control - EAWAG
  CH - 8600 DUEBENDORF

ZUERCHER, F.
  Swiss Federal Institute for
  Water Resources and Water
  Pollution Control - EAWAG
  CH - 8600 DUEBENDORF

UNITED KINGDOM

CONNOR, K.J.
  Water Research Centre
  Stevenage Laboratory
  Elder Way
  UK - STEVENAGE, Herts. SG1 1TH

CRATHORNE, B.
  Water Research Centre
  Henley Road
  UK - MEDMENHAM, Marlow,
      Bucks SL7 2HD

FIELDING, M.
  Water Research Centre
  Henley Road
  UK - MEDMENHAM, Marlow,
      Bucks SL7 2HD

FLINT, J.
  Department of the Environment
  Room A4, 19
  Romey House
  43 Marsham Street
  UK - LONDON SW1 3PY

MEEK, D.
  Water Research Centre
  Elder Way
  UK - STEVENAGE, Herts SG1 1TH

WAGGOTT, A.
  Water Research Centre
  Stevenage Laboratory
  Elder Way
  UK - STEVENAGE, Herts SG1 1TH

YUGOSLAVIA

DREVENKAR, V.
 Institute for Medical Research
 and Occupational Health
 Mose Pijade 158
 YU - 41000 ZAGREB

COMMISSION OF THE EUROPEAN COMMUNITIES

ANGELETTI, G.
 Directorate General "Science,
 Research and Development"
 200, rue de la Loi
 B - 1049 BRUSSELS

KNOEPPEL, H.
 Joint Research Centre
 I - 21020 ISPRA (Varese)

KRISTENSEN, K.
 Directorate General "Science,
 Research and Development"
 200, rue de la Loi
 B - 1049 BRUSSELS

NICOLAY, D.
 Directorate General "Information
 market and innovation"
 Jean Monnet Building B4/072
 P.O.B. 1907
 L - 2920 LUXEMBOURG

OTT, H.
 Directorate General "Science,
 Research and Development"
 200, rue de la Loi
 B - 1049 BRUSSELS

SCHAUENBURG, H.
 Joint Research Centre
 I - 21020 ISPRA (Varese)

SCHLITT, H.
 Joint Research Centre
 I - 21020 ISPRA (Varese)

STANDECKER, H.
 Directorate General "Science,
 Research and Development"
 200, rue de la Loi
 B - 1049 BRUSSELS

INDEX OF AUTHORS